A modern course in aeroelasticity

Monographs and textbooks on mechanics of solids and fluids

editor-in-chief: G. Æ. Oravas

Mechanics: Dynamical systems

editor: L. Meirovitch

1. E. H. DOWELL
   Aeroelasticity of plates and shells

2. D. G. B. EDELEN
   Lagrangian mechanics of nonconservative nonholonomic systems

3. JOHN L. JUNKINS
   An introduction to optimal estimation of dynamical systems

4. E. H. DOWELL
   A modern course in aeroelasticity

5. L. MEIROVITCH
   Computational methods in structural dynamics

A modern course in aeroelasticity

Earl H. Dowell, *editor,*
Professor of Mechanical and Aerospace Engineering
Princeton University, New Jersey, USA

Howard C. Curtiss, Jr.
Professor of Mechanical and Aerospace Engineering
Princeton University, New Jersey, USA

Robert H. Scanlan
Professor of Civil Engineering
Princeton University, New Jersey, USA

and

Fernando Sisto
Professor of Mechanical Engineering
Stevens Institute of Technology, New Jersey, USA

SIJTHOFF & NOORDHOFF   1980
Alphen aan den Rijn, The Netherlands
Rockville, Maryland, U.S.A.

ISBN 90 286 0057 4
ISBN 90 286 0737 4 (pb)

First printing 1978
Second printing 1980

Set-in-film in Northern Ireland
Printed in The Netherlands

Contents

Contents

Contents

Contents

Preface

A reader who achieves a substantial command of the material contained in this book should be able to read with understanding most of the literature in the field. Possible exceptions may be certain special aspects of the subject such as the aeroelasticity of plates and shells or the use of electronic feedback control to modify aeroelastic behavior. The first author has considered the former topic in a separate volume. The latter topic is also deserving of a separate volume.

In the first portion of the book the basic physical phenomena of divergence, control surface effectiveness, flutter and gust response of aeronautical vehicles are treated. As an indication of the expanding scope of the field, representative examples are also drawn from the non-aeronautical literature. To aid the student who is encountering these phenomena for the first time, each is introduced in the context of a simple physical model and then reconsidered systematically in more complicated models using more sophisticated mathematics.

Beyond the introductory portion of the book, there are several special features of the text. One is the treatment of unsteady aerodynamics. This crucial part of aeroelasticity is usually the most difficult for the experienced practitioner as well as the student. The discussion is developed from the basic fluid mechanics and includes a comprehensive review of the fundamental theory underlying numerical lifting surface analysis. Not only the well known results for subsonic and supersonic flow are covered; but also some of the recent developments for transonic flow, which hold promise of bringing effective solution techniques to this important regime.

Professor Sisto's chapter on Stall Flutter is an authoritative account of this important topic. A difficult and still incompletely understood phenomenon, stall flutter is discussed in terms of its fundamental aspects as well as its significance in applications. The reader will find this chapter particularly helpful as an introduction to this complex subject.

Another special feature is a series of chapters on three areas of advanced application of the fundamentals of aeroelasticity. The first of these is a discussion of Aeroelastic Problems of Civil Engineering Structures by Professor Scanlan. The next is a discussion of Aeroelasticity of Helicopters and V/STOL aircraft by Professor Curtiss. The final chapter in this series treats Aeroelasticity in Turbomachines and is by Professor Sisto. This series of chapters is unique in the aeroelasticity literature and the first author feels particularly fortunate to have the contributions of these eminent experts.

The emphasis in this book is on fundamentals because no single volume can hope to be comprehensive in terms of applications. However, the above three chapters should give the reader an appreciation for the relationship between theory and practice. One of the continual fascinations of aeroelasticity is this close interplay between fundamentals and applications. If one is to deal successfully with applications, a solid grounding in the fundamentals is essential.

For the beginning student, a first course in aeroelasticity could cover Chapters 1-3 and selected portions of 4. For a second course and the advanced student or research worker, the remaining Chapters would be appropriate. In the latter portions of the book, more comprehensive literature citations are given to permit ready access to the current literature.

The reader familiar with the standard texts by Scanlan and Rosenbaum, Fung, Bisplinghoff, Ashley and Halfman and Bisplinghoff and Ashley will appreciate readily the debt the authors owe to them. Recent books by Petre* and Forsching† should also be mentioned though these are less accessible to an english speaking audience. It is hoped the reader will find this volume a worthy successor.

---

*Petre, A., *Theory of Aeroelasticity. Vol. I Statics, Vol. II Dynamics.* In Romanian. Publishing House of the Academy of the Socialist Republic of Romania, Bucharest, 1966.
† Forsching, H. W., *Fundamentals of Aeroelasticity.* In German. Springer-Verlag, Berlin, 1974.

Short bibliography

Books

Bolotin, V. V., *Nonconservative Problems of the Elastic Theory of Stability*, Pergamon Press, 1963.

(BAH)   Bisplinghoff, R. L., Ashley, H. and Halfman, R. L., *Aeroelasticity*, Addison-Wesley Publishing Company, Cambridge, Mass., 1955.

(BA)   Bisplinghoff, R. L. and Ashley, H., *Principles of Aeroelasticity*, John Wiley and Sons, Inc., New York, N.Y., 1962. Also available in Dover Edition.

Fung, Y. C., *An Introduction to the Theory of Aeroelasticity*, John Wiley and Sons, Inc., New York, N.Y., 1955. Also available in Dover Edition.

Scanlan, R. H. and Rosenbaum, R., *Introduction to the Study of Aircraft Vibration and Flutter*, The Macmillan Company, New York, N.Y., 1951. Also available in Dover Edition.

(AGARD)   AGARD Manual on Aeroelasticity, Vols. I–VII, Beginning 1959 with continual updating.

Ashley, H., Dugundji, J. and Rainey, A. G., *Notebook for Aeroelasticity*, AIAA Professional Seminar Series, 1969.

Dowell, E. H., *Aeroelasticity of Plates and Shells*, Noordhoff International Publishing, Leyden, 1975.

Simiu, E. and Scanlan, R. H., *Wind Effects on Structures—An Introduction to Wind Engineering*, John Wiley and Sons, 1978.

In parentheses, abbreviations for the above books are indicated which are used in the text.

Survey articles

Garrick, I. E., 'Aeroelasticity—Frontiers and Beyond', 13th Von Karman Lecture, *J. of Aircraft*, Vol. 13, No. 9, 1976, pp. 641–657.

Several Authors, 'Unsteady Aerodynamics. Contribution of the Structures and Materials Panel to the Fluid Dynamics Panel Round Table Discussion on Unsteady Aerodynamics', Goettingen, May 1975, *AGARD Report* R-645, March 1976.

Rodden, W. P., *A Comparison of Methods Used in Interfering Lifting Surface Theory*, AGARD Report R-643, March 1976.

Ashley, H., 'Aeroelasticity', *Applied Mechanics Reviews*, February 1970.

Abramson, H. N., 'Hydroelasticity: A Review of Hydrofoil Flutter', *Applied Mechanics Reviews*, February 1969.

Many Authors, 'Aeroelastic Effects From a Flight Mechanics Standpoint', AGARD, Conference Proceedings No. 46, 1969.

Short bibliography

Landahl, M. T. and Stark, V. J. E., 'Numerical Lifting Surface Theory-Problems and Progress', *AIAA Journal*, No. 6, No. 11, November 1968, pp. 2049–2060.
Many Authors, 'Symposium on Fluid-Solid Interaction', *ASME Annual Winter Meeting*, November 1967.

Journals

AIAA Journal
ASCE Transactions, Engineering Mechanics Division
ASME Transactions, Journal of Applied Mechanics
International Journal of Solids and Structures
Journal of Aircraft
Journal of Sound and Vibration

Other journals will have aeroelasticity articles, of course, but these are among those with the most consistent coverage.

The impact of aeroelasticity on design is not discussed in any detail in this book. For insight into this important area the reader may consult the following volumes prepared by the National Aeronautics and Space Administration in its series on SPACE VEHICLE DESIGN CRITERIA. Although these documents focus on space vehicle applications, much of the material is relevant to aircraft as well. The depth and breadth of coverage varies considerably from one volume to the next, but each contains at least a brief State-of-the-Art review of its topic as well as a discussion of Recommended Design Practices. Further some important topics are included which have not been treated at all in the present book. These include, as already mentioned in the Preface,

Aeroelasticity of plates and shells (panel flutter) (NASA SP-8004)
Aeroelastic effects on control system dynamics (NASA SP-8016, NASA SP-8036 NASA SP-8079)

as well as

Structural response to time-dependent separated fluid flows (buffeting) (NASA SP-8001)
Fluid motions inside elastic containers (fuel sloshing) (NASA SP-8009, NASA SP-8031)
Coupled structural—propulsion instability (POGO) (NASA SP-8055)

It is intended to revise these volumes periodically to keep them up-to-date.

NASA SP-8001 1970
    Buffeting During Atmospheric Ascent
NASA SP-8002 1964
    Flight Loads Measurements During Launch and Exit
NASA SP-8003 1964
    Flutter, Buzz, and Divergence
NASA SP-8004 1972
    Panel Flutter
NASA SP-8006 1965
    Local Steady Aerodynamic Loads During Launch and Exit
NASA SP-8008 1965
    Prelaunch Ground Wind Loads
NASA SP-8012 1968
    Natural Vibration Modal Analysis

NASA SP-8016 1969
   Effects of Structural Flexibility on Spacecraft Control Systems
NASA SP-8009 1968
   Propellant Slosh Loads
NASA SP-8031 1969
   Slosh Suppression
NASA SP-8035 1970
   Wind Loads During Ascent
NASA SP-8036 1970
   Effects of Structural Flexibility on Launch Vehicle Control Systems
NASA SP-8050 1970
   Structural Vibration Prediction
NASA SP-8055 1970
   Prevention of Coupled Structure-Propulsion Instability (POGO)
NASA SP-8079 1971
   Structural Interaction with Control Systems.

Introduction

Several years ago, Collar suggested that aeroelasticity could be usefully visualized as forming a triangle of disciplines.

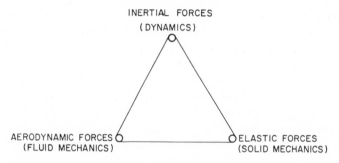

Aeroelasticity is concerned with those physical phenomena which involve significant mutual interaction among inertial, elastic and aerodynamic forces. Other important technical fields can be identified by pairing the several points of the triangle. For example,

Stability and control (flight mechanics) = dynamics + aerodynamics
Structural vibrations = dynamics + solid mechanics
Static aeroelasticity = fluid mechanics + solid mechanics

Conceptually, each of these technical fields may be thought of as a special aspect of aeroelasticity. For historical reasons only the last topic, static aeroelasticity, is normally so considered. However, the impact of aeroelasticity on stability and control (flight mechanics) has increased substantially in recent years, for example.

In modern aerospace vehicles, life can be even more complicated. For example, stresses induced by high temperature environments can be important in aeroelastic problems, hence the term
    'aerothermoelasticity'

In other applications, the dynamics of the guidance and control system may significantly affect aeroelastic problems or vice versa, hence the term 'aeroservoelasticity'

For a historical discussion of aeroelasticity including its impact on aerospace vehicle design, consult Chapter I of Bisplinghoff and Ashley (BA) and AGARD C.P. No. 46, 'Aeroelastic Effects from a Flight Mechanics Standpoint'.

We shall first concentrate on the dynamics and solid mechanics aspects of aeroelasticity with the aerodynamic forces taken as given. Subsequently, the aerodynamic aspects of aeroelasticity shall be treated from first principles. Theoretical methods will be emphasized, although these will be related to experimental methods and results where this will add to our understanding of the theory and its limitations. For simplicity, we shall begin with the special case of static aeroelasticity.

Although the technological cutting edge of the field of aeroelasticity has centered in the past on aeronautical applications, applications are found at an increasing rate in civil engineering, e.g., flows about bridges and tall buildings; mechanical engineering, e.g., flows around turbomachinery blades and fluid flows in flexible pipes; and nuclear engineering; e.g., flows about fuel elements and heat exchanger vanes. It may well be that such applications will increase in both absolute and relative number as the technology in these areas demands lighter weight structures under more severe flow conditions. Much of the fundamental theoretical and experimental developments can be applied to these areas as well and indeed it is hoped that a common language can be used in these several areas of technology. To further this hope we shall discuss subsequently in some detail several nonairfoil examples, even though our principal focus shall be on aeronautical problems. Separate chapters on civil engineering, turbomachinery and helicopter (rotor systems) applications will introduce the reader to the fascinating phenomena which arise in these fields.

Since most aeroelastic phenomena are of an undesirable character, leading to loss of design effectiveness or even sometimes spectacular structural failure as in the case of aircraft wing flutter or the Tacoma Narrows Bridge disaster, the spreading importance of aeroelastic effects will not be warmly welcomed by most design engineers. However, the mastery of the material to be discussed here will permit these effects to be better understood and dealt with if not completely overcome.

2

Static aeroelasticity

2.1 Typical section model of an airfoil

We shall find a simple, somewhat contrived, physical system useful for introducing several aeroelastic problems. This is the so-called 'typical section' which is a popular pedagogical device.* This simplified aeroelastic system consists of a rigid, flat plate airfoil mounted on a torsional spring attached to a wind tunnel wall. See Figure 2.1; the airflow over the airfoil is from left to right.

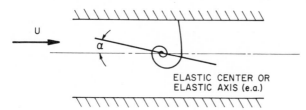

Figure 2.1 Geometry of typical section airfoil.

The principal interest in this model for the aeroelastician is the rotation of the plate (and consequent twisting of the spring), α, as a function of airspeed. If the spring were very stiff or airspeed were very slow, the rotation would be rather small; however, for flexible springs or high flow velocities the rotation may twist the spring beyond its ultimate strength and lead to structural failure. A typical plot of elastic twist, α_e, vs airspeed, U, is given in Figure 2.2. The airspeed at which the elastic twist increases rapidly to the point of failure is called the 'divergence airspeed', U_D. A major aim of any theoretical model is to accurately predict U_D. It should be emphasized that the above curve is representative not only of our typical section model but also of real aircraft wings. Indeed the

---

* See Chapter 6, BA, especially pp. 189–200.

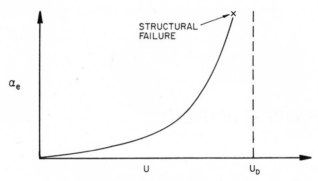

Figure 2.2 Elastic twist vs airspeed.

primary difference is not in the basic physical phenomenon of divergence but rather in the elaborateness of the theoretical analysis required to predict accurately U_D for an aircraft wing versus that required for our simple typical section model.

To determine U_D theoretically we proceed as follows. The equation of static equilibrium simply states that the sum of aerodynamic plus elastic moments about any point on the airfoil is zero. By convention, we take the point about which moments are summed as the point of spring attachment, the so-called 'elastic center' or 'elastic axis' of the airfoil.

The total aerodynamic angle of attack, α, is taken as the sum of some initial angle of attack, α_0 (with the spring untwisted), plus an additional increment due to elastic twist of the spring, α_e.

$$\alpha = \alpha_0 + \alpha_e \tag{2.1.1}$$

In addition, we define a point on the airfoil known as the 'aerodynamic center'.* This is the point on the airfoil about which the aerodynamic moment is independent of angle of attack, α. Thus, we may write the moment about the elastic axis as

$$M_y = M_{AC} + Le \tag{2.1.2}$$

where M_y moment about elastic axis or center
   M_{AC} moment about aerodynamic center,
        both moments are positive nose up
    L lift, net vertical force positive up
    e distance from aerodynamic center to
        elastic axis, positive aft.

---

* For two dimensional, incompressible flow this is at the airfoil quarter-chord; for supersonic flow it moves back to the half-chord. See Ashley and Landahl [1]. References are given at the end of each chapter.

4

From aerodynamic theory [1] (or experiment plus dimensional analysis) one has

$$L = C_L qS$$
$$M_{AC} = C_{MAC} qSc$$

(2.1.3a)

where

$$C_L = C_{L_0} + \frac{\partial C_L}{\partial \alpha} \alpha, \text{ lift coefficient}$$

(2.1.3b)

$C_{MAC} = C_{MAC_0}$, a constant, aerodynamic center moment coefficient in which

$$q = \frac{\rho U^2}{2}, \text{ dynamic pressure and}$$

ρ air density
U air velocity
$c\ \mathcal{C}$ airfoil chord
l airfoil span
S airfoil area, $c \times 1$

(2.1.3a) defines C_L and C_{MAC}. (2.1.3b) is a Taylor Series expansion of C_L for small α. C_{L_0} is the lift coefficient at $\alpha \equiv 0$. From (2.1.2), (2.1.3a) and (2.1.3b), we see the moment is also expanded in a Taylor series. The above forms are traditional in the aerodynamic literature. They are not necessarily those a nonaerodynamicist would choose.

Note that C_{L_0}, $\partial C_L/\partial \alpha$, C_{MAC_0} are nondimensional functions of airfoil shape, planform and Mach number. For a flat plate in two-dimensional incompressible flow [1]

$$\frac{\partial C_L}{\partial \alpha} = 2\pi, \qquad C_{MAC_0} = 0 = C_{L_0}$$

In what follows, we shall take $C_{L_0} \equiv 0$ for convenience and without any essential loss of information.

From (2.1.2), (2.1.3a) and (2.1.3b)

$$M_y = eqS\left[\frac{\partial C_L}{\partial \alpha}(\alpha_0 + \alpha_e)\right] + qScC_{MAC_0}$$

(2.1.4)

Now consider the elastic moment. If the spring has linear moment-twist characteristics then the elastic moment (positive nose up) is $-K_\alpha \alpha_e$ where K_α is the elastic spring constant and has units of moment (torque) per

5

angle of twist. Hence, summing moments we have

$$eqS\left[\frac{\partial C_L}{\partial \alpha}(\alpha_0 + \alpha_e)\right] + qScC_{MAC_0} - K_\alpha \alpha_e = 0 \tag{2.1.5}$$

which is the equation of static equilibrium for our 'typical section' airfoil.

Solving for the elastic twist (assuming $C_{MAC_0} = 0$ for simplicity) one obtains

$$\alpha_e = \frac{qS}{K_\alpha} \frac{e\dfrac{\partial C_L}{\partial \alpha}\alpha_0}{1 - q\dfrac{Se}{K_\alpha}\dfrac{\partial C_L}{\partial \alpha}} \tag{2.1.6}$$

This solution has several interesting properties. Perhaps most interesting is the fact that at a particular dynamic pressure the elastic twist becomes infinitely large. That is, when the denominator of the right-hand side of (2.1.6) vanishes

$$1 - q\frac{Se}{K_\alpha}\frac{\partial C_L}{\partial \alpha} = 0 \tag{2.1.7}$$

at which point $\alpha_e \to \infty$.

Equation (2.1.7) represents what is termed the 'divergence condition' and the corresponding dynamic pressure which may be obtained by solving (2.1.7) is termed the 'divergence dynamic pressure',

$$q_D \equiv \frac{K_\alpha}{Se(\partial C_L/\partial \alpha)} \tag{2.1.8}$$

Since only positive dynamic pressures are physically meaningful, note that only for $e > 0$ will divergence occur, i.e., when the aerodynamic center is ahead of the elastic axis. Using (2.1.8), (2.1.6) may be rewritten in a more concise form as

$$\alpha_e = \frac{(q/q_D)\alpha_0}{1 - q/q_D} \tag{2.1.9}$$

Of course, the elastic twist does not become infinitely large for any real airfoil; because this would require an infinitely large aerodynamic moment. Moreover, the linear relation between the elastic twist and the aerodynamic moment would be violated long before that. However, the elastic twist can become so large as to cause structural failure. For this reason, all aircraft are designed to fly below the divergence limits of all airfoil or lifting surfaces, e.g., wings, fins, control surfaces.

Now let us examine equations (2.1.5) and (2.1.9) for additional insight into our problem, again assuming $C_{MAC_0} = 0$ for simplicity. Two special cases will be informative. First, consider $\alpha_0 \equiv 0$. Then (2.1.5) may be written

$$\alpha_e \left[qS \frac{\partial C_L}{\partial \alpha} e - K_\alpha \right] = 0 \qquad (2.1.5a)$$

Excluding the trivial case $\alpha_e = 0$ we conclude from (2.1.5a) that

$$qS \frac{\partial C_L}{\partial \alpha} e - K_\alpha = 0 \qquad (2.1.7a)$$

which is the 'divergence condition'. This will be recognized as an eigenvalue problem, the vanishing of the coefficient of α_e in (2.1.5a) being the condition for nontrivial solutions of the unknown, α_e.* Hence, 'divergence' requires only a consideration of elastic deformations.

Secondly, let us consider another special case of a somewhat different type, $\alpha_0 \neq 0$, but $\alpha_e \ll \alpha_0$. Then (2.1.5a) may be written approximately as

$$eqS \frac{\partial C_L}{\partial \alpha} \alpha_0 - K_\alpha \alpha_e = 0 \qquad (2.1.10)$$

Solving

$$\alpha_e = \frac{qSe(\partial C_L/\partial \alpha)\alpha_0}{K_\alpha} \qquad (2.1.11)$$

Note this solution agrees with (2.1.6) if the denominator of (2.1.6) can be approximated by

$$1 - q \frac{Se}{K_\alpha} \frac{\partial C_L}{\partial \alpha} = 1 - \frac{q}{q_D} \approx 1$$

Hence, this approximation is equivalent to assuming that the dynamic pressure is much smaller than its divergence value. Note that the term neglected in (2.1.5) is the aerodynamic moment due to the elastic twist. Without this term, solution (2.1.11) is valid only for $q/q_D \ll 1$; it cannot predict divergence, however. This term can be usefully thought of as the 'aeroelastic feedback'.† A feedback diagram of equation (2.1.5) is given

---

* Here in static aeroelasticity q plays the role of the eigenvalue; in dynamic aeroelasticity q will be a parameter and the (complex) frequency will be the eigenvalue. This is a source of confusion for some students when they first study the subject.

† For the reader with some knowledge of feedback theory as in, for example, Savant [2].

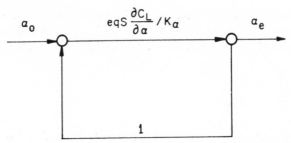

Figure 2.3   Feedback representation of aeroelastic divergence.

in Figure 2.3. Thus, when the forward loop gain exceeds unity, $qeS(\partial C_L/\partial\alpha)K_\alpha > 1$, the system is statically unstable, see equation (2.1.8). Hence, aeroelasticity can also be thought of as the study of aerodynamic + elastic feedback systems. One might also note the similarity of this divergence problem to conventional 'buckling' of structures.* Having exhausted the interpretations of this problem, we will quickly pass on to some slightly more complicated problems, but whose physical content is similar.

Typical section with control surface

We shall add a control surface to our typical section of Figure 2.1, as indicated in Figure 2.4. For simplicity, we take $\alpha_0 = C_{MAC_0} = 0$; hence, $\alpha = \alpha_e$. The aerodynamic lift is given by

$$L = qSC_L = qS\left(\frac{\partial C_L}{\partial\alpha}\alpha + \frac{\partial C_L}{\partial\delta}\delta\right) \text{ positive up} \tag{2.1.12}$$

the moment by

$$M_{AC} = qScC_{MAC} = qSc\frac{\partial C_{MAC}}{\partial\delta}\delta \text{ positive nose up} \tag{2.1.13}$$

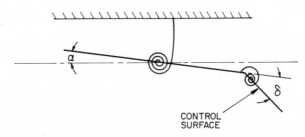

Figure 2.4   Typical section with control surface.

---

* Timoshenko and Gere [3].

8

the moment about the hinge line of the control surface by

$$H = qS_H c_H C_H = qS_H c_H \left(\frac{\partial C_H}{\partial \alpha} \alpha + \frac{\partial C_H}{\partial \delta} \delta \right) \text{ positive tail down} \quad (2.1.14)$$

where S_H is the area of control surface, c_H the chord of the control surface and C_H the (nondimensional) aerodynamic hinge moment coefficient. As before, $\frac{\partial C_L}{\partial \alpha}$, $\frac{\partial C_L}{\partial \delta}$, $\frac{\partial C_{MAC}}{\partial \delta}$, $\frac{\partial C_H}{\partial \alpha}$, $\frac{\partial C_H}{\partial \delta}$ are aerodynamic constants which vary with Mach and airfoil geometry. Note $\frac{\partial C_H}{\partial \delta}$ is typically negative.

The basic purpose of a control surface is to change the lift (or moment) on the main lifting surface. It is interesting to examine aeroelastic effects on this purpose.

To write the equations of equilibrium, we need the elastic moments about the elastic axis of the main lifting surface and about the hinge line of the control surface. These are $-K_\alpha \alpha$ (positive nose up), $-K_\delta (\delta - \delta_0)$ (positive tail down), and $\delta_e \equiv \delta - \delta_0$, where δ_e is the elastic twist of control surface in which δ_0 is the difference between the angle of zero aerodynamic control deflection and zero twist of the control surface spring.

The two equations of static moment equilibrium are

$$eqS \left(\frac{\partial C_L}{\partial \alpha} \alpha + \frac{\partial C_L}{\partial \delta} \delta \right) + qSc \frac{\partial C_{MAC}}{\partial \delta} \delta - K_\alpha \alpha = 0 \quad (2.1.15)$$

$$qS_H c_H \left(\frac{\partial C_H}{\partial \alpha} \alpha + \frac{\partial C_H}{\partial \delta} \delta \right) - K_\delta (\delta - \delta_0) = 0 \quad (2.1.16)$$

The above are two algebraic equations in two unknowns, α and δ, which can be solved by standard methods. For example, Cramer's rule gives

$$\alpha = \frac{\begin{vmatrix} 0 & eqS \dfrac{\partial C_L}{\partial \delta} + qSc \dfrac{\partial C_{MAC}}{\partial \delta} \\[2ex] -K_\delta \delta_0 & qS_H c_H \dfrac{\partial C_H}{\partial \delta} - K_\delta \end{vmatrix}}{\begin{vmatrix} eqS \dfrac{\partial C_L}{\partial \alpha} - K_\alpha & eqS \dfrac{\partial C_L}{\partial \delta} + qSc \dfrac{\partial C_{MAC}}{\partial \delta} \\[2ex] qS_H c_H \dfrac{\partial C_H}{\partial \alpha} & qS_H c_H \dfrac{\partial C_H}{\partial \delta} - K_\delta \end{vmatrix}} \quad (2.1.17)$$

9

and a similar equation for δ. To consider divergence we again set the denominator to zero. This gives a quadratic equation in the dynamic pressure q. Hence, there are two values of divergence dynamic pressure. Only the lower positive value of the two is physically significant.

In addition to the somewhat more complicated form of the divergence condition, there is a *new physical phenomenon* associated with the control surface called 'control surface reversal'. If the two springs were rigid, i.e., $K_\alpha \to \infty$ and $K_\delta \to \infty$, then $\alpha = 0$, $\delta = \delta_0$, and

$$L_r = qS \frac{\partial C_L}{\partial \delta} \delta_0 \qquad (2.1.18)$$

With flexible springs

$$L = qS \left(\frac{\partial C_L}{\partial \alpha} \alpha + \frac{\partial C_L}{\partial \delta} \delta \right) \qquad (2.1.19)$$

where α, δ are determined by solving the equilibrium equations (2.1.15), and (2.1.16). In general, the latter value of the lift will be smaller than the rigid value of lift. Indeed, the lift may actually become zero or even negative due to aeroelastic effects. Such an occurrence is called 'control surface reversal'. To simplify matters and show the essential character of control surface reversal, we will assume $K_\delta \to \infty$ and hence, $\delta \to \delta_0$ from the equilibrium condition (2.1.16). Solving the equilibrium equation (2.1.15), we obtain

$$\alpha = \delta_0 \frac{\dfrac{\partial C_L}{\partial \delta} + \dfrac{c}{e} \dfrac{\partial C_{MAC}}{\partial \delta}}{\dfrac{K_\alpha}{qSe} - \dfrac{\partial C_L}{\partial \alpha}} \qquad (2.1.20)$$

But

$$L = qS \left(\frac{\partial C_L}{\partial \delta} \delta_0 + \frac{\partial C_L}{\partial \alpha} \alpha \right)$$

$$= qS \left(\frac{\partial C_L}{\partial \delta} + \frac{\partial C_L}{\partial \alpha} \frac{\alpha}{\delta_0} \right) \delta_0 \qquad (2.1.21)$$

so that, introducing (2.1.20) into (2.1.21) and normalizing by L_r, we obtain

$$\frac{L}{L_r} = \frac{1 + q \dfrac{Sc}{K_\alpha} \dfrac{\partial C_{MAC}}{\partial \delta} \left(\dfrac{\partial C_L}{\partial \alpha} \Big/ \dfrac{\partial C_L}{\partial \delta} \right)}{1 - q \dfrac{Se}{K_\alpha} \dfrac{\partial C_L}{\partial \alpha}} \qquad (2.1.22)$$

Control surface reversal occurs when $L/L_r = 0$

$$1 + q_R \frac{Sc}{K_\alpha} \frac{\partial C_{MAC}}{\partial \delta} \left(\frac{\partial C_L}{\partial \alpha} \bigg/ \frac{\partial C_L}{\partial \delta} \right) = 0 \qquad (2.1.23)$$

where q_R is the dynamic pressure at reversal, or

$$q_R \equiv \frac{-\dfrac{K_\alpha}{Sc} \left(\dfrac{\partial C_L}{\partial \delta} \bigg/ \dfrac{\partial C_L}{\partial \alpha} \right)}{\dfrac{\partial C_{MAC}}{\partial \delta}} \qquad (2.1.24)$$

Typically, $\partial C_{MAC}/\partial \delta$ is negative, i.e., the aerodynamic moment for positive control surface rotation is nose down. Finally, (2.1.22) may be written

$$\frac{L}{L_r} = \frac{1 - q/q_R}{1 - q/q_D} \qquad (2.1.25)$$

where q_R is given by (2.1.22) and q_D by (2.1.8). It is very interesting to note that when K_δ is finite, the reversal dynamic pressure is still given by (2.1.24). However, q_D is now the lowest root of the denominator of (2.1.17). Can you reason physically why this is so?*

A graphical depiction of (2.1.25) is given in the Figure 2.5 where the two cases, $q_D > q_R$ and $q_D < q_R$, are distinguished. In the former case L/L_r decreases with increasing q and in the latter the opposite is true. Although the graphs are shown for $q > q_D$, our analysis is no longer valid when the divergence condition is exceeded without taking into account nonlinear effects.

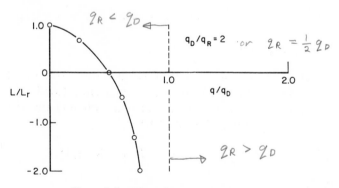

Figure 2.5 *Lift vs dynamic pressure.*

* See, BA, pp. 197–200.

11

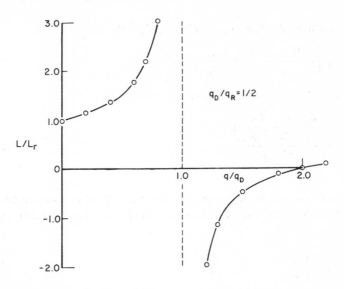

Fig 2.5 $cont'd$

Typical section—nonlinear effects

For sufficiently large twist angles, the assumption of elastic and/or aerodynamic moments proportional to twist angle becomes invalid. Typically the elastic spring becomes stiffer at larger twist angles; for example the elastic moment-twist relation might be

$$M_E = -K_\alpha \alpha_e - K_{\alpha_3} \alpha_e^3$$

where $K_\alpha > 0$, $K_{\alpha_3} > 0$. The lift-angle of attack relation might be

$$L = qS[(\partial C_L/\partial \alpha)\alpha - (\partial C_L/\partial \alpha)_3 \alpha^3]$$

where $\partial C_L/\partial \alpha$ and $(\partial C_L/\partial \alpha)_3$ are positive quantities. Note the lift decreases for large α due to flow separation from the airfoil. Combining the above in a moment equation of equilibrium and assuming for simplicity that $\alpha_0 = C_{MAC} = 0$, we obtain (recall (2.1.5))

$$eqS[(\partial C_L/\partial \alpha)\alpha_e - (\partial C_L/\partial \alpha)_3 \alpha_e^3] - [K_\alpha \alpha_e + K_{\alpha_3} \alpha_e^3] = 0$$

Rearranging,

$$\alpha_e[eq(S\,\partial C_L/\partial \alpha) - K_\alpha] - \alpha_e^3[eqS(\partial C_L/\partial \alpha)_3 + K_{\alpha_3}] = 0$$

12

Solving, we obtain the trivial solution $\alpha_e \equiv 0$, as well as

$$\alpha_e^2 = \frac{\left[eqS \dfrac{\partial C_L}{\partial \alpha} - K_\alpha \right]}{\left[eqS \left(\dfrac{\partial C_L}{\partial \alpha} \right)_3 + K_{\alpha_3} \right]}$$

To be physically meaningful α_e must be a real number; hence the right hand side of the above equation must be a positive number for the nontrivial solution $\alpha_e \neq 0$ to be possible.

For simplicity let us first assume that $e > 0$. Then we see that only for $q > q_D$ (i.e., for $eqS(\partial C_L/\partial \alpha) > K_\alpha$) are nontrivial solutions possible. See Figure 2.6. For $q < q_D$, $\alpha_e \equiv 0$ as a consequence of setting $\alpha_0 \equiv C_{MAC} \equiv 0$. Clearly for $e > 0$, $\alpha_e \equiv 0$ when $q < q_D$ where

$$q_D \equiv \frac{K_\alpha}{eqS \, \partial C_L/\partial \alpha}$$

Note that two (symmetrical) equilibrium solutions are possible for $q > q_D$. The actual choice of equilibrium position would depend upon how the airfoil is distributed (by gusts for example) or possibly upon imperfections in the spring or airfoil geometry. α_0 may be thought of as an initial imperfection and its sign would determine which of the two equilibria positions occurs. Note that for the nonlinear model α_e remains finite for any finite q.

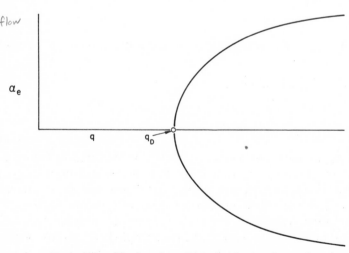

Figure 2.6a *(Nonlinear) equilibria for elastic twist: $e > 0$.*

13

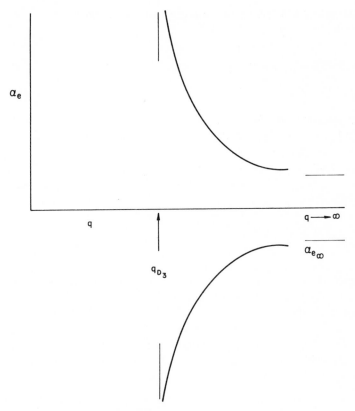

Figure 2.6b   (Nonlinear) equilibria for elastic twist: e < 0.

For $e < 0$, the equilibrium configurations would be as shown in Figure 2.6. where

$$q_{D_3} = -K_{\alpha_3}/eS(\partial C_L/\partial \alpha)_3$$

and

$$\alpha_{e_\infty}^2 = \partial C_L/\partial \alpha/(\partial C_L/\partial \alpha)_3$$

As far as the author is aware, the behavior indicated in Figure 2.6 has never been observed experimentally. Presumably structural failure would occur for $q > q_{D_3}$ even though α_{e_∞} is finite. It would be most interesting to try to achieve the above equilibrium diagram experimentally.

The above discussion does not exhaust the possible types of nonlinear behavior for the typical section model. Perhaps one of the most important nonlinearities in practice is that associated with the control surface spring and the elastic restraint of the control surface connection to the main lifting surface.*

2.2   One dimensional aeroelastic model of airfoils

Beam-rod representation of large aspect ratio wing†

We shall now turn to a more sophisticated, but more realistic beam-rod model which contains the same basic physical ingredients as the typical section. A beam-rod is here defined as a flat plate with rigid chordwise sections whose span, *l*, is substantially larger than its chord, *c*. See Figure 2.7. The airflow is in the *x* direction. The equation of static moment equilibrium for a beam-rod is

torsional ↗ *＊ misleading*

$$\frac{d}{dy}\left(GJ\frac{d\alpha_e}{dy}\right)+M_y=0 \tag{2.2.1}$$

$\alpha_e(y)$ nose up twist about the elastic axis, e.a., at station y
M_y nose up aerodynamic moment about e.a. per unit distance in the spanwise, y, direction
G shear modulus
J polar moment of inertia
   $(= ch^3/3$ for a rectangular cross-section of thickness $h, h \ll c)$
GJ torsional stiffness

Equation (2.2.1) can be derived by considering a differential element dy (see Figure 2.8) The internal elastic moment is $GJ(d\alpha_e/dy)$ from the theory of elasticity.‡ Note for $d\alpha_e/dy > 0$, $GJ(d\alpha_e/dy)$ is positive nose

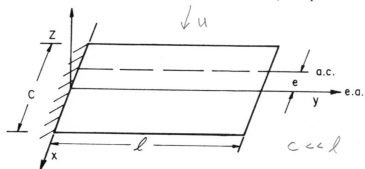

Figure 2.7   Beam-rod representation of wing.

---

* Woodcock [4].
† See Chapter 7, BA, pp. 280–295, especially pp. 288–295.
‡ Housner, and Vreeland [5].

15

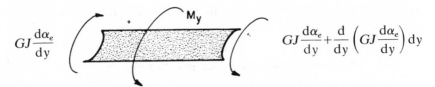

$$GJ\frac{d\alpha_e}{dy} \qquad\qquad GJ\frac{d\alpha_e}{dy}+\frac{d}{dy}\left(GJ\frac{d\alpha_e}{dy}\right)dy$$

Figure 2.8   *Differential element of beam-rod.*

down. Summing moments on the differential element, we have

$$-GJ\frac{d\alpha_e}{dy}+GJ\frac{d\alpha_e}{dy}+\frac{d}{dy}\left(GJ\frac{d\alpha_e}{dy}\right)dy+\text{H.O.T.}+M_y\,dy=0$$

In the limit, as $dy\rightarrow 0$,

$$\frac{d}{dy}\left(GJ\frac{d\alpha_e}{dy}\right)+M_y=0 \qquad\qquad (2.2.1)$$

Equation (2.2.1) is a second order differential equation in y. Associated with it are two boundary conditions. The airfoil is fixed at its root and free at its tip, so that the boundary conditions are

$$\alpha_e=0 \quad\text{at}\quad y=0 \;;\quad GJ\frac{d\alpha_e}{dy}=0 \quad\text{at}\quad y=l \qquad (2.2.2)$$

Turning now to the aerodynamic theory, we shall use the 'strip theory' approximation. That is, *we shall assume that the aerodynamic lift and moment at station y depends only on the angle of attack at station y* (and is independent of the angle of attack at other spanwise locations).

Thus moments and lift per unit span are, as before,

$$M_y=M_{AC}+Le \qquad\qquad (2.2.3a)$$

$$L\equiv qcC_L \qquad\qquad (2.2.3b)$$

where now the lift and moment coefficients are given by

$$C_L(y)=\frac{\partial C_L}{\partial\alpha}[\alpha_0(y)+\alpha_e(y)]^* \qquad\qquad (2.2.3c)$$

$$M_{AC}=qc^2C_{MAC} \quad or \quad = q\,s\,c\,C_{MAC} \qquad\qquad (2.2.3d)$$

(2.2.3b) and (2.2.3d) define C_L and C_{MAC} respectively.

---

* A more complete aerodynamic model would allow for the effect of angle of attack at one spanwise location, say η, on (nondimensional) lift at another, say y. This relation would then be replaced by $C_L(y)=\int A(y-\eta)[\alpha_0(\eta)+\alpha_e(\eta)]\,d\eta$ where A is an aerodynamic influence function which must be measured or calculated from an appropriate theory. More will be said of this later.

for   unit   span
$s = l \times c$

Using (2.2.3) in (2.2.1) and nondimensionalizing (assuming for simplicity, constant wing properties)

$$\bar{y} \equiv \frac{y}{l}$$

$$\lambda^2 \equiv \frac{ql^2}{GJ} c \frac{\partial C_L}{\partial \alpha} e$$

$$K \equiv -\frac{qcl^2}{GJ} \left(e \frac{\partial C_L}{\partial \alpha} \alpha_0 + C_{MAC_0} c \right)$$

(2.2.1) becomes

$$\frac{d^2 \alpha_e}{d\bar{y}^2} + \lambda^2 \alpha_e = K \qquad (2.2.4)$$

which is subject to boundary conditions (2.2.2). These boundary conditions have the nondimensional form

$$\alpha_e = 0 \quad \text{at} \quad \bar{y} = 0$$

$$\frac{d\alpha_e}{d\bar{y}} = 0 \quad \text{at} \quad \bar{y} = 1 \quad \rightarrow \text{Because} \quad T = 0 \qquad (2.2.5)$$

The general solution to (2.2.4) is

$$\alpha_e = A \sin \lambda \bar{y} + B \cos \lambda \bar{y} + \frac{K}{\lambda^2} \qquad (2.2.6)$$

Applying boundary conditions (2.2.5), we obtain

$$B + \frac{K}{\lambda^2} = 0, \qquad \lambda[A \cos \lambda - B \sin \lambda] = 0 \qquad (2.2.7)$$

Solving equation (2.2.7), $A = -(K/\lambda^2) \tan \lambda$, $B = -K/\lambda^2$, so that

$$\alpha_e = \frac{K}{\lambda^2} [1 - \tan \lambda \sin \lambda \bar{y} - \cos \lambda \bar{y}] \qquad (2.2.8)$$

Divergence occurs when $\alpha_e \rightarrow \infty$, i.e., $\tan \lambda \rightarrow \infty$, or $\cos \lambda \rightarrow 0$.* Thus, for $\lambda = \lambda_m = (2m - 1)\frac{\pi}{2} (m = 1, 2, 3, \ldots)$, $\alpha_e \rightarrow \infty$. The lowest of these, $\lambda_1 = \frac{\pi}{2}$, is physically significant. Using the definition of λ preceding equation

---

* Note $\lambda \equiv 0$ is not a divergence condition! Expanding (2.2.8) for $\lambda \ll 1$, we obtain $\alpha_e = \frac{K}{\lambda^2}[1 - \lambda^2 \bar{y} - \left(1 - \frac{\lambda^2 \bar{y}^2}{2}\right) + \cdots] \rightarrow K\left[\frac{\bar{y}^2}{2} - \bar{y}\right]$ as $\lambda \rightarrow 0$.

2 Static aeroelasticity

(2.2.4), the divergence dynamic pressure is

$$q = (\pi/2)^2 \frac{GJ}{2} \Big/ lce(\partial C_L/\partial \alpha) \tag{2.2.9}$$

Recognizing that $S = lc$, we see that (2.2.9) is equivalent to the typical section value, (2.1.8), with

$$K_\alpha = \left(\frac{\pi}{2}\right)^2 \frac{GJ}{l} \tag{2.2.10}$$

Eigenvalue and eigenfunction approach

One could have treated divergence from the point of view of an eigenvalue problem. Neglecting those terms which do *not* depend on the elastic twist, i.e., setting $\alpha_0 = C_{MAC_0} = 0$, we have $K = 0$ and hence

$$\frac{d^2\alpha}{d\tilde{y}^2} + \lambda^2\alpha = 0 \tag{2.2.11}$$

with

$$\alpha = 0 \quad \text{at} \quad \tilde{y} = 0$$
$$\frac{d\alpha}{d\tilde{y}} = 0 \quad \text{at} \quad \tilde{y} = 1 \tag{2.2.12}$$

The general solution is

$$\alpha = A \sin \lambda\tilde{y} + B \cos \lambda\tilde{y} \tag{2.2.13}$$

Using (2.2.12), (2.2.13)

$$B = 0$$
$$\lambda[A \cos \lambda - B \sin \lambda] = 0$$

we conclude that

$$A = 0$$

or

$$\lambda \cos \lambda = 0 \quad \text{and} \quad A \neq 0 \tag{2.2.14}$$

The latter condition, of course, is 'divergence'. Can you show that $\lambda = 0$, does not lead to divergence? What does (2.2.13) say? For each eigenvalue, $\lambda = \lambda_m = (2m-1)\frac{\pi}{2}$ there is an eigenfunction, $A \neq 0$, $B = 0$,

$$\alpha_m \sim \sin \lambda_m\tilde{y} = \sin (2m-1)\frac{\pi}{2}\,\tilde{y} \tag{2.2.15}$$

18

These eigenfunctions are of interest for a number of reasons:

(1) They give us the twist distribution at the divergence dynamic pressure as seen above in (2.2.15).

(2) They may be used to obtain a series expansion of the solution for any dynamic pressure.

(3) They are useful for developing an approximate solution for variable property wings.

Let us consider further the second of these. Now we let $\alpha_0 \neq 0$, $C_{MAC_0} \neq 0$ and begin with (2.2.4).

$$\frac{d^2\alpha_e}{d\bar{y}^2} + \lambda^2\alpha_e = K \tag{2.2.4}$$

Assume a series solution of the form

$$\alpha_e = \sum_n a_n\alpha_n(\bar{y}) \tag{2.2.16}$$

$$K = \sum_n A_n\alpha_n(\bar{y}) \tag{2.2.17}$$

where a_n, A_n are to be determined. Now it can be shown that

$$\int_0^1 \alpha_n(\bar{y})\alpha_m(\bar{y})\,d\bar{y} = \tfrac{1}{2} \quad \text{for} \quad m = n$$
$$= 0 \quad \text{for} \quad m \neq n \tag{2.2.18}$$

This is the so-called 'orthogonality condition'. We shall make use of it in what follows. First, let us determine A_n. Multiply (2.2.17) by α_m and $\int_0^1 \cdots d\bar{y}$.

$$\int_0^1 K\alpha_m(\bar{y})\,d\bar{y} = \sum_n A_n \int_0^1 \alpha_n(\bar{y})\alpha_m(\bar{y})\,d\bar{y}$$
$$= A_m\tfrac{1}{2}$$

using (2.2.18). Solving for A_m,

$$A_m = 2\int_0^1 K\alpha_m(\bar{y})\,d\bar{y} \tag{2.2.19}$$

Now let us determine a_n. Substitute (2.2.16) and (2.2.17) into (2.2.4) to obtain

$$\sum_n \left[a_n\frac{d^2\alpha_n}{d\bar{y}^2} + \lambda^2 a_n\alpha_n \right] = \sum_n A_n\alpha_n \tag{2.2.20}$$

19

Now each eigenfunction, α_n, satisfies (2.2.11).

$$\frac{d^2\alpha_n}{d\bar{y}^2} + \lambda_n^2\alpha_n = 0 \tag{2.2.11}$$

Therefore, (2.2.20) may be written

$$\sum_n a_n[-\lambda_n^2 + \lambda^2]\alpha_n = \sum A_n\alpha_n \tag{2.2.21}$$

Multiplying (2.2.21) by α_m and $\int_0^1 \cdots d\bar{y}$,

$$[\lambda^2 - \lambda_m^2]a_m\tfrac{1}{2} = A_m\tfrac{1}{2}$$

Solving for a_m,

$$a_m = \frac{A_m}{[\lambda^2 - \lambda_m^2]} \tag{2.2.22}$$

Thus,

$$\alpha_e = \sum a_n\alpha_n = \sum_n \frac{A_n}{[\lambda^2 - \lambda_n^2]}\alpha_n(\bar{y}) \tag{2.2.23}$$

where A_n is given by (2.2.19).*

Similar calculations can be carried out for airfoils whose stiffness, chord, etc., are *not* constants but vary with spanwise location. One way to do this is to first determine the eigenfunction expansion *for the variable property wing* as done above for the constant property wing. The determination of such eigenfunctions may itself be fairly complicated, however. An alternative procedure can be employed which expands the solution for *the variable property* wing in terms of the eigenfunctions of the *constant property wing*. This is the last of the reasons previously cited for examining the eigenfunctions.

Galerkin's method

The equation of equilibrium for a *variable* property wing may be obtained by substituting (2.2.3) into (2.2.1). In dimensional terms

$$\frac{d}{dy}\left(GJ\frac{d}{dy}\,\alpha_e\right) + eqc\frac{\partial C_L}{\partial\alpha}\,\alpha_e = -eqc\frac{\partial C_L}{\partial\alpha}\,\alpha_0 - qc^2C_{MAC_0} \tag{2.2.24}$$

---

*For a more detailed mathematical discussion of the above, see Hildebrand [6], pp. 224–234. This problem is one of a type known as 'Sturm–Liouville Problems'.

In nondimensional terms

$$\frac{d}{d\tilde{y}}\left(\gamma \frac{d\alpha_e}{d\tilde{y}}\right) + \lambda^2 \alpha_e \beta = K \tag{2.2.25}$$

where

$$\gamma \equiv \frac{GJ}{(GJ)_{ref}} \qquad K = -\frac{qcl^2}{(GJ)_{ref}}\left[e\frac{\partial C_L}{\partial \alpha}\alpha_0 + C_{MAC_0}c\right]$$

$$\lambda^2 \equiv \frac{ql^2 c_{ref}}{(GJ)_{ref}}\left(\frac{\partial C_L}{\partial \alpha}\right)_{ref}e_{ref} \qquad \beta = \frac{c}{c_{ref}}\frac{e}{e_{ref}}\frac{\left(\dfrac{\partial C_L}{\partial \alpha}\right)}{\left(\dfrac{\partial C_L}{\partial \alpha}\right)_{ref}}$$

Let

$$\alpha_e = \sum_n a_n \alpha_n(\tilde{y})$$

$$K = \sum_n A_n \alpha_n(\tilde{y})$$

as before. Substituting the series expansions into (2.2.25), multiplying by α_m and $\int_0^1 \cdots d\tilde{y}$,

$$\sum_n a_n\left\{\int_0^1 \frac{d}{d\tilde{y}}\left(\gamma\frac{d\alpha_n}{d\tilde{y}}\right)\alpha_m\, d\tilde{y} + \lambda^2\int_0^1 \beta\alpha_n\alpha_m\, d\tilde{y}\right\}$$

$$= \sum_n A_n\int_0^1 \alpha_n\alpha_m\, d\tilde{y} = \frac{A_m}{2} \tag{2.2.26}$$

The first and second terms cannot be simplified further unless the eigenfunctions or 'modes' employed are eigenfunctions for the variable property wing. Hence, a_n is not as simply related to A_n as in the constant property wing example. (2.2.26) represents a system of equations for the a_n. In matrix notation

$$[C_{mn}]\{a_n\} = \{A_m\}\tfrac{1}{2} \tag{2.2.27}$$

where

$$C_{mn} \equiv \int_0^1 \frac{d}{d\tilde{y}}\left(\gamma\frac{d\alpha_n}{d\tilde{y}}\right)\alpha_m\, d\tilde{y} + \lambda^2\int_0^1 \beta\alpha_n\alpha_m\, d\tilde{y}$$

By truncating the series to a finite number of terms, we may formally solve for the a_n,

$$\{a_n\} = \tfrac{1}{2}[C_{mn}]^{-1}\{A_m\} \tag{2.2.28}$$

Illustrated by Solution to
 # 3 on pp 423-28

21

2 Static aeroelasticity

The divergence condition is simply that the determinant of C_{mn} vanish (and hence $a_n \to \infty$)

$$|C_{mn}| = 0 \qquad (2.2.29)$$

which is a polynomial in λ^2. It should be emphasized that for an 'exact' solution, (2.2.27), (2.2.28) etc., are infinite systems of equations (in an infinite number of unknowns). In practice, some large but finite number of equations is used to obtain an accurate approximation. By systematically increasing the terms in the series, the convergence of the method can be assessed. This procedure is usually referred to as Galerkin's method or as a 'modal' method.* The modes, α_n, used are called 'primitive modes' to distinguish them from eigenfunctions, i.e., they are 'primitive functions' for a variable property wing even though eigenfunctions for a constant property wing.

2.3 Rolling of a straight wing

We shall now consider a more complex physical and mathematical variation on our earlier static aeroelastic lifting surface (wing) studies. For variety, we treat a new physical situation, the rolling of a wing (rotation about the root axis). Nevertheless, we shall meet again our old friends, 'divergence' and 'control surface effectiveness' or 'reversal'.

The present analysis differs from the previous one as follows:

(a) integral equation formulation vs. differential equation formulation

(b) aerodynamic induction effects vs. 'strip' theory

(c) 'lumped element' method of solution vs. modal (or eigenfunction) solution.

The geometry of the problem is shown in Figure 2.9.

Integral equation of equilibrium

The integral equation of equilibrium is

$$\alpha(y) = \int_0^l C^{\alpha\alpha}(y, \eta) M_y(\eta) \, d\eta \qquad (2.3.1)\dagger$$

compare w/(2.3.20)

---

* Duncan [7].

† For simplicity, $\alpha_0 \equiv 0$ in what follows.

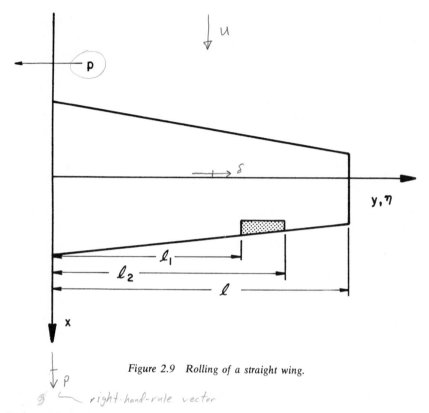

Figure 2.9   *Rolling of a straight wing.*

Before deriving the above equation, let us first consider the physical interpretation of $C^{\alpha\alpha}$:

Apply a unit point moment at some point, say $y = \gamma$, i.e.,

$$M_y(\eta) = \delta(\eta - \gamma).$$

Then (2.3.1) becomes

$$\alpha(y) = \int_0^l C^{\alpha\alpha}(y, \eta)\, \delta(\eta - \gamma)\, d\eta$$

$$= C^{\alpha\alpha}(y, \gamma) \tag{2.3.2}$$

Thus, $C^{\alpha\alpha}(y, \gamma)$ is the twist at y due to a unit moment at γ, or alternatively, $C^{\alpha\alpha}(y, \eta)$ is the twist at y due to a unit moment at η. $C^{\alpha\alpha}$ is called a *structural influence function.*

23

Also note that (2.3.1) states that to obtain the total twist, one multiplies the actual distributed torque, M_y, by $C^{\alpha\alpha}$ and sums (integrates) over the span. This is physically plausible.

$C^{\alpha\alpha}$ plays a central role in the integral equation formulation.* The physical interpretation of $C^{\alpha\alpha}$ suggests a convenient means of measuring $C^{\alpha\alpha}$ in a laboratory experiment. By successively placing unit couples at various locations along the wing and measuring the twists of all such stations for each loading position we can determine $C^{\alpha\alpha}$. This capability for measuring $C^{\alpha\alpha}$ gives the integral equation a preferred place in aeroelastic analysis where $C^{\alpha\alpha}$ and/or GJ are not always easily determinable from purely theoretical considerations.

Derivation of equation of equilibrium. Now consider a derivation of (2.3.1) taking as our starting point the differential equation of equilibrium. We have, you may recall,

$$\frac{d}{dy}\left(GJ\frac{d\alpha}{dy}\right) = -M_y \tag{2.3.3}$$

with

$$\alpha(0) = 0 \quad \text{and} \quad \frac{d\alpha}{dy}(l) = 0 \tag{2.3.4}$$

as boundary conditions.

As a special case of (2.3.3) and (2.3.4) we have for a unit torque applied at $y = \eta$,

$$\frac{d}{dy}GJ\frac{dC^{\alpha\alpha}}{dy} = -\delta(y-\eta) \tag{2.3.5}$$

with

$$C^{\alpha\alpha}(0, \eta) = 0 \quad \text{and} \quad \frac{dC^{\alpha\alpha}}{dy}(l, \eta) = 0 \tag{2.3.6}$$

Multiply (2.3.5) by $\alpha(y)$ and integrate over the span,

$$\int_0^l \alpha(y)\frac{d}{dy}\left(GJ\frac{dC^{\alpha\alpha}}{dy}\right)dy = -\int_0^l \delta(y-\eta)\alpha(y)\,dy$$

$$= -\alpha(\eta) \tag{2.3.7}$$

---

*For additional discussion, see the following selected references: Hildebrand [6] pp. 388–394 and BAH, pp. 39–44.

Integrate LHS of (2.3.7) by parts,

$$\alpha GJ \frac{dC^{\alpha\alpha}}{dy}\Bigg|_0^l - GJ\frac{d\alpha}{dy}C^{\alpha\alpha}\Bigg|_0^l + \int_0^l C^{\alpha\alpha}\frac{d}{dy}\left(GJ\frac{d\alpha}{dy}\right)dy = -\alpha(\eta) \quad (2.3.8)$$

Using boundary conditions, (2.3.4) and (2.3.6), the first two terms of LHS of (2.3.8) vanish. Using (2.3.3) the integral term may be simplified and we obtain,

$$\alpha(\eta) = \int_0^l C^{\alpha\alpha}(y, \eta)M_y(y)\,dy \quad (2.3.9)$$

Interchanging y and η,

$$\alpha(y) = \int_0^l C^{\alpha\alpha}(\eta, y)M_y(\eta)\,d\eta \quad (2.3.10)$$

(2.3.10) is identical to (1), *if*

$$C^{\alpha\alpha}(\eta, y) = C^{\alpha\alpha}(y, \eta) \quad (2.3.11)$$

We shall prove (2.3.11) subsequently.

· *Calculation of $C^{\alpha\alpha}$.* We shall calculate $C^{\alpha\alpha}$ from (2.3.5) using (2.3.6). Integrating (2.3.5) with respect to y from 0 to y_1,

$$GJ(y_1)\frac{dC^{\alpha\alpha}}{dy}(y_1, \eta) - GJ(0)\frac{dC^{\alpha\alpha}}{dy}(0, \eta) \quad = \quad \int_0^{y_1} -\delta(y-\eta)\,dy$$

$$\begin{aligned} &= -1 \quad \text{if} \quad y_1 > \eta \\ &= 0 \quad \text{if} \quad y_1 < \eta \end{aligned} \equiv S(y_1, \eta) \Big\} \text{ by def. of } DD \text{ fn} \quad (2.3.12)$$

Sketch of function $S(y_1, \eta)$.

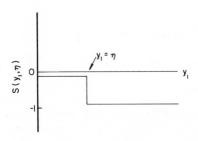

Dividing (2.3.12) by $GJ(y_1)$ and integrating with respect to y_1 from 0 to y_2,

25

$$C^{\alpha\alpha}(y_2, \eta) - C^{\alpha\alpha}(0, \eta) - GJ(0)\frac{\mathrm{d}C^{\alpha\alpha}}{\mathrm{d}y}(0, \eta)\int_0^{y_2}\frac{1}{GJ}\mathrm{d}y_1$$

$$= \int_0^{y_2}\frac{S(y_1, \eta)}{GJ(y_1)}\mathrm{d}y_1$$

$$= -\int_\eta^{y_2}\frac{1}{GJ(y_1)}\mathrm{d}y_1 \quad \text{for} \quad y_2 > \eta$$

$$= 0 \qquad\qquad \text{for} \quad y_2 < \eta \qquad (2.3.13)$$

From boundary conditions, (2.3.6),

(a) $C^{\alpha\alpha}(0, \eta) = 0$

(b) $\dfrac{\mathrm{d}C^{\alpha\alpha}}{\mathrm{d}y}(l, \eta) = 0$

These may be used to evaluate the unknown terms in (2.3.12) and (2.3.13). Evaluating (2.3.12) at $y_1 = l$

(c) $GJ(l)\dfrac{\mathrm{d}C^{\alpha\alpha}}{\mathrm{d}y}^{\!\!\!\!0}(l, \eta) - GJ\dfrac{\mathrm{d}C^{\alpha\alpha}}{\mathrm{d}y}(0, \eta) = -1$

Using (a) and (c), (2.3.13) may be written,

$$C^{\alpha\alpha}(y_2, \eta) = \int_0^{y_2}\frac{1}{GJ}\mathrm{d}y_1 - \int_\eta^{y_2}\frac{1}{GJ}\mathrm{d}y_1$$

$$= \int_0^{\eta}\frac{1}{GJ}\mathrm{d}y_1 \quad \text{for} \quad y_2 > \eta$$

$$= \int_0^{y_2}\frac{1}{GJ}\mathrm{d}y_1 \quad \text{for} \quad y_2 < \eta$$

One may drop the dummy subscript on y_2, of course. Thus

$$C^{\alpha\alpha}(y, \eta) = \int_0^{y}\frac{1}{GJ}\mathrm{d}y_1 \quad \text{for} \quad y < \eta$$

$$= \int_0^{\eta}\frac{1}{GJ}\mathrm{d}y_1 \quad \text{for} \quad y > \eta \qquad (2.3.14)$$

Note from the above result we may conclude by interchanging y and η that

$$C^{\alpha\alpha}(y, \eta) = C^{\alpha\alpha}(\eta, y)$$

This is a particular example of a more general principle known as Maxwell's Reciprocity Theorem* which says that all structural influence

---

* Bisplinghoff, Mar, and Pian [8], p. 247.

functions for linear elastic bodies are symmetric in their arguments. In the case of $C^{\alpha\alpha}$ these are y and η, of course.

Aerodynamic forces (including spanwise induction)

First, let us identify the aerodynamic angle of attack; i.e., the angle between the airfoil chord and relative airflow. See Figure 2.10. Hence, the total angle of attack due to twisting and rolling is

$$\alpha_{\text{Total}} = \alpha(y) - \frac{py}{U}$$

The control surface will be assumed rigid and its rotation is given by

$$\delta(y) = \delta_R \quad \text{for} \quad l_1 < y < l_2$$
$$= 0 \quad \text{otherwise}$$

From aerodynamic theory or experiment

$$C_L \equiv \frac{L}{qc} = \int_0^l A^{L\alpha}(y, \eta)\alpha_T(\eta)\frac{d\eta}{l} + \int_0^l A^{L\delta}(y, \eta)\,\delta(\eta)\frac{d\eta}{l} \tag{2.3.15}$$

Here $A^{L\alpha}$, $A^{L\delta}$ are aerodynamic influence functions; as written, they are nondimensional. Thus, $A^{L\alpha}$ is nondimensional lift at y due to unit angle of attack at η. Substituting for α_T and δ, (2.3.15) becomes,

$$C_L = \int_0^l A^{L\alpha}\,\alpha\,\frac{d\eta}{l} - \frac{pl}{U}\int_0^l A^{L\alpha}\,\frac{\eta}{l}\frac{d\eta}{l} + \delta_R\int_{l_1}^{l_2} A^{L\delta}\,\frac{d\eta}{l}$$

$$C_L = \int_0^l A^{L\alpha}\alpha\,\frac{d\eta}{l} + \frac{pl}{U}\frac{\partial C_L}{\partial\left(\dfrac{pl}{U}\right)} + \delta_R\frac{\partial C_L}{\partial\delta_R} \tag{2.3.16}$$

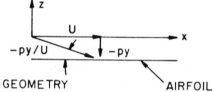

GEOMETRY   AIRFOIL

Figure 2.10

27

where

$$\frac{\partial C_L}{\partial \left(\frac{pl}{U}\right)}(y) \equiv -\int_0^l A^{L\alpha} \frac{\eta}{l} \frac{d\eta}{l}$$

and

$$\frac{\partial C_L}{\partial \delta_R}(y) \equiv \int_{l_1}^{l_2} A^{L\delta} \frac{d\eta}{l}$$

Physical Interpretation of $A^{L\alpha}$ and $A^{L\delta}$: $A^{L\alpha}$ is the lift coefficient at y due to unit angle of attack at η. $A^{L\delta}$ is the lift coefficient at y due to unit rotation of control surface at η.

Physical Interpretation of $\partial C_L/\partial(pl/U)$ and $\partial C_L/\partial\delta_R$: $\partial C_L/\partial(pl/U)$ is the lift coefficient at y due to unit rolling velocity, pl/U. $\partial C_L/\partial\delta_R$ is the lift coefficient at y due to unit control surface rotation, δ_R.

As usual

$$C_{MAC} \equiv \frac{M_{AC}}{qc^2} = \frac{\partial C_{MAC}}{\partial \delta_R} \delta_R \tag{2.3.17}$$

is the aerodynamic coefficient moment (about a.c.) at y due to control surface rotation. Note

$$\partial C_{MAC}/\partial \alpha_T \equiv 0$$

by definition of the aerodynamic center. Finally the total moment loading about the elastic axis is

$$M_y = M_{AC} + Le$$
$$= qc[C_{MAC}c + C_L e] \tag{2.3.18}$$

Using (2.3.16) and (2.3.17), the above becomes

$$M_y = qc\left[c\frac{\partial C_{MAC}}{\partial \delta_R}\delta_R + e\left\{\int_0^l A^{L\alpha}\alpha\frac{d\eta}{l} + \frac{\partial C_L}{\partial\left(\frac{pl}{U}\right)}\left(\frac{pl}{U}\right) + \frac{\partial C_L}{\partial\delta_R}\delta_R\right\}\right]$$

$$\tag{2.3.19}$$

Note that $A^{L\alpha}$, $A^{L\delta}$ are more difficult to measure than their structural counterpart, $C^{\alpha\alpha}$. One requires an experimental model to which one can apply unit angles of attack at various discrete points along the span of the wing. This requires a rather sophisticated model and also introduces experimental difficulties in establishing and maintaining a smooth flow over the airfoil. Conversely

$$\frac{\partial C_L}{\partial \frac{pl}{U}}, \frac{\partial C_L}{\partial \delta_R} \quad \text{and} \quad \frac{\partial C_{MAC}}{\partial \delta_R}$$

are relatively easy to measure since they only require a rolling or control surface rotation of *a rigid wing with the same geometry* as the flexible airfoil of interest.

Aeroelastic equations of equilibrium and lumped element solution method

The key relations are (2.3.1) and (2.3.19). The former describes the twist due to an aerodynamic moment load, the latter the aerodynamic moment due to twist as well as rolling and control surface rotation.

By substituting (2.3.19) in (2.3.1), one could obtain a single equation for α. However, this equation is not easily solved analytically except for some simple cases, which are more readily handled by the differential equation approach. Hence, we seek an approximate solution technique. Perhaps the most obvious and convenient method is to approximate the integrals in (2.3.1) and (2.3.19) by sums, i.e., the wing is broken into various spanwise segments or 'lumped elements'. For example, (2.3.1) would be approximated as:

$$\alpha(y_i) \cong \sum_{j=1}^{N} C^{\alpha\alpha}(y_i, \eta_j) M_y(\eta_j) \Delta\eta \qquad i = 1, \ldots, N \qquad (2.3.20)$$

where $\Delta\eta$ is the segment width and N the total number of segments. Similarly, (2.3.19) may be written:

$$M_y(y_i) \cong qc \left\{ \left[c \frac{\partial C_{MAC}}{\partial \delta_R} \delta_R + e \frac{\partial C_L}{\partial \left(\frac{pl}{U}\right)} \frac{pl}{U} + e \frac{\partial C_L}{\partial \delta_R} \delta_R \right] \right.$$

$$\left. + e \sum_{j=1}^{N} A^{L\alpha}(y_i, \eta_j) \alpha(\eta_j) \frac{\Delta\eta}{l} \right\} \qquad i = 1, \ldots, N \qquad (2.3.21)$$

To further manipulate (2.3.20) and (2.3.21), it is convenient to use matrix notation. That is,

$$\{\alpha\} = \Delta\eta [C^{\alpha\alpha}]\{M_y\} \qquad (2.3.20)$$

29

and

$$\{M_y\} = q\left[\begin{smallmatrix} \diagdown c^2 \diagdown \end{smallmatrix}\right]\left\{\frac{\partial C_{MAC}}{\partial \delta_R}\right\}\delta_R$$

$$+ q\left[\begin{smallmatrix} \diagdown ce \diagdown \end{smallmatrix}\right]\left\{\frac{\partial C_L}{\partial\left(\frac{pl}{U}\right)}\right\}\frac{pl}{U}$$

$$+ q\left[\begin{smallmatrix} \diagdown ce \diagdown \end{smallmatrix}\right]\left\{\frac{\partial C_L}{\partial \delta_R}\right\}\delta_R$$

$$+ q\left[\begin{smallmatrix} \diagdown ce \diagdown \end{smallmatrix}\right][A^{L\alpha}]\{\alpha\}\frac{\Delta\eta}{l} \qquad (2.3.21)$$

All full matrices are of order $N \times N$ and row or column matrices of order N. Substituting (2.3.21) into (2.3.20), and rearranging terms gives,

$$\left[\left[\diagdown 1 \diagdown\right] - q\frac{(\Delta\eta)^2}{l}[E][A^{L\alpha}]\right]\{\alpha\} = \{f\} \qquad (2.3.22)$$

where the following definitions apply

$$\{f\} \equiv q[E]\left\{\left\{\frac{\partial C_L}{\partial \delta_R}\right\}\delta_R + \left\{\frac{\partial C_L}{\partial\left(\frac{pl}{U}\right)}\right\}\frac{pl}{U}\right\}\Delta\eta$$

$$+ q[F]\left\{\frac{\partial C_{MAC}}{\partial \delta_R}\right\}\delta_R\,\Delta\eta$$

$$[E] \equiv [C^{\alpha\alpha}]\left[\begin{smallmatrix} \diagdown ce \diagdown \end{smallmatrix}\right]$$

$$[F] \equiv [C^{\alpha\alpha}]\left[\begin{smallmatrix} \diagdown c^2 \diagdown \end{smallmatrix}\right]$$

Further defining

$$[D] \equiv \left[\diagdown 1 \diagdown\right] - q\frac{(\Delta\eta)^2}{l}[E][A^{L\alpha}]$$

we may formally solve (2.2.22) as

$$\{\alpha\} = [D]^{-1}\{f\} \qquad (2.3.23)$$

Now let us interpret this solution.

Divergence

Recall that the inverse does not exist if

$$|D| = 0 \qquad (2.3.24)$$

and hence,

$$\{\alpha\} \to \{\infty\}.$$

(2.3.24) gives rise to an eigenvalue problem for the divergence dynamic pressure, q_D. Note (2.3.24) is a polynomial in q.

The lowest positive root (eigenvalue) of (2.3.24) gives the q of physical interest, i.e., $q_{\text{Divergence}}$. Rather than seeking the roots of the polynomial we might more simply plot $|D|$ versus q to determine the values of dynamic pressure for which the determinant is zero. A schematic of such results for various choices of N is shown below in Figure 2.11. From the above results we may plot q_D (the lowest positive q for which $|D| = 0$) vs. N as shown below in Figure 2.12. The 'exact' value of q_D is obtained as $N \to \infty$. Usually reasonably accurate results can be obtained for small values of N, say 10 or so. The divergence speed calculated above does not depend upon the rolling of the wing, i.e., p is considered prescribed, e.g., $p = 0$.

Reversal and rolling effectiveness

In the above we have taken pl/U as known; however, in reality it is a function of δ_R and the problem parameters through the requirement that the wing be in static rolling equilibrium, i.e., it is an additional degree of

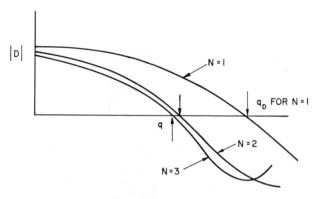

Figure 2.11 Characteristic determinant vs dynamic pressure.

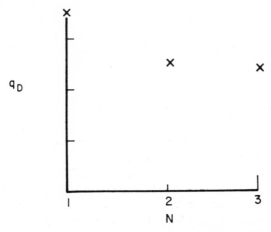

Figure 2.12   *Convergence of divergence dynamic pressure with modal number.*

freedom. For rolling equilibrium at a steady roll rate, p, the rolling moment about the x-axis is zero.

$$M_{\text{Rolling}} \equiv 2 \int_0^l Ly \, dy = 0 \qquad (2.3.25)$$

Approximating (2.3.25),

$$\sum_i L_i y_i \, \Delta y = 0 \qquad (2.3.26)$$

or, in matrix notation,

$$2 \lfloor y \rfloor \{L\} \, \Delta y = 0 \qquad (2.3.27)$$

or

$$2q \lfloor cy \rfloor \{C_L\} \, \Delta y = 0$$

From (2.3.16), using the 'lumped element' approximation and matrix notation,

$$\{C_L\} = \frac{\Delta \eta}{l} [A^{L\alpha}]\{\alpha\} + \left\{\frac{\partial C_L}{\partial \delta_R}\right\} \delta_R + \left\{\frac{\partial C_L}{\partial \left(\frac{pl}{U}\right)}\right\} \frac{pl}{U} \qquad (2.3.16)$$

Substitution of (2.3.16) into (2.3.27) gives

$$\lfloor cy \rfloor \left\{ \frac{\Delta \eta}{l} [A^{L\alpha}]\{\alpha\} + \left\{\frac{\partial C_L}{\partial \delta_R}\right\} \delta_R + \left\{\frac{\partial C_L}{\partial \left(\frac{pl}{U}\right)}\right\} \frac{pl}{U} \right\} = 0 \qquad (2.3.28)$$

Note that (2.3.28) is a single algebraic equation. (2.3.28) plus (2.3.20) and (2.3.21) are $2N+1$ linear algebraic equations in the $N(\alpha)$ plus $N(M_y)$ plus $1(p)$ unknowns. As before $\{M_y\}$ is normally eliminated using (2.3.21) in (2.3.20) to obtain N, equation (2.3.22), plus 1, equation (2.3.28), equations in $N(\alpha)$ plus $1(p)$ unknowns. In either case the divergence condition may be determined by setting the determinant of coefficients to zero and determining the smallest positive eigenvalue, $q = q_D$.

For $q < q_D$, pl/U (and $\{\alpha\}$) may be determined from (2.3.22) and (2.3.28). Since our mathematical model is linear

$$pl/U \sim \delta_R$$

and hence a convenient plot of the results is as shown in Figure 2.13. As

$$q \to q_D, \frac{pl}{U} \, (\text{and} \, \{\alpha\}) \to \infty.$$

Another qualitatively different type of result may sometimes occur. See Figure 2.14. If

$$\frac{pl}{U/\delta_R} \to 0 \quad \text{for} \quad q \to q_R < q_D$$

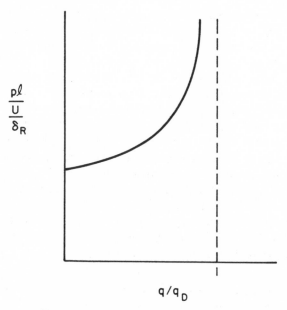

Figure 2.13 *Roll rate vs dynamic pressure.*

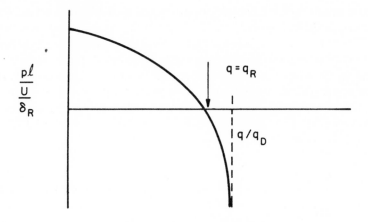

Figure 2.14   Roll rate vs dynamic pressure.

then 'rolling reversal' is said to have occurred and the corresponding $q = q_R$ is called the 'reversal dynamic pressure'. The basic phenomenon is the same as that encountered previously as 'control surface reversal'. Figures 2.13 and 2.14 should be compared to Figures 2.5a,b.

It is worth emphasizing that the divergence condition obtained above by permitting p to be determined by (static) rolling equilibrium will be different from that obtained previously by assuming $p = 0$. The latter physically corresponds to an aircraft constrained not to roll, as might be the case for some wind tunnel models. The former corresponds to no restraint with respect to roll, either structural or pilot induced.*

The above analysis has introduced the simple yet powerful idea of structural and aerodynamic influence functions. While the utility of the concept has been illustrated for a one-dimensional aeroelastic model, not the least advantage of such an approach is the conceptual ease with which the basic notion can be extended to two-dimensional models, e.g., plate-like structures, or even three-dimensional ones (though the latter is rarely needed for aeroelastic problems).

In a subsequent section we briefly outline the generalization to two-dimensional models. Later this subject will be considered in more depth in the context of dynamic aeroelasticity.

---

* This distinction between the two ways in which the aircraft may be restrained has recently received new emphasis in the context of the oblique wing concept. Weisshaar and Ashley [9].

Integral equation eigenvalue problem and the experimental determination of influence functions

For the special case of a constant section wing with 'strip theory' aerodynamics one may formulate a standard integral equation eigenvalue problem for the determination of divergence. In itself this problem is of little interest. However, it does lead to some interesting results with respect to the determination of the structural and aerodynamic influence functions by experimental means.

For such a wing,

$$M_y = Le + M_{AC}$$

$$= eqc\frac{\partial C_L}{\partial \alpha}\alpha + \cdots$$

where the omitted terms are independent of twist and may therefore be ignored for the divergence (eigenvalue) problem. Also the coefficients of α may be taken as constants for a constant section wing. Substituting the above expression into the integral equation of structural equilibrium we have

$$\alpha(y) = eqc\frac{\partial C_L}{\partial \alpha}\int_0^l C^{\alpha\alpha}(y, \eta)\alpha(\eta)\,\mathrm{d}\eta$$

This is an eigenvalue problem in integral form where the eigenvalue is

$$\lambda \equiv eqc\frac{\partial C_L}{\partial \alpha}$$

One may solve this problem for the corresponding eigenvalues and eigenfunctions which satisfy the equation

$$\alpha_n(y) = \lambda_n \int_0^l C^{\alpha\alpha}(y, \eta)\alpha_n(\eta)\,\mathrm{d}\eta$$

Incidentally, the restriction to a constant section wing was unnecessary and with a moderate amount of effort one could even use a more sophisticated aerodynamic model. Such complications are not warranted here.

These eigenfunctions or similar functions may be usefully employed to determine by experimental means the structural, $C^{\alpha\alpha}$, and aerodynamic, $A^{L\alpha}$, influence functions. The former is not as attractive as the use of point unit structural loads as we shall see; however, the

35

procedure outlined below for the determination of $A^{L\alpha}$ probably deserves more attention than it has previously received.

Assume the structural influence function can be expanded in terms of the eigenfunctions

$$C^{\alpha\alpha}(y, \eta) = \sum_n C_n(y)\alpha_n(\eta) \tag{2.3.29}$$

where the C_n are to be determined. Also recall that

$$\alpha_n(y) = \lambda_n \int_0^l C^{\alpha\alpha}(y, \eta)\alpha_n(\eta)\, d\eta \tag{2.3.30}$$

and the α_n are the eigenfunctions and λ_n the eigenvalues of $C^{\alpha\alpha}$ satisfying (2.3.30) and an orthogonality condition

$$\int \alpha_n\alpha_m\, dy = 0 \quad \text{for} \quad m \neq n$$

Then multiply (2.3.29) by $\alpha_m(\eta)$ and integrate over the span of the wing; the result is

$$C_m(y) = \frac{\displaystyle\int_0^l C^{\alpha\alpha}(y, \eta)\alpha_m(\eta)\, d\eta}{\displaystyle\int_0^l \alpha_m^2(\eta)\, d\eta}$$

from (2.3.30)

$$= \frac{\alpha_m(y)}{\lambda_m \displaystyle\int_0^l \alpha_m^2(\eta)\, d\eta} \tag{2.3.31}$$

Hence (2.3.31) in (2.3.29) gives

$$C^{\alpha\alpha}(y, \eta) = \sum_n \frac{\alpha_n(y)\alpha_n(\eta)}{\lambda_n \displaystyle\int_0^l \alpha_n^2(\eta)\, d\eta} \tag{2.3.32}$$

Thus if the eigenfunctions are known then the Green's function is readily determined from (2.3.32). Normally this holds no special advantage since the determination of the α_n, theoretically or experimentally, is at least as difficult as determining the Green's function, $C^{\alpha\alpha}$, directly. Indeed as discussed previously if we apply unit moments at various points along the span the resulting twist distribution is a direct measure of $C^{\alpha\alpha}$. A somewhat less direct way of measuring $C^{\alpha\alpha}$ is also possible which makes use of the expansion of the Green's (influence) function. Again using (2.3.29)

$$C^{\alpha\alpha}(y, \eta) = \sum_n C_n \alpha_n(\eta) \qquad C_n(y)$$

(2.3.29)

and assuming the α_n are orthogonal (although not necessarily eigenfunctions of the problem at hand) we have

$$C_n(y) = \frac{\displaystyle\int_0^l C^{\alpha\alpha}(y, \eta)\alpha_n(\eta)\,\mathrm{d}\eta}{\displaystyle\int_0^l \alpha_n^2(\eta)\,\mathrm{d}\eta}$$

(2.3.33)

Now we have the relation between twist and moment

$$\alpha(y) = \int_0^l C^{\alpha\alpha}(y, \eta)M_y(\eta)\,\mathrm{d}\eta$$

(2.3.34)

Clearly if we use a moment distribution

$$M_y(\eta) = \alpha_n(\eta)$$

the resulting twist distribution will be (from 2.3.33)

$$\alpha(y) = C_n(y)\int_0^l \alpha_n^2(\eta)\,\mathrm{d}\eta$$

(2.3.35)

Hence we may determine the expansion of the Green's function by successively applying moment distributions in the form of the expansion functions and measuring the resultant twist distribution. For the structural influence function this offers no advantage in practice since it is easier to apply point moments rather than moment distributions.

However, for the aerodynamic Green's functions the situation is different. In the latter case we are applying a certain twist to the wing and measuring the resulting aerodynamic moment distribution. It is generally desirable to maintain a smooth (if twisted) aerodynamic surface to avoid complications of flow separation and roughness and hence the application of a point twist distribution is less desirable than a distributed one. We quickly summarize the key relations for the aerodynamic influence function. Assume

$$A^{L\alpha}(y, \eta) = \sum_n A_n^{L\alpha}(y)\alpha_n(\eta)$$

(2.3.36)

37

2 Static aeroelasticity

We know that

$$C_L(y) = \int_0^l A^{L\alpha}(y, \eta)\alpha(\eta)\, d\eta \qquad (2.3.37)$$

For orthogonal functions, α_n, we determine from (2.3.36) that

$$A_n^{L\alpha}(y) = \frac{\displaystyle\int_0^l A^{L\alpha}(y, \eta)\alpha_n(\eta)\, d\eta}{\displaystyle\int_0^l \alpha_n^2(\eta)\, d\eta} \qquad (2.3.38)$$

Applying the twist distribution $\alpha = \alpha_n(\eta)$ to the wing, we see from (2.3.37) and (2.3.38) that the resulting lift distribution is

$$C_L(y) = A_n^{L\alpha}(y) \int_0^l \alpha_n^2(\eta)\, d\eta \qquad (2.3.39)$$

Hence, by measuring the lift distributions on 'warped wings' with twist distributions $\alpha_n(\eta)$ we may completely determine the aerodynamic influence function in terms of its expansion (2.3.36). This technique or a similar one has been used occasionally,* but not as frequently as one might expect, possibly because of the cost and expense of testing the number of wings sufficient to establish the convergence of the series. In this regard, if one uses the α_n for a Galerkin or modal expansion solution for the complete aeroelastic problem one can show that the number of C_n, $A_n^{L\alpha}$ required is equal to the number of modes, α_n, employed in the twist expansion.

2.4 Two dimensional aeroelastic model of lifting surfaces

We consider in turn, structural modeling, aerodynamic modeling, the combining of the two into an aeroelastic model, and its solution.

Two dimensional structures—integral representation

The two dimensional or plate analog to the one-dimensional or beam-rod model is

$$w(x, y) = \iint C^{wp}(x, y; \xi, \eta)p(\xi, \eta)\, d\xi\, d\eta \qquad (2.4.1)$$

---

* Covert [10].

where w vertical deflection at a point, x, y, on plate

$\quad\ p$ force/area (pressure) at point ξ, η on plate

$\quad C^{wp}$ deflection at x, y due to unit pressure at ξ, η

Note that w and p are taken as positive in the same direction. For the special case where

$$w(x, y) = h(y) + x\alpha(y) \tag{2.4.2}$$

and

$$C^{wp}(x, y; \xi, \eta) = C^{hF}(y, \eta) + xC^{\alpha F}(y, \eta) + \xi C^{hM}(y, \eta) + x\xi C^{\alpha M}(y, \eta) \tag{2.4.3}$$

C^{hM} *is the deflection of* y *axis at* y *due to a unit torque at* η

with the definitions,

C^{hF} is the deflection of y axis at y due to unit force F *@* η

$C^{\alpha F}$ is the twist about y axis at y due to unit force F ~~etc.~~, *@* η

$C^{\alpha M}$ *" " " " " " " " torque @* η

we may retrieve our beam-rod result. Note that (2.4.2) and (2.4.3) may be thought of as polynomial (Taylor Series) expansions of deflections.

Substituting (2.4.2), (2.4.3) into (2.4.1), we have

$$
\begin{aligned}
h(y) + x\alpha(y) = &\left[\iint C^{hF}\left(\int p(\xi, \eta)\, d\xi\right) d\eta \right. \\
&+ \left. \int C^{hM}\left(\int \xi p(\xi, \eta)\, d\xi\right) d\eta\right] \\
&+ x\left[\iint C^{\alpha F}\left(\int p(\xi, \eta)\, d\xi\right) d\eta \right. \\
&+ \left. \int C^{\alpha M}\left(\int \xi p(\xi, \eta)\, d\xi\right) d\eta\right]
\end{aligned}
\tag{2.4.4}
$$

If y, η lie along an elastic axis, then $C^{hM} = C^{\alpha F} = 0$. Equating coefficients of like powers of x, we obtain

$$h(y) = \int C^{hF}(y, \eta) F(\eta)\, d\eta \tag{2.4.5a}$$

$$\alpha(y) = \int C^{\alpha M}(y, \eta) M(\eta)\, d\eta \tag{2.4.5b}$$

same as (2.3.1)

where

$$F \equiv \int p\, d\xi, \qquad M \equiv \int p\xi\, d\xi$$

39

✳ sketch h

(2.4.5b) is our previous result. Since for static aeroelastic problems, M is only a function of α (and not of h), (2.4.5b) may be solved independently of (2.4.5a). Subsequently (2.4.5b) may be solved to determine h if desired. (2.4.5a) has no effect on divergence or control surface reversal, of course, and hence we were justified in neglecting it in our previous discussion.

Two dimensional aerodynamic surfaces—integral representation

In a similar manner (for simplicity we only include deformation dependent aerodynamic forces to illustrate the method),

$$\frac{p(x, y)}{q} = \iint A^{p w_x}(x, y; \xi, \eta) \frac{\partial w}{\partial \xi}(\xi, \eta) \frac{\mathrm{d}\xi \, \mathrm{d}\eta}{c_r \, l} \qquad (2.4.6)$$

where $A^{p w_x}$ nondimensional aerodynamic pressure at x, y due to unit $\partial w / \partial \xi$ at point ξ, η

c_r reference chord, l reference span

For the special case

$$w = h + x\alpha$$

and, hence,

$$\frac{\partial w}{\partial x} = \alpha$$

we may retrieve our beam-rod aerodynamic result.

For example, we may compute the lift as

$$L \equiv \int p \, \mathrm{d}x = q c_r \int_0^l A^{L\alpha}(y, \eta) \alpha(\eta) \frac{\mathrm{d}\eta}{l} \qquad (2.4.7)$$

where

$$A^{L\alpha} \equiv \iint A^{p w_x}(x, y; \xi, \eta) \frac{\mathrm{d}\xi \, \mathrm{d}x}{c_r \, c_r}$$

Solution by matrix-lumped element approach

Approximating the integrals by sums and using matrix notation, (2.4.1) becomes

$$\{w\} = \Delta \xi \, \Delta \eta [C^{wp}] \{p\} \qquad (2.4.8)$$

and (2.4.6) becomes

$$\{p\} = q \frac{\Delta\xi}{c_r} \frac{\Delta\eta}{l} [A^{pw_x}] \left(\frac{\partial w}{\partial \xi}\right) \tag{2.4.9}$$

Now

$$\left(\frac{\partial w}{\partial \xi}\right)_i \cong \frac{w_{i+1} - w_{i-1}}{2\,\Delta\xi}$$

is a difference representation of the surface slope. Hence

$$\left(\frac{\partial w}{\partial \xi}\right) = \frac{1}{2\,\Delta\xi} [\overline{W}]\{w\} = \frac{1}{2\,\Delta\xi} \begin{bmatrix} [W][0][0][0] \\ [W][0][0] \\ [W][0] \\ [W] \end{bmatrix} \{w\}^* \tag{2.4.10}$$

is the result shown for *four* spanwise locations, where

$$[W] = \begin{bmatrix} 0 & 1 & 0 & 0 & \cdot \\ -1 & 0 & 1 & 0 & \cdot \\ 0 & -1 & 0 & 1 & \cdot \\ & & \cdot & \cdot & \cdot \\ \cdots & 0 & 0 & -1 & 0 \end{bmatrix}$$

$$\underbrace{\qquad\qquad}_{\text{number of chordwise locations}}$$

is a numerical weighting matrix. From (2.4.8), (2.4.9), (2.4.10), we obtain an equation for w,

$$[D]\{w\} \equiv \left[\left[\diagdown 1 \right] - q \frac{(\Delta\xi)^2}{c_r} \frac{(\Delta\eta)^2}{l} \frac{1}{2\,\Delta\xi} [C^{wp}][A^{pw_x}][W] \right] \{w\} = \{0\} \tag{2.4.11}$$

For divergence

$$|D| = 0$$

which permits the determination of q_D.

---

* For definiteness consider a rectangular wing divided up into small (rectangular) finite difference boxes. The weighting matrix $[(W)]$ is for a given spanwise location and various chordwise boxes. The elements in the matrices, $\{\partial w/\partial\xi\}$ and $\{w\}$, are ordered according to fixed spanwise location and then over all chordwise locations. This numerical scheme is only illustrative and not necessarily that which one might choose to use in practice.

2.5   Nonairfoil physical problems

Fluid flow through a flexible pipe

Another static aeroelastic problem exhibiting divergence is encountered in long slender pipes with a flowing fluid.* See Figure 2.15. We shall assume the fluid is incompressible and has no significant variation across the cross-section of the pipe. Thus, the aerodynamic loading per unit length along the pipe is (invoking the concept of an equivalent fluid added mass moving with the pipe and including the effects of convection velocity),† U,

$$-L = \rho A \left[\frac{\partial}{\partial t} + U \frac{\partial}{\partial x} \right]^2 w = \rho A \left[\frac{\partial^2 w}{\partial t^2} + 2 U \frac{\partial^2 w}{\partial x \, \partial t} + U^2 \frac{\partial^2 w}{\partial x^2} \right] \qquad (2.5.1)$$

where $A \equiv \pi R^2$, open area for circular pipe
  ρ, U fluid density, axial velocity
  w transverse deflection of the pipe
  x axial coordinate
  t time

The equation for the beam-like slender pipe is

$$EI \frac{\partial^4 w}{\partial x^4} + m_p \frac{\partial^2 w}{\partial t^2} = L \qquad (2.5.2)$$

where $m_p \equiv \rho_p \, 2\pi R h$ for a thin hollow circular pipe of thickness h, mass per unit length
  EI beam bending stiffness

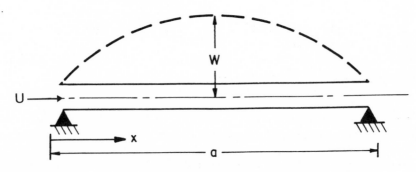

Figure 2.15   Fluid flow through a flexible pipe.

---

* Housner [11].
† See Section 3.4.

Both static and dynamic aeroelastic phenomena are possible for this physical model but for the moment we shall only consider the former. Further we shall consider for simplicity simply supported or pinned boundary conditions, i.e.,

$$w = 0$$

and

$$M \equiv EI\frac{\partial^2 w}{\partial x^2} = 0 \quad \text{at} \quad x = 0, a \tag{2.5.3}$$

where M is the elastic bending moment and, a, the pipe length.

Substituting (2.5.1) into (2.5.2) and dropping time derivatives consistent with limiting our concern to static phenomena, we have

$$EI\frac{\partial^4 w}{\partial x^4} + \rho A U^2 \frac{\partial^2 w}{\partial x^2} = 0 \tag{2.5.4}$$

subject to boundary conditions

$$w = \frac{\partial^2 w}{\partial x^2} = 0 \quad \text{at} \quad x = 0, a \tag{2.5.5}$$

The above equations can be recognized as the same as those governing the buckling of a beam under a compressive load of magnitude,* P. The equivalence is

$$P = \rho U^2 A$$

Formally we may compute the buckling or divergence dynamic pressure by assuming†

$$w = \sum_{i=1}^{4} A_i e^{p_i x}$$

where the p_i are the four roots of the characteristic equation associated with (2.5.4),

$$EI p^4 + \rho U^2 A p^2 = 0$$

Thus

$$p_{1,2} = 0$$

$$p_3, p_4 = \pm i\left(\frac{\rho U^2 A}{EI}\right)^{\frac{1}{2}}$$

---

* Timoshenko and Gere [3].

† Alternatively one could use Galerkin's method for (2.5.4) and (2.5.5) or convert them into an integral equation to be solved by the 'lumped element' method.

and

$$w = A_1 + A_2 x + A_3 \sin \frac{\lambda x}{a} + A_4 \cos \frac{\lambda x}{a} \qquad (2.5.6)$$

where

$$\lambda^2 \equiv \left(\frac{\rho U^2 A}{EI} \right) a^2$$

Using the boundary conditions (2.5.5) with (2.5.6) we may determine that

$$A_1 = A_2 = A_4 = 0$$

and either $A_3 = 0$ or $\sin \lambda = 0$
For nontrivial solutions

$$A_3 \neq 0$$

and

$$\sin \lambda = 0$$

or

$$\lambda = \pi, 2\pi, 3\pi, \text{etc.} \qquad (2.5.7)$$

Note that $\lambda = 0$ is a trivial solution, e.g., $w \equiv 0$.

Of the several eigenvalue solutions the smallest nontrivial one is of greatest physical interest, i.e.,

$$\lambda = \pi$$

The corresponding divergence or buckling dynamic pressure is

$$\rho U^2 = \frac{EI}{Aa^2} \pi^2 \qquad (2.5.8)$$

Note that λ^2 is a nondimensional ratio of aerodynamic to elastic stiffness; we shall call it and similar numbers we shall encounter an 'aeroelastic stiffness number'. It is as basic to aeroelasticity as Mach number and Reynolds number are to fluid mechanics. Recall that in our typical section study we also encountered an 'aeroelastic stiffness number', namely,

$$\frac{qS \dfrac{\partial C_L}{\partial \alpha}}{K_\alpha} e \qquad or \qquad \frac{q S \dfrac{\partial C_L}{\partial \alpha}}{K_\alpha / e}$$

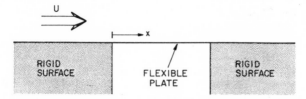

Figure 2.16   Fluid flow over a flexible wall.

as well as in the (uniform) beam-rod wing model,

$$\frac{q(lc)e\frac{\partial C_L}{\partial \alpha}}{\frac{GJ}{l}}$$

(Low speed) fluid flow over a flexible wall

A mathematically similar problem arises when a flexible plate is embedded in an otherwise rigid surface. See Figure 2.16. This is a simplified model of a physical situation which arises in nuclear reactor heat exchangers, for example. Aeronautical applications may be found in the local skin deformations on aircraft and missiles. Early airships may have encountered aeroelastic skin buckling.*

For a one dimensional (beam) structural representation of the wall, the equation of equilibrium is, as in our previous example,

$$EI\frac{\partial^4 w}{\partial x^4} = L$$

Also, as a rough approximation, it has been shown that the aerodynamic loading may be written†

$$L \sim \rho U^2 \frac{\partial^2 w}{\partial x^2}$$

Hence using this aerodynamic model, there is a formal mathematical analogy to the previous example and the aeroelastic calculation is the same. For more details and a more accurate aerodynamic model, the cited references should be consulted.

---

* Shute [12], p. 95.
† Dowell [13], p. 19, Kornecki [14], Kornecki, Dowell and O'Brien [15].

2 Static aeroelasticity

References for Chapter 2

[1] Ashley, H. and Landahl, M., *Aerodynamics of Wings and Bodies*, Addison-Wesley, 1965.

[2] Savant, Jr., C. J., *Basic Feedback Control System Design*, McGraw-Hill, 1958.

[3] Timoshenko, S. P., and Gere, J., *Theory of Elastic Stability*, McGraw-Hill, 1961.

[4] Woodcock, D. L., 'Structural Non-linearities', Vol. I, Chapter 6, *AGARD Manual on Aeroelasticity*.

[5] Housner, G. W. and Vreeland, T., Jr., *The Analysis of Stress and Deformation*, The MacMillan Co., 1966.

[6] Hildebrand, F. B., *Advance Calculus for Engineers*, Prentice-Hall, Inc. 1962.

[7] Duncan, W. J. 'Galerkin's Methods in Mechanics and Differential Equations', Br. A.R.C., R&M., 1798, 1937.

[8] Bisplinghoff, R. L., Mar, J. W. and Pian, T. H. H., *Statics of Deformable Solids*, Addison-Wesley, 1965.

[9] Weisshaar, T. A. and Ashley, H., 'Static Aeroelasticity and the Flying Wing, Revisited', *J. Aircraft*, Vol. 11 (Nov. 1974) pp. 718–720.

[10] Covert, E. E., 'The Aerodynamics of Distorted Surfaces', *Proceedings of Symposium on Aerothermoelasticity*. ASD TR 61–645, 1961, pp. 369–398.

[11] Housner, G. W., 'Bending Vibrations of a Pipe Line Containing Flowing Fluid', *Journal of Applied Mechanics*, Vol. 19 (June 1952) p. 205.

[12] Shute, N., *Slide Rule*, Wm. Morrow & Co., Inc., New York, N.Y. 10016.

[13] Dowell, E. H. *Aeroelasticity of Plates and Shells*, Noordhoff International Publishing, 1974.

[14] Kornecki, A., 'Static and Dynamic Instability of Panels and Cylindrical Shells in Subsonic Potential Flow', *J. Sound Vibration*, Vol. 32 (1974) pp. 251–263.

[15] Kornecki, A., Dowell, E. H., and O'Brien, J., 'On the Aeroelastic Instability of Two-Dimensional Panels in Uniform Incompressible Flow', *J. Sound Vibration*, Vol. 47 (1976) pp. 163–178.

3

Dynamic aeroelasticity

In static aeroelasticity we have considered various mathematical models of aeroelastic systems. In all of these, however, the fundamental physical content consisted of two distinct phenomena, 'divergence' or static instability, and loss of aerodynamic effectiveness as typified by 'control surface reversal'. Turning to dynamic aeroelasticity we shall again be concerned with only a few distinct fundamental physical phenomena. However, they will appear in various theoretical models of increasing sophistication. The principal phenomena of interest are (1) 'flutter' or dynamic instability and (2) response to various dynamic loadings as modified by aeroelastic effects. In the latter category primary attention will be devoted to (external) aerodynamic loadings such as atmospheric turbulence or 'gusts'. These loadings are essentially random in nature and must be treated accordingly. Other loadings of interest may be impulsive or discrete in nature such as the sudden loading due to maneuvering of a flight vehicle as a result of control surface rotation.

To discuss these phenomena we must first develop the dynamic theoretical models. This naturally leads us to a discussion of how one obtains the equations of motion for a given aeroelastic system including the requisite aerodynamic forces. Our initial discussion of aerodynamic forces will be conceptual rather than detailed. Later, in Chapter 4, these forces are developed from the fundamentals of fluid mechanics. We shall begin by using the 'typical section' as a pedagogical device for illustrating the physical content of dynamic aeroelasticity. Subsequently using the concepts of structural and aerodynamic influence and impulse functions, we shall discuss a rather general model of an aeroelastic system. The solution techniques for our aeroelastic models are for the most part standard for the modern treatment of the dynamics of linear systems and again we use the typical section to introduce these methods.

We now turn to a discussion of energy and work methods which have proven very useful for the development of structural equations of motion.

3 Dynamic aeroelasticity

In principle, one may use Newton's Second Law (plus Hooke's Law) to obtain the equations of motion for any elastic body. However, normally an alternative procedure based on Hamilton's Principle or Lagrange's Equations is used.* For systems with many degrees of freedom, the latter are more economical and systematic.

We shall briefly review these methods here by first deriving them from Newton's Second Law for a single particle and then generalizing them for many particles and/or a continuous body. One of the major advantages over the Newtonian formulation is that we will deal with work and energy (scalars) as contrasted with accelerations and forces (vectors).

3.1 Hamilton's principle

Single particle

Newton's Law states

$$\vec{F} = m \frac{d^2 \vec{r}}{dt^2} \tag{3.1.1}$$

where $\vec{F}$ is the force vector and $\vec{r}$ is the displacement vector, representing the actual path of particle.

Consider an adjacent path, $\vec{r} + \delta\vec{r}$, where $\delta\vec{r}$ is a 'virtual displacement' which is small in some appropriate sense. If the time interval of interest is $t = t_1 \rightarrow t_2$ then we shall require that

$$\delta\vec{r} = 0 \quad \text{at} \quad t = t_1, t_2$$

although this can be generalized. Thus, the actual and adjacent paths coincide at $t = t_1$ or t_2. $(3.1.1)$

Now form the dot product of (1) with $\delta\vec{r}$ and $\int_{t_1}^{t_2} \cdots dt$. The result is

$$\int_{t_1}^{t_2} \left(m \frac{d^2 \vec{r}}{dt^2} \cdot \delta\vec{r} - \vec{F} \cdot \delta\vec{r} \right) dt = 0 \tag{3.1.2}$$

The second term in brackets can be identified as work or more precisely the 'virtual work'. The 'virtual work' is defined as the work done by the actual forces being moved through the virtual displacement. We assume that the force remains fixed during the virtual displacement or, equivalently, the virtual displacement occurs instantaneously, i.e., $\delta t = 0$.

It follows that the first term must also have the dimensions of work

---

* See, for example, Meirovitch [1].

48

(or energy). To see this more explicitly, we manipulate the first term by an integration by parts as follows:

$$m \int_{t_1}^{t_2} \frac{d^2\vec{r}}{dt^2} \cdot \delta\vec{r}\, dt = m\frac{d\vec{r}}{dt} \cdot \delta\vec{r}\bigg|_{t_1}^{t_2}$$

$$- m \int_{t_1}^{t_2} \frac{d\vec{r}}{dt} \cdot \frac{d}{dt}(\delta r)\, dt$$

$$= -m \int_{t_1}^{t_2} \frac{d\vec{r}}{dt} \cdot \delta\frac{d\vec{r}}{dt}\, dt$$

$$= -\frac{m}{2} \int_{t_1}^{t_2} \delta\left(\frac{d\vec{r}}{dt} \cdot \frac{d\vec{r}}{dt}\right) dt \tag{3.1.3}$$

Hence (3.1.2) becomes

$$\int_{t_1}^{t_2} \left[\frac{1}{2} m\delta\left(\frac{d\vec{r}}{dt} \cdot \frac{d\vec{r}}{dt}\right) + F \cdot \delta\vec{r}\right] dt = 0$$

or

$$\int_{t_1}^{t_2} \delta[T + W]\, dt = 0 \tag{3.1.4}$$

where

$$\delta T \equiv \delta\frac{1}{2} m\frac{d\vec{r}}{dt} \cdot \frac{d\vec{r}}{dt} \tag{3.1.5}$$

is defined as the 'virtual kinetic energy' and

$$\delta W \equiv \vec{F} \cdot \delta\vec{r} \tag{3.1.6}$$

is the 'virtual work'. Hence, the problem is cast in the form of scalar quantities, work and energy. (3.1.4) is Hamilton's Principle. It is equivalent to Newton's Law.

Before proceeding further it is desirable to pause to consider whether we can reverse our procedure, i.e., starting from (3.1.4), can we proceed to (3.1.1)? It is not immediately obvious that this is possible. After all, Hamilton's Principle represent an integrated statement over the time interval of interest while Newton's Second Law holds at every instant in time. By formally reversing our mathematical steps however, we may proceed from (3.1.4) to (3.1.2). To take the final step from (3.1.2) to (3.1.1) we must recognize that our choice of $\delta\vec{r}$ is arbitrary. Hence, if (3.1.2) is to hold for any possible choice of $\delta\vec{r}$, (3.1.2) must follow. To demonstrate this we note that, if $\delta\vec{r}$ is arbitrary and (3.1.1) were not true,

49

then it would be possible select $\delta\vec{r}$ such that (3.1.2) would not be true. Hence (3.1.2) implies (3.1.1) if $\delta\vec{r}$ is arbitrary.

Many particles

The previous development is readily generalized to many particles. Indeed, the basic principle remains the same and only the work and energy expressions are changed as follows:

$$\delta T = \sum_i \frac{m_i}{2} \delta\left(\frac{\mathrm{d}\vec{r}_i}{\mathrm{d}t} \cdot \frac{\mathrm{d}\vec{r}_i}{\mathrm{d}t}\right) \tag{3.1.7}$$

$$\delta W = \sum_i \vec{F}_i \cdot \delta\vec{r}_i \tag{3.1.8}$$

where m_i is the mass of ith particle,
$\vec{r}_i$ is the displacement of ith particle, $\hspace{2cm}$ (3.1.9)
and $\hspace{1cm}$ $\vec{F}_i$ is the force acting on ith particle.

Continuous body

For a continuous body (3.1.7) and (3.1.8) are replaced by (3.1.10) and (3.1.11).

$$\delta T = \iiint_{\text{volume}} \frac{\rho}{2} \delta \frac{\mathrm{d}\vec{r}}{\mathrm{d}t} \cdot \frac{\mathrm{d}\vec{r}}{\mathrm{d}t} \mathrm{d}V \tag{3.1.10}$$

where ρ is the density (mass per unit volume), V is the volume, and δW is the virtual work done by external applied forces and internal elastic forces. For example, if $\vec{f}$ is the ~~vector~~ vector body force per unit volume and $\vec{p}$ the surface force per unit area then

$$\delta W = \iiint_{\text{volume}} \vec{f} \cdot \delta\vec{r} \, \mathrm{d}V + \iint_{\text{surface area}} \vec{p} \cdot \delta\vec{r} \, \mathrm{d}A \tag{3.1.11}$$

Potential energy

In a course on elasticity* it would be shown that the work done by internal elastic forces is the negative of the virtual elastic potential

---

* Bisplinghoff, Mar, and Pian [2], Timoshenko and Goodier [3].

energy. The simplest example is that of an elastic spring. See sketch below.

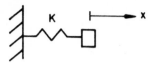

The force in the spring is

$-Kx$

where the minus sign arises from the fact that the force of the spring on the mass opposes the displacement, x. The virtual work is

$$\delta W = -Kx\,\delta x$$

$$= -\delta\frac{Kx^2}{2}$$

The virtual change in potential energy is

δU

$$\delta W \equiv -\delta W$$

$$= \delta\frac{Kx^2}{2} = \delta\left(\frac{Fx}{2}\right) \qquad (3.1.12)$$

$$where \quad F = Kx$$

Considering the other extreme, the most complete description of the potential energy of an elastic body which satisfies Hooke's Law is (see Bisplinghoff, Mar and Pian [2])

$$U = \tfrac{1}{2}\iiint_V [\sigma_{xx}\varepsilon_{xx} + \sigma_{xy}\varepsilon_{xy} + \sigma_{yx}\varepsilon_{yx} + \cdots]\,dV \qquad (3.1.13)$$

where σ_{xx} is the stress component (analogous to F) and ε_{xx} is the strain component (analogous to x).

From this general expression for potential (strain) energy of an elastic body we may derive some useful results for the bending and twisting of beams and plates. For the bending of a beam, the usual assumption of plane sections over the beam cross-section remaining plane leads to a strain-displacement relation of the form

$$\varepsilon_{yy} = -z\frac{\partial^2 w}{\partial y^2}$$

where z is the vertical coordinate through beam, w is the vertical

51

3 Dynamic aeroelasticity

displacement of beam, Hooke's Law reads,

$$\sigma_{yy} = E\varepsilon_{yy} = -Ez\frac{\partial^2 w}{\partial y^2}$$

and we assume all other stresses are negligible

$$\sigma_{yz} = \sigma_{xy} = \sigma_{xz} = \sigma_{xx} = \sigma_{zz} = 0$$

If we further assume $w(x, y, z) = h(y)$ where y is the lengthwise coordinate axis of the beam, then

$$U = \frac{1}{2}\int EI\left(\frac{\partial^2 h}{\partial y^2}\right)^2 dy \qquad *$$

where

$$I \equiv \int z^2\, dz \int dx$$

For the <u>twisting of a thin beam</u>, analogous reasoning leads to similar results.

$$\varepsilon_{xy} = -z\frac{\partial^2 w}{\partial x\, \partial y}$$

$$\sigma_{xy} = \frac{E}{(1+\nu)}\varepsilon_{xy} = -\frac{E}{(1+\nu)}z\frac{\partial^2 w}{\partial x\, \partial y}$$

Thus

$$U = \frac{1}{2}\int GJ\left(\frac{\partial\alpha}{\partial y}\right)^2 dy \qquad \nu - \text{Poisson's Ratio}$$

where

$$G \equiv \frac{E}{2(1+\nu)}, \qquad J \equiv 4\int z^2\, dz \int dx \quad \text{and} \quad w = x\alpha(y)$$

The above can be generalized to the <u>bending of a plate</u> in two dimensions.

$$\varepsilon_{yy} = -z\left[\frac{\partial^2 w}{\partial y^2} + \nu\frac{\partial^2 w}{\partial x^2}\right]$$

$$\varepsilon_{xx} = -z\left[\frac{\partial^2 w}{\partial x^2} + \nu\frac{\partial^2 w}{\partial y^2}\right]$$

$$\varepsilon_{xy} = -z\frac{\partial^2 w}{\partial x\, \partial y}$$

52

$$\sigma_{xx} = \frac{E}{(1-\nu^2)} \left[\varepsilon_{xx} + \nu\varepsilon_{yy} \right]$$

$$\sigma_{yy} = \frac{E}{(1-\nu^2)} \left[\varepsilon_{yy} + \nu\varepsilon_{xx} \right]$$

$$\sigma_{xy} = \frac{E}{(1+\nu)} \varepsilon_{xy}$$

and

$$U = \frac{1}{2} \iint D \left[\left(\frac{\partial^2 w}{\partial x^2} \right)^2 + \left(\frac{\partial^2 w}{\partial y^2} \right)^2 + 2\nu \frac{\partial^2 w}{\partial x^2} \frac{\partial^2 w}{\partial y^2} \right.$$ *derived pp 435-6*

$$\left. + 2(1-\nu) \left(\frac{\partial^2 w}{\partial x \, \partial y} \right)^2 \right] \mathrm{d}x \, \mathrm{d}y$$

where

$$D \equiv \frac{E}{(1-\nu^2)} \int_{-h/2}^{+h/2} z^2 \, \mathrm{d}y, \text{ plate bending stiffness}$$
dz

and   *h , thickness*

$$w = w(x, y)$$

Nonpotential forces

Now, if one divides the virtual work into potential and nonpotential contributions, one has Hamilton's Principle in the form

$$\int \left[(\delta T - \delta U) + \underbrace{\overrightarrow{F_{NC}} \cdot \overrightarrow{\delta r}}_{\delta W_{NC}} \right] \mathrm{d}t = 0 \tag{3.1.14}$$

where F_{NC} includes only the nonpotential (or nonconservative) forces.

In our aeroelastic problems the nonconservative virtual work is a result of aerodynamic loading. For example, the virtual work due to the aerodynamic pressure (force per unit area) on a two-dimensional plate is clearly

$$\delta W_{NC} = \iint p \, \delta w \, \mathrm{d}x \, \mathrm{d}y$$

Note that if the deflection is taken to be a consequence of a chordwise rigid rotation about and bending of a spanwise elastic axis located at, say $x = 0$, then

$$w = h(y) + x\alpha(y)$$

53

3 Dynamic aeroelasticity

and hence

$$\delta W = \int \left[\int p \, dx \right] \delta h \, dy + \int [\int px \, dx] \, \delta\alpha \, dy$$

where $L = \int p \, dx$ net vertical force/per unit span

$M_y \equiv \int px \, dx$ net moment about y axis per unit span

Thus, for this special case,

$$\delta W = \int L \, \delta h \, dy + \int M_y \, \delta\alpha \, dy$$

3.2 Lagrange's equations

Lagrange's equations may be obtained by reversing the process by which we obtained Hamilton's Principle. However to obtain a more general result than simply a retrieval of Newton's Second Law we introduce the notion of 'generalized' coordinates. A 'generalized' coordinate is one which is arbitrary and independent (of other coordinates). A set of 'generalized' coordinates is sufficient* to describe the motion of a dynamical system. That is, the displacement of a particle or point in a continuous body may be written

$$\vec{r} = \vec{r}(q_1, q_2, q_3, \ldots, t) \tag{3.2.1}$$

where q_i is the ith generalized coordinate. From (3.2.1) it follows that

$$T = T(\dot{q}_i, q_i, t) \tag{3.2.2}$$
$$U = U(\dot{q}_i, q_i, t)$$

Thus Hamilton's Principle may be written

$$\int_{t_1}^{t_2} [\delta(T - U) + \delta W_{NC}] \, dt = 0 \tag{3.1.14}$$

Using (3.2.2) in (3.1.14)

$$\sum_i \int_{t_1}^{t_2} \left[\frac{\partial(T - U)}{\partial \dot{q}_i} \delta\dot{q}_i + \frac{\partial(T - U)}{\partial q_i} \delta q_i + Q_i \, \delta q_i \right] dt = 0 \tag{3.2.3}$$

---

* and necessary, i.e., they are independent.

where the generalized forces, Q_i, are known from

$$\delta W_{NC} \equiv \sum_i Q_i \, \delta q_i \tag{3.2.4}$$

As we will see (3.2.4) *defines the* Q_i as the coefficients of δq_i in an expression for δW_{NC} which must be obtained independently of (3.2.4). Integrating the first term of (3.2.3) by parts (noting that $\delta q_i = 0$ $t = t_1, t_2$) we have

$$\sum_i \overset{0}{\cancel{\frac{\partial(T-U)}{\partial q_i} \delta q_i}} \Big|_{t_1}^{t_2} + \int_{t_1}^{t_2} \left[-\frac{\mathrm{d}}{\mathrm{d}t} \frac{\partial(T-U)}{\partial \dot{q}_i} \delta q_i \right.$$

$$\left. + \frac{\partial(T-U)}{\partial q_i} \delta q_i + Q_i \, \delta q_i \right] \mathrm{d}t = 0 \tag{3.2.5}$$

Collecting terms

$$\sum_i \int_{t_1}^{t_2} \left[-\frac{\mathrm{d}}{\mathrm{d}t} \frac{\partial(T-U)}{\partial \dot{q}_i} + \frac{\partial(T-U)}{\partial q_i} + Q_i \right] \delta q_i \, \mathrm{d}t = 0 \tag{3.2.6}$$

Since the δq_i are independent and arbitrary it follows that each bracketed quantity must be zero, i.e.,

$$-\frac{\mathrm{d}}{\mathrm{d}t} \frac{\partial(T-U)}{\partial \dot{q}_i} + \frac{\partial(T-U)}{\partial q_i} + Q_i = 0 \qquad i = 1, 2, \ldots \tag{3.2.7}$$

These are Lagrange's equations.   *see p 120 condensed form*

Example — Typical section equations of motion

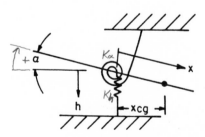

Figure 3.1   *Geometry of typical section airfoil*

x is measured along chord from e.a.; note that x is *not* a generalized coordinate, e.g., it cannot undergo a virtual change.

$$\text{generalized} \atop \text{coordinates} \quad \begin{cases} q_1 = h, \\ q_2 = \alpha, \end{cases}$$

3 Dynamic aeroelasticity

The displacement of any point on the airfoil is

$$\vec{r} = u\vec{i} + w\vec{k} \tag{3.2.8}$$

where u is the horizontal displacement component, w is the vertical displacement component, and $\vec{i}, \vec{k}$ are the unit, cartesian vectors.

From geometry

$$\left. \begin{aligned} u &= x[\cos \alpha - 1] \simeq 0 \\ w &= -h - x \sin \alpha \simeq -h - x\alpha \end{aligned} \right\} \text{for } \alpha \ll 1 \tag{3.2.9}$$

Hence,

$$\begin{aligned} T &= \frac{1}{2} \int \left[\left(\frac{dw}{dt} \right)^2 + \left(\frac{du}{dt} \right)^2 \right] \rho \, dx \\ &\simeq \frac{1}{2} \int \left(\frac{dw}{dt} \right)^2 \rho \, dx \\ &= \tfrac{1}{2} \int (-\dot{h} - \dot{\alpha}x)^2 \rho \, dx \quad \longrightarrow \quad \tfrac{1}{2} m \bar{v}^2 \\ &= \tfrac{1}{2}\dot{h}^2 \int \rho \, dx + \tfrac{1}{2} 2\dot{h}\dot{\alpha} \int x\rho \, dx + \tfrac{1}{2}\dot{\alpha}^2 \int x^2 \rho \, dx \\ &= \tfrac{1}{2}\dot{h}^2 m + \tfrac{1}{2} 2\dot{h}\dot{\alpha}S_\alpha + \tfrac{1}{2}\dot{\alpha}^2 I_\alpha \end{aligned} \tag{3.2.10}$$

$$m \equiv \int \rho \, dx$$

$$S_\alpha \equiv \int \rho x \, dx \equiv x_{c.g.} m$$

$$I_\alpha \equiv \int \rho x^2 \, dx$$

The potential energy is

$$U = \tfrac{1}{2} K_h h^2 + \tfrac{1}{2} K_\alpha \alpha^2 \tag{3.2.11}$$

For our system, Lagrange's equations are

$$-\frac{d}{dt} \left(\frac{\partial(T-U)}{\partial \dot{h}} \right) + \frac{\partial(T-U)}{\partial h} + Q_h = 0$$

$$-\frac{d}{dt} \left(\frac{\partial(T-U)}{\partial \dot{\alpha}} \right) + \frac{\partial(T-U)}{\partial \alpha} + Q_\alpha = 0 \tag{3.2.12}$$

where

$$\delta W_{NC} = Q_h \, \delta h + Q_\alpha \, \delta \alpha \tag{3.2.13}$$

Now let us evaluate the terms in (3.2.12) and (3.2.13). Except for Q_h, Q_α these are readily obtained by using (3.2.10) and (3.2.11) in (3.2.12). Hence, let us first consider the determination of Q_h, Q_α. To do this we calculate *independently* the work done by the aerodynamic forces.

$$
\begin{aligned}
\delta W_{NC} &= \int p \, \delta w \, dx \\
&= \int p(-\delta h - x \, \delta \alpha) \, dx \\
&= \delta h \left(-\int p \, dx \right) + \delta \alpha \left(-\int px \, dx \right) \\
&= \delta h(-L) + \delta \alpha (M_y)
\end{aligned}
\qquad (3.2.14)
$$

where we identify from (3.2.13) and (3.2.14)

$$
L \equiv \int p \, dx = -Q_h
$$

$$
M_y \equiv -\int px \, dx = Q_\alpha
$$

Note the sign convention is that p is positive up, L is positive up and M_y is positive nose up. Putting it all together, noting that

$$
\frac{\partial(T-U)}{\partial h} = -K_h h \quad \text{etc.}
$$

we have from Lagrange's equations

$$
-\frac{d}{dt}(m\dot{h} + S_\alpha \dot{\alpha}) - K_h h - L = 0
$$

$$
-\frac{d}{dt}(S_\alpha \dot{h} + I_\alpha \dot{\alpha}) - K_\alpha \alpha + M_y = 0
\qquad (3.2.15)
$$

These are the equations of motion for the 'typical section' in terms of the particular coordinates h and α.

Other choices of generalized coordinates are possible; indeed, one of the principal advantages of Lagrange's equations is this freedom to make various choices of generalized coordinates. The choice used above simplifies the potential energy but not the kinetic energy. If the generalized coordinates were chosen to be the translation of and rotation about the center of mass the kinetic energy would be simplified, viz.

$$
T = \frac{m}{2} \dot{h}_{cm}^2 + \frac{I_{cm}}{2} \dot{\alpha}_{cm}^2
$$

See solution (a) App. 2
pp 443-445

but the potential energy would be more complicated. Also the relevant aerodynamic moment would be that about the center of mass axis rather than that about the elastic axis (spring attachment point).

Another choice might be the translation of and rotation about the aerodynamic center axis though this choice is much less often used than those discussed above.

Finally we note that there is a particular choice of coordinates which leads to a maximum simplification of the inertial and elastic terms (though not necessarily the aerodynamic terms). These may be determined by making some arbitrary initial choice of coordinates, e.g., h and α, and then determining the 'normal modes' of the system in terms of these.* These 'normal modes' provide us with a coordinate transformation from the initial coordinates, h and α, to the coordinates of maximum simplicity. We shall consider this matter further subsequently.

3.3   Dynamics of the typical section model of an airfoil

To study the dynamics of aeroelastic systems, we shall use the 'typical section'† as a device for exploring mathematical tools and the physical content associated with such systems. To simplify matters, we begin by assuming the aerodynamic forces *are given* where $p(x, t)$ is the aerodynamic pressure, L, the resultant (lift) force and M_y the resultant moment about the elastic axis. See Figure 3.2. The equations of motion are

$$m\ddot{h} + K_h h + S_\alpha \ddot{\alpha} = -L \tag{3.3.1}$$

$$S_\alpha \ddot{h} + I_\alpha \ddot{\alpha} + K_\alpha \alpha = M_y \tag{3.3.2}$$

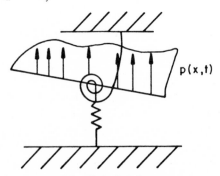

Figure 3.2   Typical section geometry

* Meirovitch [4].
† BA, pp. 201–246.

where

$$L \equiv \int p \, dx$$

$$M_y \equiv \int px \, dx$$

We will find it convenient also to define the 'uncoupled natural frequencies',

$$\omega_h^2 \equiv K_h/m, \qquad \omega_\alpha^2 \equiv K_\alpha/I_\alpha \tag{3.3.3}$$

These are 'natural frequencies' of the system for $S_\alpha \equiv 0$ as we shall see in a moment.

Sinusoidal motion

This is the simplest type of motion; however, as we shall see, we can exploit it systematically to study more complicated motions.
Let

$$L = \bar{L}e^{i\omega t}, \qquad M_y = \bar{M}_y e^{i\omega t}$$
$$h = \bar{h}e^{i\omega t}, \qquad \alpha = \bar{\alpha}e^{i\omega t} \tag{3.3.4}$$

Substituting (3.3.4) and (3.3.3) into (3.3.2) we have in matrix notation

$$\begin{bmatrix} m(-\omega^2 + \omega_h^2) & -S_\alpha \omega^2 \\ -S_\alpha \omega^2 & I_\alpha(-\omega^2 + \omega_\alpha^2) \end{bmatrix} \begin{Bmatrix} \bar{h}e^{i\omega t} \\ \bar{\alpha}e^{i\omega t} \end{Bmatrix} = \begin{Bmatrix} -\bar{L}e^{i\omega t} \\ \bar{M}_y e^{i\omega t} \end{Bmatrix} \tag{3.3.5}$$

Solving for $\bar{h}, \bar{\alpha}$ we have

$$\frac{\bar{h}}{\bar{L}} = \frac{-[1 - (\omega/\omega_\alpha)^2] + d/b \dfrac{x_\alpha}{r_\alpha^2}\left(\dfrac{\omega}{\omega_\alpha}\right)^2}{K_h\left\{[1 - (\omega/\omega_\alpha)^2][1 - (\omega/\omega_h)^2] - \dfrac{x_\alpha^2}{r_\alpha^2}\left(\dfrac{\omega}{\omega_\alpha}\right)^2\left(\dfrac{\omega}{\omega_h}\right)^2\right\}}$$

$$\equiv H_{hL}\left(\omega/\omega_\alpha; \frac{\omega_h}{\omega_\alpha}, d/b, x_\alpha, r_\alpha\right) \tag{3.3.6}$$

where

$$d \equiv \bar{M}_y/\bar{L}$$

where b is the reference length (usually selected as half-chord by tradition),

$$x_\alpha \equiv \frac{S_\alpha}{mb} = \frac{x_{c.g.}}{b}$$

59

and

$$r_\alpha^2 \equiv \frac{I_\alpha}{mb^2}$$

A plot of H_{hL} is shown below in Figure 3.3. $\dfrac{\omega_1}{\omega_\alpha}$, $\dfrac{\omega_2}{\omega_\alpha}$ are the roots of the denominator, the system 'natural frequencies'.

$$\frac{\omega_1^2}{\omega_h \omega_\alpha}, \frac{\omega_2^2}{\omega_h \omega_\alpha} = \frac{\left[\frac{\omega_h}{\omega_\alpha} + \frac{\omega_\alpha}{\omega_h} \right] \pm \left\{ \left[\frac{\omega_h}{\omega_\alpha} + \frac{\omega_\alpha}{\omega_h} \right]^2 - 4 \left[1 - \frac{x_\alpha^2}{r_\alpha^2} \right] \right\}^{\frac{1}{2}}}{2[1 - x_\alpha^2/r_\alpha^2]} \tag{3.3.7}$$

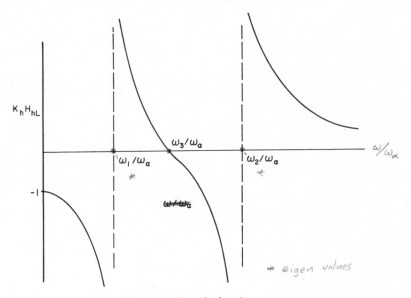

Figure 3.3   Transfer function.

A similar equation may be derived for

$$\frac{\bar{\alpha}}{L} \equiv H_{\alpha L}\left(\omega/\omega_\alpha; \frac{\omega_h}{\omega_\alpha}, d/b, x_\alpha, r_\alpha \right) \tag{3.3.8}$$

ω_1 and ω_2 are again the natural frequencies. H_{hL}, $H_{\alpha L}$ are so-called 'transfer functions'; they are 'mechanical' or 'structural transfer functions' as they describe the motion of the structural system under specified loading. Later on we shall have occasion to consider 'aerodynamic transfer functions' and also 'aeroelastic transfer functions'. ω_3/ω_α is the root of the numerator of H_{hL} (but not in general of $H_{\alpha L}$ which will vanish at a different frequency),

$$\left(\frac{\omega_3}{\omega_\alpha}\right)^2 = \frac{1}{1+(d/b)x_\alpha/r_\alpha^2} \qquad (3.3.9)$$

set numerator of (3.3.6) to zero

Note that infinite response occurs at the natural frequencies, ω_1 and ω_2, for both H_{hL} and $H_{\alpha L}$. This is not an instability; it is 'resonance' with the infinite response due to the absence of any damping in the system. Had structural or aerodynamic damping been included as will be done in later examples, then the transfer functions would become complex numbers which is a mathematical complication. However, the magnitude of the transfer functions would remain finite though large at $\omega = \omega_1$, ω_2 which is an improvement in the realism of the physical model. With L and M assumed given, which admittedly is somewhat artificial, the question of instability does not arise, We will elaborate on this point later when we discuss the notion of instability in a more precise way.

From sinusoidal motion we may proceed to periodic (but not necessarily sinusoidal) motion.

Periodic motion

The above analysis can be generalized to any periodic motion by expanding the motion into a Fourier (sinusoidal) series. Define:

$T_0 \equiv$ basic period

$\omega_0 \equiv 2\pi/T_0$, fundamental frequency

Then a periodic force, $L(t)$, may be written as

$$L(t) = \sum_{n=-\infty}^{\infty} L_n e^{+in\omega_0 t} \qquad (3.3.10)$$

where

$$L_n = \frac{1}{T_0} \int_{-T_0/2}^{T_0/2} L(t) e^{-in\omega_0 t}\, dt \qquad (3.3.11)$$

Using (3.3.10) and (3.3.6),

$$h(t) = \sum_n H_{hL}\left(\frac{\omega_0 n}{\omega_\alpha}\right) L_n e^{in\omega_0 t} \qquad (3.3.12)$$

From periodic motion we may proceed to arbitrary time dependent motion.

61

3 Dynamic aeroelasticity

By taking the limit as the basic period becomes infinitely long, $T_0 \to \infty$, we obtain results for nonperiodic motion.

Define

$$\omega \equiv n\omega_0$$

$$\Delta\omega \equiv \Delta n\omega_0{}^\dagger = \omega_0 = 2\pi/T_0 \quad \text{frequency increment}$$

$$L^*(\omega) \equiv \frac{L_n}{\Delta\omega} = \frac{L_n T_0}{2\pi} \quad \text{force per frequency increment}$$

Then (3.3.10) becomes

$$L(t) = \int_{-\infty}^{\infty} L^*(\omega)e^{+i\omega t}\,d\omega \tag{3.3.10}$$

(3.3.11) becomes

$$L^*(\omega) = \frac{1}{2\pi}\int_{-\infty}^{\infty} L(t)e^{-i\omega t}\,dt \tag{3.3.11}$$

(3.3.12) becomes

$$h(t) = \int_{-\infty}^{\infty} H_{hL}(\omega/\omega_\alpha)L^*(\omega)e^{i\omega t}\,d\omega \tag{3.3.12}$$

An interesting alternate form of (3.3.12) can be obtained by substituting (3.3.11) into (3.3.12). Using a dummy time variable, τ, in (3.3.11) and interchanging order of integration in (3.3.12), gives

$$h(t) = \int_{-\infty}^{\infty} I_{hL}(t-\tau)L(\tau)\,d\tau \tag{3.3.13}$$

where

$$I_{hL}(t) \equiv \frac{1}{2\pi}\int_{-\infty}^{\infty} H_{hL}(\omega/\omega_\alpha)e^{i\omega t}\,d\omega \tag{3.3.14}$$

Comparing (3.3.12) and (3.3.14), note that I_{hL} is response to $L^*(\omega) = \dfrac{1}{2\pi}$ or from (3.3.10) and (3.3.11), $L(t) = \delta(t)$. Hence, I is the response to an impulse force and is thus called the impulse function.

(3.3.10)–(3.3.12) are Fourier transform relations and (3.3.13) is a so-called convolution integral.

---

† Note $\Delta n = 1$ since any n is an integer.

Note (3.3.13) is suitable for treating transient motion; however, a special case of the Fourier transform is often used for transient motion. This is the Laplace transform.

Laplace transform. Consider

$$L(\tau) = 0 \quad \text{for} \quad \tau < 0$$

also

$$I_{hL}(t - \tau) = 0 \quad \text{for} \quad t - \tau < 0$$

The latter will be true for any physically realizable system since the system cannot respond before the force is applied.
Define

$$p \equiv i\omega; \quad \text{thus} \quad \omega = -ip$$

and

$$L^\dagger \equiv 2\pi L^*(-ip)$$

then (3.3.10) becomes

$$L(t) = \frac{1}{2\pi i} \int_{-i\infty}^{i\infty} L^\dagger e^{pt} \, dp \tag{3.3.15}$$

(3.3.11) becomes

$$L^\dagger = \int_0^\infty L(t) e^{-pt} dt$$

(3.3.13) becomes

$$h(t) = \int_0^t I_{hL}(t - \tau) L(\tau) \, d\tau$$

where

$$I_{hL}(t) = \frac{1}{2\pi i} \int_{-i\infty}^{i\infty} H_{hL}\left(\frac{-ip}{\omega_\alpha}\right) e^{pt} \, dp$$

Utilization of transform integral approach for arbitrary motion. There are several complementary approaches in practice. In one the transfer function, H_{hL}, is first determined through consideration of simple sinusoidal motion. Then the impulse function is evaluated from

$$I_{hL}(t) = \frac{1}{2\pi} \int_{-\infty}^\infty H_{hL}(\omega) e^{i\omega t} \, d\omega \tag{3.3.14}$$

63

3 Dynamic aeroelasticity

and the response is obtained from

$$h(t) = \int_0^t I_{hL}(t-\tau)L(\tau)\,\mathrm{d}\tau \tag{3.3.13}$$

Alternatively, knowing the transfer function, $H_{hL}(\omega)$, the transform of the input force is determined from

$$L^*(\omega) = \frac{1}{2\pi}\int_{-\infty}^{\infty} L(t)\mathrm{e}^{-i\omega t}\,\mathrm{d}\omega \tag{3.3.11}$$

and the response from

$$h(t) = \int_{-\infty}^{\infty} H_{hL}(\omega)L^*(\omega)\mathrm{e}^{i\omega t}\,\mathrm{d}\omega \tag{3.3.12}$$

Both approaches give the same result, of course.

example

As a simple example we consider the translation of our typical section for $S_\alpha \equiv 0$, i.e., the center of mass coincides with the elastic axis or spring attachment point. This uncouples the rotation from translation and we need only consider

$$m\ddot{h} + K_h h = -L \tag{3.3.1}$$

We assume a force of the form

$$L = \mathrm{e}^{-at} \quad \text{for} \quad t > 0$$
$$= 0 \quad \text{for} \quad t < 0 \tag{3.3.16}$$

From our equation of motion (or (3.3.6) for $S_\alpha = x_\alpha = 0$) we determine the transfer function as

$$H_{hL}(\omega) = \frac{-1}{m[\omega_h^2 - \omega^2]}, \qquad \omega_h^2 \equiv K_h/m \tag{3.3.6}$$

From (3.3.14), using the above and evaluating the integral, we have

$$I_{hL}(t) = -\frac{1}{m\omega_h}\sin \omega_h t \quad \text{for} \quad t > 0 \tag{3.3.17}$$
$$= 0 \quad \text{for} \quad t < 0$$

From (3.3.13), using above (3.3.17) for I_{hL} and given L, we obtain

$$h(t) = -\frac{1}{m\omega_h}\left\{\frac{\omega_h\mathrm{e}^{-at} - \omega_h\cos\omega_h t + a\sin\omega_h t}{a^2 + \omega_h^2}\right\} \tag{3.3.18}$$

64

We can obtain the same result using our alternative method. Calculating L^* from (3.3.11) for our given L, we have

$$L = \frac{1}{2\pi} \frac{1}{a + i\omega}$$

Using above and the previously obtained transfer function in (3.3.12) we obtain the response. The result is, of course, the same as that determined before. Note that in accordance with our assumption of a system initially at rest, $h = \dot{h} = 0$ at $t = 0$. Examining our solution, (3.3.18), for large time we see that

$$h \rightarrow -\frac{1}{m\omega_h} \left\{ \frac{-\omega_h \cos \omega_h t + a \sin \omega_h t}{a^2 + \omega_h^2} \right\} \quad \text{as} \quad t \rightarrow \infty$$

This indicates that the system continues to respond even though the force L, approaches zero for large time! This result is quite unrealistic physically and is a consequence of our ignoring structural damping in our model. Had we included this effect in our equation of motion using a conventional analytical damping model[†]

$$m[\ddot{h} + 2\zeta_n \omega_h \dot{h}] + K_h h = -L \tag{3.3.1}$$

the response would have been

$$h = -\frac{1}{m\omega_h} \left\{ \frac{\omega_h e^{-at} + [-\omega_h \cos \omega_h t + a \sin \omega_h t] e^{-\zeta_h \omega_h t}}{a^2 + \omega_h^2} \right\} \tag{3.3.19}$$

for small damping, $\zeta_h \ll 1$, which is the usual situation. Now $h \rightarrow 0$, for $t \rightarrow \infty$. Furthermore, if the force persists for a long time, i.e., $a \rightarrow 0$, then

$$h(t) \rightarrow -\frac{1}{m\omega_h} \left\{ \frac{\omega_h}{\omega_h^2} \right\} = -\frac{1}{K_h}$$

which is the usual static or steady state response to a force of unit amplitude. The terms which approach zero for large time due to structural damping are usually termed the transient part of the solution. If

$$a \ll \zeta_h \omega_h$$

the transient solution dies out rapidly compared to the force and we usually are interested in the steady state response. If

$$a \gg \zeta_h \omega_h$$

the 'impulsive' force dies out rapidly and we are normally interested in

---

[†] Meirovitch [4].

the transient response. Frequently the maximum response is of greatest interest. A well known result is that the peak dynamic response is approximately twice the static response if the force persists for a long time and the damping is small. That is, if

$$\zeta_h \ll 1$$

$$a \ll \omega_h$$

then $h_{\max}$ occurs when (see (3.3.19))

$$\cos \omega_n t \cong -1 \quad \text{or} \quad t = \frac{\pi}{\omega_h}$$

$$\sin \omega_n t \cong 0$$

and

$$h_{\max} = -\frac{1}{m\omega_h} \frac{\omega_h}{\omega_h^2} [1 - (-1)]$$

$$= -\frac{2}{K_h}$$

The reader may wish to consider other special combinations of the relative sizes of

$\quad a \quad$ force time constant

$\quad \omega_h \quad$ system natural time constant

$\quad \zeta_h \omega_h \quad$ damping time constant

A great deal of insight into the dynamics of linear systems can be gained thereby.

The question arises which of the two approaches is to be preferred. The answer depends upon a number of factors, including the computational efficiency and physical insight desired. Roughly speaking the second approach, which is essentially a frequency domain approach, is to be preferred when analytical solutions are to be attempted or physical insight based on the degree of frequency 'matching' or 'mis-matching' of H_{hL} and L^* is desired. In this respect larger response obviously will be obtained if the maxima of H and L^* occur near the same frequencies, i.e., they are 'matched', and lesser response otherwise, 'mismatched'. The first approach, which is essentially a time approach, is generally to be preferred when numerical methods are attempted and quantitative accuracy is of prime importance.

Other variations on these methods are possible. For example the transfer function, H_{hL}, and impulse function, I_{hL}, may be determined experimentally. Also the impulse function may be determined directly from the equation of motion, bypassing any consideration of the transfer function. To illustrate this latter remark, consider our simple example

$$m\ddot{h} + K_h h = -L \tag{3.3.1}$$

The impulse function is the response for h due to $L(t) = \delta(t)$. Hence, it must satisfy

$$m\ddot{I}_{hL} + K_h I_{hL} = -\delta(t) \tag{3.3.20}$$

Let us integrate the above from $t = 0$ to ε.

$$\int_0^\varepsilon [m\ddot{I}_{hL} + K_h I_{hL}]\, \mathrm{d}t = -\int_0^\varepsilon \delta(t)\, \mathrm{d}t$$

or

$$m\dot{I}_{hL} \Big|_0^\varepsilon + K_h \int_0^\varepsilon I_{hL}\, \mathrm{d}t = -1$$

In the limit as $\varepsilon \to 0^+$, we obtain the 'initial condition',

$$\dot{I}_{hL}(0^+) = -\frac{1}{m} \tag{3.3.21}$$

and also

$$I_{hL}(0^+) = 0$$

Hence, solving (3.3.20) and using the initial velocity condition, (3.3.21), we obtain

$$I_{hL} = -\frac{1}{m\omega_h} \sin \omega_h t \quad \text{for} \quad t > 0 \tag{3.3.17}$$

which is the same result obtained previously.

Finally, all of these ideas can be generalized to many degrees of freedom. In particular using the concept of 'normal modes' any multi-degree-of-freedom system can be reduced to a system of uncoupled single-degree-of-freedom systems.† As will become clear, when aerodynamic forces are present the concept of normal modes which decouple the various degrees of freedom is not as easily applied and one must usually deal with all the degrees of freedom which are of interest simultaneously.

---

† Meirovitch [4].

3   Dynamic aeroelasticity

Random motion

A random motion is by definition one whose detailed behavior is neither repeatable nor of great interest. Hence attention is focused on certain averages, usually the mean value and also the mean square value. The mean value may be treated as a static loading and response problem and hence we shall concentrate on the mean square relations which are the simplest characterization of *random, dynamic* response.

Relationship between mean values. To see the equivalence between mean value dynamic response and static response, consider

$$h(t) = \int_{-\infty}^{\infty} I_{hL}(t-\tau)L(\tau)\,d\tau \tag{3.3.13}$$

and take the mean of both sides (here a bar above the quantity denotes its mean, which should not be confused with that symbol's previous use in our discussion of sinusoidal motion)

$$\bar{h} \equiv \lim_{T\to\infty} \frac{1}{2T} \int_{-T}^{T} h(t)\,dt$$

$$= \lim_{T\to\infty} \frac{1}{2T} \int_{-T}^{T} \int_{-\infty}^{\infty} I_{hL}(t-\tau)L(\tau)\,d\tau\,dt$$

Interchanging the order of integration and making a change of variables, the right hand side becomes

$$= \int_{-\infty}^{\infty} \left\{ \lim_{T\to\infty} \frac{1}{2T} \int_{-T}^{T} L(t-\tau)\,dt \right\} I_{hL}(\tau)\,d\tau$$

$$= \bar{L} \int_{-\infty}^{\infty} I_{hL}(\tau)\,d\tau$$

$$= \bar{L} H_{hL}(\omega = 0)$$

$$= -\frac{\bar{L}}{K_h}$$

which is just the usual static relation. Unfortunately, no such simple relation exists between the *mean square* values. Instead all frequency components of the transfer function, H_{hL}, contribute. Because of this it proves useful to generalize the definition of a mean square.

Relationship between mean square values. A more general and informative quantity than the mean square, the correlation function, ϕ, can be

68

defined as

$$\phi_{LL}(\tau) \equiv \lim_{T \to \infty} \frac{1}{2T} \int_{-T}^{T} L(t)L(t+\tau) \, dt \tag{3.3.22}$$

The mean square of L, $\overline{L^2}$, is given by

$$\overline{L^2} = \phi_{LL}(\tau = 0) \tag{3.3.23}$$

As $\tau \to \infty$, $\phi_{LL} \to 0$ if L is truly a random function since $L(t)$ and $L(t+\tau)$ will be 'uncorrelated'. Indeed, a useful check on the randomness of L is to examine ϕ for large τ. Analogous to (3.3.22), we may define

$$\phi_{hh}(\tau) \equiv \lim_{T \to \infty} \frac{1}{2T} \int_{-T}^{T} h(t)h(t+\tau) \, dt$$

$$\phi_{hL}(\tau) \equiv \lim_{T \to \infty} \frac{1}{2T} \int_{-T}^{T} h(t)L(t+\tau) \, dt \tag{3.3.24}$$

ϕ_{hL} is the 'cross-correlation' between h and L. ϕ_{hh} and ϕ_{LL} are 'autocorrelations'. The Fourier Transform of the correlation function is also a quantity of considerable interest, the 'power spectra',

$$\Phi_{LL}(\omega) \equiv \frac{1}{\pi} \int_{-\infty}^{\infty} \phi_{LL}(\tau)e^{-i\omega\tau} \, d\tau \tag{3.3.25}$$

(Note that a factor of two difference exists in (3.3.25) from the usual Fourier transform definition. This is by tradition.) From (3.3.25), we have

$$\phi_{LL}(\tau) = \frac{1}{2} \int_{-\infty}^{\infty} \Phi_{LL}(\omega)e^{i\omega\tau} \, d\omega$$

$$= \int_{0}^{\infty} \Phi_{LL}(\omega) \cos \omega\tau \, d\omega \tag{3.3.26}$$

The latter follows since $\Phi_{LL}(\omega)$ is a real even function of ω. Note

$$\overline{L^2} = \phi_{LL}(0) = \int_{0}^{\infty} \Phi_{LL}(\omega) \, d\omega \tag{3.3.27}$$

Hence a knowledge of Φ_{LL} is sufficient to determine the mean square. It turns out to be most convenient to relate the power spectra of L to that of h and use (3.3.27) or its counterpart for h to determine the mean square values.

3 Dynamic aeroelasticity

To relate the power spectra, it is useful to start with a substitution of (3.3.13) into the first of (3.3.24).

$$\phi_{hh}(\tau) = \lim_{T \to \infty} \frac{1}{2T} \int_{-T}^{T} \left\{ \int_{-\infty}^{\infty} L(\tau_1) I_{hL}(t - \tau_1)\, d\tau_1 \right\}$$

$$\times \left\{ \int_{-\infty}^{\infty} L(\tau_2) I_{hL}(t + \tau - \tau_2)\, d\tau_2 \right\} dt$$

Interchanging order of integrations and using a change of integration variables

$$t' \equiv t - \tau_1; \qquad \tau_1 = t - t'$$
$$t'' \equiv t + \tau - \tau_2; \qquad \tau_2 = t + \tau - t''$$

we have

$$\phi_{hh} = \int_{-\infty}^{\infty} \int_{-\infty}^{\infty} I_{hL}(t') I_{hL}(t'') \phi_{LL}(\tau + t' - t'')\, dt'\, dt'' \qquad (3.3.28)$$

One could determine $\overline{h^2}$ from (3.3.28)

$$\overline{h^2} = \phi_{hh}(\tau = 0) = \int_{-\infty}^{\infty} \int_{-\infty}^{+\infty} I_{hL}(t') I_{hL}(t'') \phi_{LL}(t' - t'')\, dt'\, dt'' \qquad (3.3.29)$$

However we shall proceed by taking the Fourier Transform of (3.3.28).

$$\Phi_{hh} \equiv \frac{1}{\pi} \int_{-\infty}^{\infty} \phi_{hh}(\tau) e^{-i\omega\tau}\, d\tau$$

$$= \frac{1}{\pi} \int\int\int^{\infty}_{-\infty} I_{hL}(t') I_{hL}(t'') \phi_{LL}(\tau + t' - t'') e^{-i\omega\tau}\, dt'\, dt''\, d\tau$$

$$= \frac{1}{\pi} \int\int\int^{\infty}_{-\infty} I_{hL}(t') e^{+i\omega t'} I_{hL}(t'') e^{-i\omega t''}$$

$$\times \phi_{LL}(\tau + t' - t'') \exp -i\omega(\tau + t' - t'')\, dt'\, dt''\, d\tau$$

Defining a new variable

$$\tau' \equiv \tau + t' - t''$$
$$d\tau' = d\tau$$

we see that

$$\Phi_{hh}(\omega) = H_{hL}(\omega)H_{hL}(-\omega)\Phi_{LL}(\omega)$$

(3.3.30)

One can also determine that

$$\Phi_{hL}(\omega) = H_{hL}(\omega)\Phi_{LL}(\omega)$$

$$\Phi_{hh}(\omega) = H_{hL}(-\omega)\Phi_{hL}(\omega)$$

(3.3.31)

(3.3.23)[30] is a powerful and well-known relation.* The basic procedure is to determine Φ_{LL} by analysis or measurement, compute Φ_{hh} from (3.3.30) and $\overline{h^2}$ from an equation analogous to (3.3.26)[27]

$$\overline{h^2} = \int_0^\infty \Phi_{hh}(\omega)\, d\omega = \int_0^\infty |H_{hL}(\omega)|^2 \, \Phi_{LL}(\omega)\, d\omega$$

(3.3.32)

Let us illustrate the utility of the foregoing discussion by an example.

Example: airfoil response to a gust.   Again for simplicity consider translation only.

$$m\ddot{h} + K_h h = -L$$

(3.3.1)

Also for simplicity assume quasi-steady aerodynamics.†

$$L = qS\frac{\partial C_L}{\partial \alpha}\left[\frac{\dot{h}}{U} + \frac{w_G}{U}\right]$$

(3.3.33)

w_G taken as positive up, is a vertical fluid 'gust' velocity, which varies randomly with time but is assumed here to be uniformly distributed spatially over the airfoil chord. Various transfer functions may be defined and calculated. For example

$$\frac{\bar{h}}{\bar{L}} \equiv H_{hL} = \frac{-1}{m[-\omega^2 + \omega_h^2]}, \qquad \omega_h^2 \equiv K_h/m$$

(3.3.34)

is the structural (and inertial) transfer function‡ (motion due to lift) (cf. (3.3.6))

$$\frac{\bar{L}}{\bar{h}} \equiv H_{Lh} = qS\frac{\partial C_L}{\partial \alpha}\frac{i\omega}{U}$$

(3.3.35)

---

* Crandall and Mark [5].

† $\dfrac{\dot{h}}{U} + \dfrac{w_G}{U}$ is an effective angle of attack, α.

‡ Here we choose to use a dimensional rather than a dimensionless transfer function.

71

3 Dynamic aeroelasticity

is the aerodynamic transfer function (lift due to motion)

$$\frac{\bar{L}}{\bar{w}_G} \equiv H_{Lw_G} = qS\frac{\partial C_L}{\partial \alpha}\frac{1}{U} \tag{3.3.36}$$

is the aerodynamic transfer function* (lift due to gust velocity field)

$$H_{hw_G} \equiv \frac{\bar{h}}{\bar{w}_G} = \frac{-H_{Lw_G}}{\left[-\dfrac{1}{H_{hL}}+H_{Lh}\right]} \tag{3.3.37}$$

is the aeroelastic transfer function (motion due to gust velocity field).

The most general of these is the aeroelastic transfer function which may be expressed in terms of the structural and aerodynamic transfer functions, (3.3.37). Using our random force-response relations, we have from (3.3.32)

$$\bar{h}^2 = \int_0^\infty |H_{hw_G}|^2\, \Phi_{w_Gw_G}\, d\omega$$

$$= \int_0^\infty \frac{\left[qS\dfrac{\partial C_L}{\partial \alpha}\dfrac{1}{U}\right]^2}{[-m\omega^2+K_h]^2+\left[qS\dfrac{\partial C_L}{\partial \alpha}\dfrac{\omega}{U}\right]^2}\, \Phi_{w_Gw_G}\, d\omega$$

If we Define an effective damping constant as

$$\zeta \equiv \frac{qS\dfrac{\partial C_L}{\partial \alpha}.\dfrac{1}{U}}{2\sqrt{mK_h}} \tag{3.3.38}$$

then

$$\bar{h}^2 = \frac{\left[qS\dfrac{\partial C_L}{\partial \alpha}\dfrac{1}{U}\right]^2}{m^2}\int_0^\infty \frac{\Phi_{w_Gw_G}\, d\omega}{[-\omega^2+\omega_h^2]^2+4\zeta^2\omega_h^2\omega^2}$$

which, for small ζ, may be evaluated as[†]

$$\bar{h}^2 \cong \frac{qS\dfrac{\partial C_L}{\partial \alpha}}{K_h}\frac{\pi}{U}\Phi_{w_Gw_G}(\omega=\omega_h) \tag{3.3.39}$$

---

* We ignore a subtlety here in the interest of brevity. For a 'frozen gust', we must take $w_G = \bar{w}_G \exp i\omega(t-x/U_\infty)$ in determining this transfer function. See later discussion in Sections 3.6, 4.2 and 4.3.

† Crandall and Mark; the essence of the approximation is that for small ζ, $\Phi_{w_Gw_G}(\omega) \cong \Phi_{w_Gw_G}(\omega_h)$ and may be taken outside the integral. See subsequent discussion of graphical analysis.

3.3 Dynamics of the typical section model of an airfoil

Typically, the gust power spectral density is given by this

$$\Phi_{w_G w_G}(\omega) = \bar{w}_G^2 \frac{L_G}{\pi U} \frac{1 + 3\left(\frac{\omega L_G}{U}\right)^{2*}}{\left[1 + \left(\frac{\omega L_G}{U}\right)^2\right]^2} \tag{3.3.40}$$

as determined from experiment or considerations of the statistical theory of atmospheric turbulence. Here, L_G is the 'scale length of turbulence'; which is not to be confused with the lift force. Nondimensionalizing and using (3.3.39) and (3.3.40), we obtain

$$\frac{\bar{h}^2/b^2}{w_G^2/U^2} = qS \frac{\frac{\partial C_L}{\partial \alpha}}{K_n b} \frac{\frac{\omega_h L_G}{U}}{\frac{\omega_h b}{U}} \left\{ \frac{1 + 3\left(\frac{\omega_h L_G}{U}\right)^2}{\left[1 + \left(\frac{\omega_h L_G}{U}\right)^2\right]^2} \right\} \tag{3.3.41}$$

Note as $\frac{\omega_h L_G}{U} \to 0$ or ∞, $\bar{h}^2/b^2 \to 0$. Recall L_G is the characteristic length associated with the random gust field. Hence, for very large or very small characteristic lengths the airfoil is unresponsive to the gust. For what $\frac{\omega_h L_G}{b}$ does the largest response occur?

As an alternative to the above discussion, a correlation function approach could be taken where one uses the time domain and the aeroelastic impulse function,

$$\frac{I_{hw_G}}{b} = -\frac{qS \frac{\partial C_L}{\partial \alpha} \frac{1}{U} e^{-\zeta \omega_h t}}{m b \omega_h^2 \sqrt{1 - \zeta^2}} \sin \sqrt{1 - \zeta^2} \omega_h t \tag{3.3.42}$$

but we shall not pursue this here. Instead the frequency domain analysis is pursued further.

It is useful to consider the preceding calculation in graphical form for a moment. The (square of the) transfer function is plotted in Figure 3.4, and the gust power spectral density in Figure 3.5. (aeroelastic)

We note that the power spectral density is slowly varying with ω relative to the square of the transfer function which peaks sharply near $\omega = \omega_h$. Hence one may, to a close approximation, take the power spectral density as a constant with its value determined at $\omega = \omega_h$ in

---

* Houbolt, Steiner and Pratt [6]. Also see later discussion in Section 3.6.

73

3 Dynamic aeroelasticity

square of
(3,3,37)

$$\left| H_{hw_G} \right|^2$$

see p.101
for more
complicated example

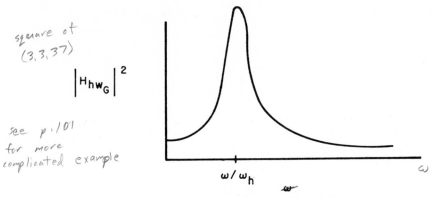

ω/ω_h

ω

Figure 3.4   Aeroelastic transfer function.

computing the mean square response. This is a simple but powerful idea which carries over to many degrees-of-freedom, and hence many resonances, provided the resonant frequencies of the transfer function are known. For some aeroelastic systems, locating the resonances may prove difficult.

There are other difficulties with the approach which should be pointed out. First of all we note that including the (aerodynamic) damping due to motion is necessary to obtain a physically meanful result. Without it the computed response would be infinite! Hence, an accurate evaluation of the effective damping for an aeroelastic system is essential in random response studies. It is known that in general the available aerodynamic theories are less reliable for evaluating the (out-of-phase with displacement) damping forces than those forces in-phase with displacement.* Another difficulty may arise if instead of evaluating the mean square displacement response we instead seek to determine the mean

eq 3.3.40

see p.100

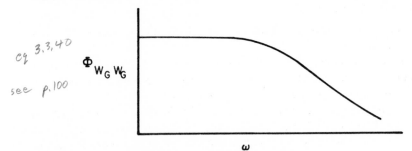

$\Phi_{W_G W_G}$

ω

Figure 3.5   Gust (auto) power spectral density.

* Acum [7].

square of acceleration. The latter quantity is frequently of greater interest from the standpoint of design. The relevant transfer function is given by

$$H_{\ddot{h}w_G} = (i\omega)^2 H_{hw_G} \tag{3.3.43}$$

and the mean square is therefore

$$\ddot{h}^2 = \frac{\int_0^\infty \omega^4 \left[qS \frac{\partial C_L}{\partial \alpha} \frac{1}{U} \right]^2 \Phi_{w_G w_G} \, d\omega}{[-m\omega^2 + K_n]^2 + \left[qS \frac{\partial C_L}{\partial \alpha} \frac{\omega}{U} \right]^2} \tag{3.3.44}$$

If we make the same approximation as before that $\Phi_{w_G w_G}$ is a constant, we are in difficulty because $|H_{hw_G}|^2$ does not approach zero as $\omega \to \infty$ and hence the integral formally diverges. This means greater care must be exercised in evaluating the integral and in particular considering the high frequency behavior of the gust power spectral density. Also, one may need to use a more elaborate aerodynamic theory. In the present example we have used a quasi-steady aerodynamic theory which is reasonably accurate for low frequencies;* however, to evaluate the acceleration response it will frequently be necessary to use a full unsteady aerodynamic theory in order to obtain accurate results at high frequencies in (3.3.44).

Measurement of power spectra. We briefly digress to consider an important application of (3.3.27) to the experimental determination of power spectra. For definiteness consider the measurement of gust power spectra. Analogous to (3.3.27) we have

$$\bar{w}_G^2 = \int_0^\infty \Phi_{w_G w_G}(\omega) \, d\omega \tag{3.3.45}$$

It is assumed that a device is available to measure w_G over a useful range of frequencies. The electronic signal from this device is then sent to an electronic 'filter'. The latter, in its most ideal form, has a transfer function given by

$$H_{Fw_G} = 1 \quad \text{for} \quad \omega_c - \frac{\Delta\omega}{2} \, \omega < \omega_c + \frac{\Delta\omega}{2} \tag{3.3.46}$$

$$= 0 \quad \text{otherwise}$$

where $\omega_c \equiv$ center frequency of the filter

$\Delta\omega \equiv$ frequency bandwidth of the filter

---

*Acum [7].

Now if we assume that the power spectrum varies slowly with ω and we choose a filter with $\Delta\omega \ll \omega_c$, then (3.3.45) may be approximated by taking $\Phi_{w_G w_G}(\omega) \cong \Phi_{w_G w_G}(\omega_c)$ and moving it outside the integral. The result is

$$\overline{w_G^2} \cong \Phi_{w_G w_G}(\omega_c)\,\Delta\omega$$

Solving for the power spectrum,

$$\Phi_{w_G w_G}(\omega_c) = \frac{\overline{w_G^2}}{\Delta\omega} \tag{3.3.47}$$

By systematically changing the filter center frequency, the power spectrum may be determined over the desired range of frequency. The frequency bandwidth, $\Delta\omega$, and the time length over which $\overline{w_G^2}$ is calculated must be chosen with care. For a discussion of these matters, the reader may consult Crandall and Mark [5], and references cited therein.

Extension of discussion on random motion to two-dimensional plate-like structures with many degrees of freedom: This extension is considered briefly in Appendix I, 'A Primer for Structural Response to Random Pressure Fluctuations'.

Flutter — an introduction to dynamic aeroelastic instability

The most dramatic physical phenomenon in the field of aeroelasticity is flutter, a dynamic instability which often leads to catastrophic structural failure. One of the difficulties in studying this phenomenon is that it is not one but many. Here we shall introduce one type of flutter using the typical section structural model and a *steady flow* aerodynamic model. The latter is a highly simplifying assumption whose accuracy we shall discuss in more detail later. From (3.3.1) and with a steady aerodynamic model, $L = qS\dfrac{\partial C_L}{\partial \alpha}\,\alpha$, $M_y = eL$, the equations of motion are

$$m\ddot{h} + S_\alpha \ddot{\alpha} + K_h h + qS\frac{\partial C_L}{\partial \alpha}\alpha = 0 \tag{3.3.48}$$

$$I_\alpha \ddot{\alpha} + S_\alpha \ddot{h} + K_\alpha \alpha - qSe\frac{\partial C_L}{\partial \alpha}\alpha = 0$$

To investigate the stability of this system we assume solutions of the form

$$h = \bar{h}e^{pt} \tag{3.3.49}$$

$$\alpha = \bar{\alpha}e^{pt}$$

and determine the possible values of p, which are in general complex numbers. If the real part of any value of p is positive, then the motion diverges exponentially with time, cf. (3.3.49), and the typical section is unstable.

To determine p, substitute (3.3.49) into (3.3.48) and use matrix notation to obtain

$$\begin{bmatrix} [mp^2 + K_h] & S_\alpha p^2 + qS\dfrac{\partial C_L}{\partial \alpha} \\ S_\alpha p^2 & I_\alpha p^2 + K_\alpha - qSe\dfrac{\partial C_L}{\partial \alpha} \end{bmatrix} \begin{Bmatrix} \bar{h}e^{pt} \\ \bar{\alpha}e^{pt} \end{Bmatrix} = \begin{Bmatrix} 0 \\ 0 \end{Bmatrix} \tag{3.3.50}$$

For nontrivial solutions the determinant of coefficients is set to zero which determines p, viz.

$$Ap^4 + Bp^2 + C = 0 \tag{3.3.51}$$

where

$$A \equiv mI_\alpha - S_\alpha^2$$

static imbalance

$$S_\alpha = \int \rho x \, dx = x_{cg}\, m$$

$$B \equiv m\left[K_\alpha - qSe\frac{\partial C_L}{\partial \alpha} \right] + K_h I_\alpha - S_\alpha qS\frac{\partial C_L}{\partial \alpha}$$

$$C \equiv K_h\left[K_\alpha - qSe\frac{\partial C_L}{\partial \alpha} \right]$$

$$I_\alpha = \int \rho x^2 \, dx \qquad \text{referred to the e.a.}$$

Solving (3.3.51),

$$p^2 = \frac{-B \pm [B^2 - 4AC]^{\frac{1}{2}}}{2A} \tag{3.3.52}$$

see p 7

$$qSe\frac{\partial C_L}{\partial \alpha} - K_\alpha = 0$$

divergence cond.

and taking the square root of (3.3.52) determines p.

The signs of A, B and C determine the nature of the solution. A is always positive for any distribution of mass; C is positive as long as q is less than its divergence value, i.e.

$$\left[K_\alpha - qSe\frac{\partial C_L}{\partial \alpha} \right] > 0$$

which is the only case of interest as far as flutter is concerned. B may be either positive or negative; re-writing

$$B = mK_\alpha + K_h I_\alpha - [me + S_\alpha]qS\frac{\partial C_L}{\partial \alpha} \tag{3.3.53}$$

If $[me + S_\alpha] < 0$ then $B > 0$ for all q.

3 Dynamic aeroelasticity

Otherwise $B<0$ when

$$K_\alpha + \frac{K_h I_\alpha}{m} - \left[1 + \frac{S_\alpha}{me}\right] qSe \frac{\partial C_L}{\partial \alpha} < 0$$

Consider in turn the two possibilities, $B>0$ and $B<0$.

$B>0$: Then the values of p^2 from (3.3.52) are real and negative provided

$$B^2 - 4AC > 0$$

and hence the values of p are purely imaginary, representing neutrally stable oscillations. On the other hand if

$$B^2 - 4AC < 0$$

the values of p^2 are complex and hence at least one value of p will have a positive real part indicating an unstable motion. Thus

$$B^2 - 4AC = 0 \tag{3.3.54}$$

gives the boundary between neutrally stable and unstable motion. From (3.3.54) one may compute an explicit value of q at which the dynamic stability, 'flutter', occurs.

From (3.3.54) we have

$$Dq_F^2 + Eq_F + F = 0$$

$$q_F = \frac{-E \pm [E^2 - 4DF]^{\frac{1}{2}}}{2D} \tag{3.3.55}$$

where

$$D \equiv \left\{ [me + S_\alpha] S \frac{\partial C_L}{\partial \alpha} \right\}^2$$

$$E \equiv \left\{ -2[me + S_\alpha][mK_\alpha + K_h I_\alpha] + 4[mI_\alpha - S_\alpha^2] e K_h \right\} S \frac{\partial C_L}{\partial \alpha}$$

$$F = [mK_\alpha + K_h I_\alpha]^2 - 4[mI_\alpha - S_\alpha^2] K_h K_\alpha$$

In order for flutter to occur at least one of the q_F determined by (3.3.55) must be real and positive. If both are, the smaller of the two is the more critical; if neither are, flutter does not occur. Pines* has studied this example in some detail and derived a number of interesting results.

---

* Pines [8].

78

Perhaps the most important of these is that for

$$S_\alpha \leq 0$$

i.e., the center of gravity is ahead of the elastic axis, no flutter occurs. Conversely as S_α increases in a positive sense the dynamic pressure at which flutter occurs, q_F, is decreased. In practice, mass is often added to a flutter prone structure so as to decrease S_α and raise q_F. Such a structure is said to have been 'mass balanced'. Now consider the other possibility for B.

$B < 0$: B is positive for $q \equiv 0$ (cf. (3.3.51) et. seq.) and will only become negative for sufficiently large q. However,

$$B^2 - 4AC = 0$$

will occur before

$$B = 0$$

since $A > 0$, $C > 0$. Hence, to determine when flutter occurs, only $B > 0$ need be considered.

Quasi-steady, aerodynamic theory

Often it is necessary to determine p by numerical methods as a function of q in order to evaluate flutter. For example, if one uses the slightly more complex 'quasi-steady' aerodynamic theory which includes the effective angle of attack contribution, $\dot{h}/U$, so that

$$qS \frac{\partial C_L}{\partial \alpha} \alpha$$

becomes

$$qS \frac{\partial C_L}{\partial \alpha} \left[\alpha + \frac{\dot{h}}{U} \right] = \rho \frac{US}{2} \frac{\partial C_L}{\partial \alpha} [U\alpha + \dot{h}]$$

then (3.3.51) will contain terms proportional to p and p^3 and the values of p must be determined numerically. An example of such a calculation is given in Figure 6.30 of B.A. which is reproduced below as Figure 3.6.
     Denote

$$p = p_R + i\omega$$
$$\omega_h^2 \equiv K_h/m, \quad \omega_\alpha^2 \equiv K_\alpha/I_\alpha$$
$$x_\alpha \equiv S_\alpha/mb, \quad r_\alpha^2 \equiv I_\alpha/mb^2$$
$$b = \text{a reference length}$$

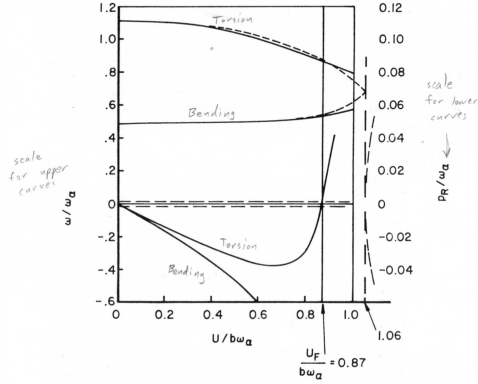

Figure 3.6   *Dimensionless frequency ω/ω_2 and damping p_R/ω_α of the aeroelastic modes of the typical section, estimated using steady-state aerodynamic operators and plotted vs. reduced airspeed $U/b\omega_\alpha$. System parameters are $x_\alpha = 0.05,\ 0.5$ $r_\alpha = 0.5,\ \omega_h/\omega_\alpha = 0.5,\ (2m/\pi\rho_\infty bS) = 10,$ $e/b = 0.4, \dfrac{\partial C_L}{\partial \alpha} = 2\pi.$ Solid curves — with aerodynamic damping. Dashed curves — without aerodynamic damping.*

Since the values of p are complex conjugate only half of them are shown. The solid lines are for the $\dot{h}/U$ or aerodynamic damping effect included and the dash lines for it omitted. There are several interesting points to be made.   (dashed lines)

(1) With aerodynamic damping omitted, the typical section model is neutrally stable until $U = U_F$. For $U = U_F$ the bending and torsion frequencies merge and for $U > U_F$ the system is unstable.

(2) With aerodynamic damping included, for small U all values of p are stable and flutter occurs at sufficiently large U where p_R changes sign from negative to positive. There is a tendency for the frequencies to merge but complete merging does not occur.

(3) The addition of aerodynamic damping reduces, *in this example for this approximate aerodynamic theory*, the flutter velocity U_F. This last result has been a source of consternation (and research papers). Whether it occurs in the real physical problem or whether it is a consequence of our simplified theoretical model is not known. No experiment has yet been performed where the aerodynamic (or structural) damping has been systematically varied to verify or refute this result.

Finally we mention one further general complication which commonly occurs in analysis. When even more elaborate, fully unsteady aerodynamic theories are employed, the aerodynamic forces are usually only conveniently known for neutrally stable motion, i.e.,

$$p = i\omega, \qquad p_R \equiv 0$$

Hence, indirect or iterative methods are usually required to effect a solution for $U = U_F$ and often no information is obtained for $U < U_F$ or $U > U_F$. We shall return to this issue later.

3.4 Aerodynamic forces for airfoils — an introduction and summary

Having developed the mathematical tools for treating the dynamics of our aeroelastic system, we now turn to a topic previously deferred, the determination of the aerodynamic forces. Usually, we wish to relate the aerodynamic lift and moment to the motion of the airfoil. In order not to break unduly the continuity of our discussion of aeroelastic phenomena, we give a brief summary of known results here and defer a discussion of the aerodynamic theory from first principles until Chapter 4.

From aerodynamic theory we know that the motion appears in the aerodynamic force relations through the 'downwash',

$$w_a \equiv \frac{\partial z_a}{\partial t} + U_\infty \frac{\partial z_a}{\partial x} \tag{3.4.1}$$

where z_a is vertical displacement of airfoil at point x, y at time t. We shall not give a formal derivation of (3.4.1) here but shall indicate the physical basis from which it follows. For an inviscid fluid the boundary condition at a fluid-solid interface, e.g., at the surface of an airfoil, requires that the fluid velocity component normal to the surface be equal to the normal velocity of the surface on the instantaneous position of the surface. (If we have a nearly planar solid surface undergoing small motions relative to its own dimensions we may apply the boundary condition on some average position of the body, say $z = 0$, rather than on the instantaneous position of the surface, $z = z_a$.) In a coordinate system fixed with respect to the fluid the boundary condition would read

$$w_a = \frac{\partial z_a}{\partial t}$$

where w_a is the normal fluid velocity component, the so-called 'downwash', and $\frac{\partial z_a}{\partial t}$ is the normal velocity of the body surface. In a coordinate system fixed with respect to the body there is an additional convection term as given in (3.4.1). This may be derived by a formal transformation from fixed fluid to fixed body axes.

Finally if in addition to the mean flow velocity, U, we also have a vertical gust velocity, w_G, then the boundary condition is that the total normal fluid velocity at the body surface be equal to the normal body velocity, i.e.,

$$w_{\text{total}} \equiv w_a + w_G = \frac{\partial z_a}{\partial x} + U_\infty \frac{\partial z_a}{\partial x}$$

where w_a is the additional fluid downwash due to the presence of the airfoil beyond that given by the prescribed gust downwash w_G. The pressure loading on the airfoil is

$$p + p_G$$

where p is the pressure due to

$$w_a = - w_G(x, t) + \frac{\partial z_a}{\partial t} + U \frac{\partial z_a}{\partial x}$$

and p_G is the prescribed pressure corresponding to the given w_G. Note, however, that p_G is continuous through $z = 0$ and hence gives no net pressure loading on the airfoil. Thus, only the pressure p due to downwash w_a is of interest in most applications.

For the typical section

$$z_a = - h - \alpha x \tag{3.4.2}$$

and

$$w_a = - w_G - \dot{h} - \dot{\alpha} x - U_\infty \alpha$$

From the first and last terms we note that $\dfrac{w_G}{U_\infty}$ is in some sense equivalent to an angle of attack, although it is an angle of attack which varies with position along the airfoil, $w_G = w_G(x, t)$!

Using the concept of aerodynamic impulse functions, we may now relate lift and moment to h, α and w_G. For simplicity let us neglect w_G for the present.

The aerodynamic force and moment can be written

$$
\begin{aligned}
L(t) \sim & \int_{-\infty}^{\infty} I_{L\dot{h}}(t-\tau)[\dot{h}(\tau) + U_\infty \alpha(\tau)] \, d\tau \\
& + \int_{-\infty}^{\infty} I_{L\dot{\alpha}}(t-\tau)\dot{\alpha}(\tau) \, d\tau
\end{aligned}
\tag{3.4.3}
$$

(3.4.3) is the aerodynamic analog to (3.3.13). Note that $\dot{h} + U_\infty \alpha$ always appear in the same combination in w_a from (3.4.2). It is conventional to express (3.4.3) in nondimensional form. Thus,

$$
\begin{aligned}
\frac{L}{qb} = & \int_{-\infty}^{\infty} I_{L\dot{h}}(s-\sigma)\left[\frac{d\frac{h}{b}(\sigma)}{d\sigma} + \alpha(\sigma)\right] d\sigma \\
& + \int_{-\infty}^{\infty} I_{L\dot{\alpha}}(s-\sigma)\left[\frac{d\alpha(\sigma)}{d\sigma}\right] d\sigma
\end{aligned}
\tag{3.4.4}
$$

and

$$
\begin{aligned}
\frac{M_y}{qb^2} = & \int_{-\infty}^{\infty} I_{M\dot{h}}(s-\sigma)\left[\frac{d\frac{h}{b}(\sigma)}{d\sigma} + \alpha(\sigma)\right] d\sigma \\
& + \int_{-\infty}^{\infty} I_{M\dot{\alpha}}(s-\sigma)\left[\frac{d\alpha(\sigma)}{d\sigma}\right] d\sigma
\end{aligned}
$$

where

$$
s \equiv \frac{tU_\infty}{b}, \qquad \sigma \equiv \frac{\tau U_\infty}{b}
$$

For the typical section, the 'aerodynamic impulse functions', $I_{L\dot{h}}$, etc., depend also upon Mach number. More generally, for a wing they vary with wing geometry as well.

(3.4.4) may be used to develop relations for sinusoidal motion by reversing the mathematical process which led to (3.3.13). Taking the Fourier Transform of (3.4.4),

$$
\frac{\bar{L}(k)}{qb} \equiv \int_{-\infty}^{\infty} \frac{L(s)}{qb} e^{-iks} \, ds = \int_{-\infty}^{\infty} \int_{-\infty}^{\infty} I_{L\dot{h}}(s-\sigma)\left[\frac{d\frac{h}{b}}{d\sigma} + \alpha\right] e^{-iks} \, d\sigma \, ds + \cdots
\tag{3.4.5}
$$

3 Dynamic aeroelasticity

where the *reduced frequency* is given by

$$k \equiv \frac{\omega b}{U_\infty}$$

Defining

$$\gamma \equiv s - \sigma, \qquad d\gamma = ds$$

$$\frac{\bar{L}(k)}{qb} = \int_{-\infty}^{\infty} \int I_{L\dot{h}}(\gamma) \left[\frac{d\frac{h}{b}}{d\sigma} + \alpha \right] e^{-ik\gamma} e^{-ik\sigma} \, d\sigma \, d\gamma + \cdots$$

$$= H_{L\dot{h}}(k) \left[ik\frac{\bar{h}}{b} + \bar{\alpha} \right] + \cdots \qquad (3.4.6)$$

where

$$H_{L\dot{h}}(k) \equiv \int_{-\infty}^{\infty} I_{L\dot{h}}(\gamma) e^{-ik\gamma} \, d\gamma$$

$$\frac{\bar{h}}{b} \equiv \int_{-\infty}^{\infty} \frac{h(\sigma)}{b} e^{-ik\sigma} \, d\sigma$$

$$\bar{\alpha} \equiv \int_{-\infty}^{\infty} \alpha(\sigma) e^{-ik\sigma} \, d\sigma$$

$H_{L\dot{h}}$, etc., are 'aerodynamic transfer functions'. From (3.4.4), (3.4.6) we may write

$$\frac{\bar{L}}{qb} = H_{L\dot{h}} \left[ik\frac{\bar{h}}{b} + \bar{\alpha} \right] + H_{L\dot{\alpha}} ik\bar{\alpha}$$

$$\frac{\bar{M}_y}{qb^2} = H_{M\dot{h}} \left[ik\frac{\bar{h}}{b} + \bar{\alpha} \right] + H_{M\dot{\alpha}} ik\bar{\alpha} \qquad (3.4.7)$$

Remember that 'transfer functions', aerodynamic or otherwise, may be determined from a consideration of sinusoidal motion only. Also note that (3.4.2), (3.4.3) and (3.4.7) are written for pitching about an axis $x = 0$. That is, the origin of the coordinate system is taken at the pitch axis. By convention, in aerodynamic analyses the axis is usually taken at mid-chord. Hence

$$z_a = -h - \alpha(x - x_{e.a.})$$

$$w = -\dot{h} - \dot{\alpha}(x - x_{e.a.}) - U_\infty \alpha$$

$$= (-\dot{h} - U_\infty \alpha) - \dot{\alpha}(x - x_{e.a.})$$

$$= (-\dot{h} - U_\infty \alpha + \dot{\alpha} x_{e.a.}) - \dot{\alpha} x \qquad (3.4.2)$$

where

$x_{e.a.}$ = distance from mid-chord to e.a.

(3.4.4) and (3.4.7) should be modified accordingly, i.e.,

$$\frac{\mathrm{d}\,\dfrac{h}{b}}{\mathrm{d}\sigma} + \alpha$$

is replaced by

$$\frac{\mathrm{d}\,\dfrac{h}{b}}{\mathrm{d}\sigma} + \alpha - \dot{\alpha}a \quad \text{where} \quad a \equiv \frac{x_{e.a.}}{b}$$

In the following table we summarize general state-of-the-art as far as available aerodynamic theories in terms of Mach number range and geometry. All assume inviscid, linearized flow models. The transonic range, $M \approx 1$, is a currently active area of research.

Aerodynamic theories available

Mach number	Geometry	
	Two dimensional	Three dimensional
$M \ll 1$	Available	Rather elaborate numerical methods available for determing transfer functions.
$M \approx 1$	Available but of limited utility because of inherent three dimensionality of flow	Rather elaborate numerical methods available for determining (linear, inviscid) transfer functions; nonlinear and/or viscous effects may be important, however.
$M \gg 1$	Available and simple because of weak memory effect.	Available and simple because of weak three dimensional effects.

The results for high speed ($M \gg 1$) flow are particularly simple. In the limit of large Mach number the (perturbation) pressure loading on an airfoil is given by

$$p = \rho \frac{U_\infty^2}{M} \left[\frac{\dfrac{\partial z_a}{\partial t} + U_\infty \dfrac{\partial z_a}{\partial x}}{U_\infty} \right]$$

85

3 Dynamic aeroelasticity

or

$$p = \rho a_\infty \left[\frac{\partial z_a}{\partial t} + U_\infty \frac{\partial z_a}{\partial x} \right]$$

This is a local, zero memory relation in that the pressure at position x, y at time t depends only on the motion at the same position and time and does *not* depend upon the motion at other positions (local effect) or at previous times (zero memory effect). This is sometimes referred to as aerodynamic 'piston theory'* since the pressure is that on a piston in a tube with velocity

$$w_a = \frac{\partial z_a}{\partial t} + U \frac{\partial z_a}{\partial x}$$

This pressure-velocity relation has been widely used in recent years in aeroelasticity and is also well known in one-dimensional plane wave acoustic theory. Impulse and transfer functions are readily derivable using aerodynamic 'piston theory'.

The 'aerodynamic impulse functions' and 'aerodynamic transfer functions' for two-dimensional, incompressible flow, although not as simple as those for $M \gg 1$, are well-known.† The forms normally employed are somewhat different from the notation of (3.4.4) and (3.4.7). For example, the lift due to transient motion is normally written

$$\frac{L}{qb} = 2\pi \left[\frac{d^2 \frac{h}{b}}{ds^2} + \frac{d\alpha}{ds} - a \frac{d^2\alpha}{ds^2} \right]$$

$$+ 4\pi \left\{ \phi(0) \left[\frac{d \frac{h}{b}}{ds} + \alpha + \left(\frac{1}{2} - a \right) \frac{d\alpha}{ds} \right] \right.$$

$$+ \left. \int_0^s \left(\frac{d \frac{h}{b}}{d\sigma} + \alpha + \left(\frac{1}{2} - a \right) \frac{d\alpha}{d\sigma} \right) \dot{\phi}(s - \sigma) \, d\sigma \right\} \tag{3.4.8}$$

One can put (3.4.8) into the form of (3.4.4) where

$$
\begin{aligned}
I_{L\dot{h}} &= 2\pi D + 4\pi \dot{\phi} + 4\pi \phi(0) \, \delta \\
I_{L\dot{\alpha}} &= 2\pi \dot{\phi} + 2\pi \phi(0) \, \delta
\end{aligned}
\tag{3.4.9}
$$

---

* Ashley, and Zartarian [9]. Also see Chapter 4.
† See Chapter 4.

Here δ is the delta function and D the doublet function, the latter being the derivative of a delta function. In practice, one would use (3.4.8) rather than (3.4.4) since delta and doublet functions are not suitable for numerical integration, etc. However, (3.4.8) and (3.4.4) are formally equivalent using (3.4.9) Note that (3.4.8) is more amenable to physical interpretation also. The terms outside the integral involving $\ddot{h}$ and $\ddot{\alpha}$ may be identified as inertial terms, sometimes called 'virtual mass' terms. These are usually negligible compared to the inertial terms of the airfoil itself if the fluid is air.* The quantity

$$-\left[\frac{\mathrm{d}\frac{h}{b}}{\mathrm{d}s}+\alpha+\left(\frac{1}{2}-a\right)\frac{\mathrm{d}\alpha}{\mathrm{d}s}\right]$$

may be identified as the downwash at the $\frac{3}{4}$ chord, Hence, the $\frac{3}{4}$ chord has been given a special place for two-dimensional, incompressible flow. Finally, note that the 'aerodynamic impulse functions', I_{Lh}, I_{Li}, can be expressed entirely in terms of a single function ϕ, the so-called Wagner function.† This function is given below in Figure 3.7. A useful approximate formulae is

$$\phi(s) = 1 - 0.165 e^{-0.0455s} - 0.335 e^{-0.3s} \tag{3.4.10}$$

For Mach numbers greater than zero, the compressibility of the flow smooths out the delta and doublet functions of (3.4.9) and no such simple

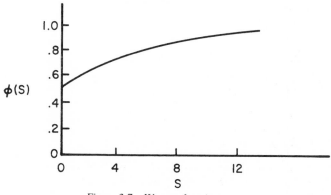

Figure 3.7   Wagner function.

---

* For light bodies or heavy fluids, e.g., lighter-than-airships or submarines, they may be important.

† For a clear, concise discussion of *transient*, two-dimensional, incompressible aerodynamics, see Sears [10], and the discussion of Sears' work in BAH, pp. 288–293.

form as (3.4.8) exists. Hence, only for incompressible flow is the form, (3.4.8), particularly useful.

Finally, we should mention that analogous impulse functions exist for gust loading due to gust vertical velocity, w_G.

$$\frac{L_G}{qb} = \oint_{-\infty}^{\infty} I_{LG}(s-\sigma)\frac{w_G(\sigma)}{U}\,d\sigma$$

$$\frac{M_{yG}}{qb^2} = \int_{-\infty}^{\infty} I_{MG}(s-\sigma)\frac{w_G(\sigma)}{U}\,d\sigma$$

(3.4.11)

For incompressible flow

$$I_{LG} = 4\pi\dot{\psi}$$

$$I_{MG} = I_{LG}(\tfrac{1}{2}+a)$$

where ψ, the Kussner function, is given by (see Figure 3.8)

$$\psi(s) = 1 - 0.5e^{-0.13s} - 0.5e^{-s}$$

(3.4.12)

The Wagner and Kussner functions have been widely employed for transient aerodynamic loading of airfoils. Even for compressible, subsonic flow they are frequently used with empirical corrections for Mach number effects. Relatively simple, exact formulae exist for two-dimensional, supersonic flow also.* However, for subsonic and/or three-dimensional flow the aerodynamic impulse functions must be determined by fairly elaborate numerical means.

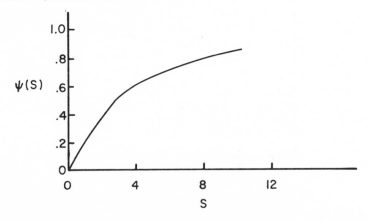

Figure 3.8   Kussner function.

---

* See BAH, pp. 367–375, for a traditional approach and Chapter 4 for an approach via Laplace and Fourier Transforms.

Finally we note that (3.4.11) may be written in the frequency domain as

$$\frac{\bar{L}_G}{qb} = H_{LG}(\omega)\frac{\bar{w}_G}{U}$$

$$\frac{\bar{M}_y}{qb^2} = H_{MG}(\omega)\frac{\bar{w}_G}{U} \qquad (3.4.13)$$

(3.4.7) and (3.4.13) will be useful when we treat the gust problem as a random process and make use of power spectral techniques. For further discussion of gust aerodynamics, see Sections 4.2 and 4.3.

General approximations

Frequently simplifying assumptions are made with respect to the spatial or temporal dependence of the aerodynamic forces. Here we discuss three widely used approximations.

'*Strip theory*' approximation.   In this approximation, one employs the known results for two-dimensional flow (infinite span airfoil) to calculate the aerodynamic forces on a lifting surface of finite span. The essence of the approximation is to consider each spanwise station as though it were a portion of an infinite span wing with uniform spanwise properties. Therefore the lift (or, more generally, chordwise pressure distribution) at any spanwise station is assumed to depend only on the downwash at that station as given by two-dimensional aerodynamic theory and to be independent of the downwash at any other spanwise station.

'*Quasisteady*' *approximation*:
The strip theory approximation discussed above is unambiguous and its meaning is generally accepted. Unfortunately, this is not true for the quasi-steady approximation. Its qualitative meaning is generally accepted, i.e., one ignores the temporal memory effect in the aerodynamic model and assumes the aerodynamic forces at any time depend only on the motion of the airfoil at that same time and are independent of the motion at earlier times. That is, the history of the motion is neglected as far as determining aerodynamic forces. For example, the piston theory aerodynamic approximation is inherently a quasi-steady approximation.

As an example of the ambiguity that can develop in constructing a quasi-steady approximation, consider the aerodynamic forces for two-dimensional, incompressible flow, e.g., see (3.4.8). One such approximation which is sometimes used is to approximate the Wagner function by

$$\phi = 1$$

and hence

$$\phi(0) = 1, \qquad \dot{\phi} = 0$$

This is clearly a quasi-steady model since the convolution integral in (3.4.8) may now be evaluated in terms of the airfoil motion at the present time, $s \equiv \dfrac{tU_\infty}{b}$, and thus the aerodynamic forces are independent of the history of the airfoil motion. An alternate quasi-steady approximation which is used on occasion is to first obtain the aerodynamic forces for steady motion, e.g., only those terms which involve α in (3.4.8) and then to define an equivalent unsteady angle of attack

$$\alpha + \frac{dh}{dt} \frac{1}{U_\infty}$$

to replace α everywhere in the steady aerodynamic theory. Clearly this second quasi-steady approximation is different from the first. (An interesting and relatively short exercise for the reader is to work out and compare these two approximations in detail using (3.4.8).) However, both are used in practice and the reader should be careful to determine exactly what a given author means by 'quasi-steady approximation'.

The ambiguity could be removed if there were general agreement that what is meant by the quasi-steady approximation is an expansion in reduced frequency for sinusoidal airfoil motion. However, even then, there would have to be agreement as to the number of terms to be retained in the expansion. (Recall that powers of frequency formally correspond to time derivatives.)

Slender body or slender (low aspect ratio) wing approximation. Another approximation based upon spatial considerations is possible when the lifting surface is of low aspect ratio or one is dealing with a slender body. In such cases the chordwise spatial rates of change (derivatives) may be neglected compared to spanwise rates of change and hence the chordwise coordinate effectively becomes a parameter rather than an independent coordinate. This approach is generally attributed to R. T. Jones.[*] It is useful as an asymptotic check on numerical methods for slender bodies and low aspect ratio wings. However it is useful for quantitative predictions for only a modest range of practical lifting surfaces.

A particularly interesting result is available for the external flow about a slender body when the body has rigid cross-sections and deforms

---

[*] Jones [11].

only in the flow direction, i.e.,

$$z_a(x, y, t) = z_a(x, t)$$

The lift force per unit chordwise distance is given by*

$$L = -\rho_\infty \frac{\mathrm{d}S}{\mathrm{d}x} U \left[U \frac{\partial z_a}{\partial x} + \frac{\partial z_a}{\partial t} \right]$$

$$- \rho_\infty S \left[U^2 \frac{\partial^2 z_a}{\partial x^2} + 2U \frac{\partial^2 z_a}{\partial x \, \partial t} + \frac{\partial^2 z_a}{\partial t^2} \right] \tag{3.4.14}$$

For a cylinder of constant, circular cross-section

$$S = \pi R^2, \qquad \frac{\mathrm{d}S}{\mathrm{d}x} = 0$$

and (3.4.14) becomes

$$L = -\rho_\infty S \left[U^2 \frac{\partial^2 z_a}{\partial x^2} + 2U \frac{\partial^2 z_a}{\partial x \, \partial t} + \frac{\partial^2 z_a}{\partial t^2} \right] \tag{3.4.15}$$

It is interesting to note that (3.4.15) is the form of the lift force used by Paidoussis and others for *internal* flows. Recall Section 2.5, equation (2.5.2). Dowell and Widnall, among others, have shown under what circumstances (3.4.15) is a rational approximation for external and internal flows.†

3.5 Solutions to the aeroelastic equations of motion

With the development of the aerodynamic relations, we may now turn to the question of solving the aeroelastic equations of motion. Substituting (3.4.4) into (3.3.1) and (3.3.2), these equations become:

$$m\ddot{h} + S_\alpha \ddot{\alpha} + K_h h = -L = \left\{ -\int_0^s I_{L\dot{h}}(s-\sigma) \left[\frac{\mathrm{d} \frac{h}{b}}{\mathrm{d}\sigma} + \alpha \right] \mathrm{d}\sigma \right.$$

$$-\int_0^s I_{L\dot{\alpha}}(s-\sigma) \frac{\mathrm{d}\alpha}{\mathrm{d}\sigma} \mathrm{d}\sigma$$

$$\left. -\int_0^s I_{LG}(s-\sigma) \frac{w_G}{U} \mathrm{d}\sigma \right\} qb$$

---

* BAH, p. 418.

† Dowell and Widnall [12], Widnall and Dowell [13], Dowell [14].

3 Dynamic aeroelasticity

and

$$I_\alpha \ddot{\alpha} + S_\alpha \ddot{h} + K_\alpha \alpha = M_y$$

$$= \left\{ \int_0^s I_{M\dot{h}}(s-\sigma) \left[\frac{d\dfrac{h}{b}}{d\sigma} + \alpha \right] d\sigma \right.$$

$$+ \int_0^s I_{M\dot{\alpha}}(s-\sigma) \frac{d\alpha}{d\sigma} d\sigma$$

$$\left. + \int_0^s I_{MG}(s-\sigma) \frac{w_G}{U} d\sigma \right\} qb^2 \qquad (3.5.1)$$

$$s \equiv \frac{tU_\infty}{b}$$

$I_{L\dot{h}}$, etc., nondimensional impulse functions. (3.5.1) are linear, differential-integral equations for h and α. They may be solved in several ways, all of which involve a moderate amount of numerical work. Basically, we may distinguish between those methods which treat the problem in the time domain and those which work in the frequency domain. The possibilities are numerous and we shall discuss representative examples of solution techniques rather than attempt to be exhaustive.

Time domain solutions

In this day and (computer) age, perhaps the most straightforward way of solving (3.5.1) (and similar equations which arise for more complicated aeroelastic systems) is by numerical integration using finite differences. Such integration is normally done on a digital computer. A simplified version of the procedure follows:

Basically, we seek a step by step solution for the time history of the motion. In particular, given the motion at some time, t, we wish to be able to obtain the motion at some later time, $t+\Delta t$. In general Δt must be sufficiently small; just how small we will discuss in a moment. In relating the solution at time, $t+\Delta t$, to that at time, t, we use the idea of a Taylor series, i.e.,

$$h(t+\Delta t) = h(t) + \frac{dh(t)}{dt} \Delta t + \frac{1}{2} \frac{d^2 h(t)}{dt^2} (\Delta t)^2 + \cdots$$

$$\alpha(t+\Delta t) = \alpha(t) + \frac{d\alpha(t)}{dt} \Delta t + \frac{1}{2} \frac{d^2\alpha(t)}{dt^2} (\Delta t) + \cdots \qquad (3.5.2)$$

If we think of starting the solution at the initial instant, $t = 0$, we see that normally $h(0)$, $dh(0)/dt$, $\alpha(0)$, $d\alpha(0)/dt$, are given as initial conditions

92

since we are dealing with (two) second order equations for h and α. However, in general, $d^2h(0)/dt^2$, $d^2\alpha(0)/dt^2$ and all higher order derivatives are *not* specified. They can be determined though from equations of motion themselves, (3.5.1). (3.5.1) are two *algebraic* equations for d^2h/dt^2, $d^2\alpha/dt^2$, in terms of lower order derivatives. Hence, they may readily be solved for d^2h/dt^2, $d^2\alpha/dt^2$. Moreover, by differentiating (3.5.1) successively the higher order derivatives may also be determined, e.g., d^3h/dt^3, etc. Hence, by using the equations of motion themselves the terms in the Taylor Series may be evaluated, (3.5.2), and h at $t = \Delta t$ determined. Repeating this procedure, the time history may be determined at $t = 2 \, \Delta t$, $3 \, \Delta t$, $4 \, \Delta t$, etc.

The above is the essence of the procedure. However, there are many variations on this basic theme and there are almost as many numerical integration schemes as there are people using them.* This is perhaps for two reasons: (1) an efficient scheme is desired (this involves essentially a trade-off between the size of Δt and the number of terms retained in the series, (3.5.2) or more generally a trade-off between Δt and the complexity of the algorithm relating $h(t + \Delta t)$ to $h(t)$); (2) some schemes including the one outlined above, are numerically unstable (i.e., numerical errors grow exponentially) if Δt is too large. This has led to a stability theory for difference schemes to determine the critical Δt and also the development of difference schemes which are inherently stable for all Δt. Generally speaking, a simple difference scheme such as the one described here will be stable if Δt is small compared to the shortest natural period of the system, say one-tenth or so. A popular method which is inherently stable for all Δt is due to Houbolt.†

An alternative but somewhat similar method to stepwise numerical integration is based upon the use of an analog computer. In such a device one again solves for the highest derivatives which are integrated by electrical devices and fed back to form the differential equation. The difficulty with such a device in the present context is having electrical components to perform the aerodynamic integrations of (3.5.1) using electronic function generators to obtain I_{Lh}, etc. There are also hybrid computers, combination analog-digital, which have their devotees for problems of this type. Generally, the digital computer is simplest to program and most flexible. However, if a large number of computations are contemplated then there may be economic advantages to using analog or hybrid computers.‡

---

* Hamming [15].

† Houbolt [16].

‡ Hausner [17].

Finally, analytical solutions or semi-analytical solutions may be obtained under certain special circumstances given sufficient simplification of the system dynamics and aerodynamics. These are usually obtained via a Laplace Transform. Since the Laplace Transform is a special case of the Fourier Transform, we defer a discussion of this topic to the following section on frequency domain solutions.

Frequency domain solutions

An alternative procedure to the time domain approach is to treat the problem in the frequency domain. This approach is more popular and widely used today than the time domain approach. Perhaps the most important reason for this is the fact that the aerodynamic theory is much more completely developed for simple harmonic motion than for arbitrary time dependent motion. That is, the unsteady aerodynamicist normally provides $H_{L\dot{h}}$, for example, rather than $I_{L\dot{h}}$. Of course, these two quantities form a Fourier Transform pair,

$$H_{L\dot{h}}(k) = \int_{-\infty}^{\infty} I_{L\dot{h}}(s)e^{-iks}\, ds$$

$$I_{L\dot{h}}(s) = \frac{1}{2\pi} \int_{-\infty}^{\infty} H_{L\dot{h}}(k)e^{iks}\, dk$$

(3.5.3)

where

$$k \equiv \frac{\omega b}{U}, \qquad s \equiv \frac{tU}{b}$$

and, in principle, given $H_{L\dot{h}}$ one can compute $I_{L\dot{h}}(s)$. However, for the more complex (and more accurate) aerodynamic theories $H_{L\dot{h}}$ is a highly oscillatory function which is frequently only known numerically at a relatively small number of frequencies, k. Hence, although there have been attempts to obtain $I_{L\dot{h}}$ by a numerical integration of $H_{L\dot{h}}$ over all frequency, they have not been conspicuously successful. Fortunately, for a determination of the stability characteristics of a system, e.g., flutter speed, one need only consider the frequency characteristics of the system dynamics, per se, and may avoid such integrations.

Another reason for the popularity of the frequency domain method is the powerful power spectral description of random loads such as gust loads, landing loads (over randomly rough surfaces), etc. These require a frequency domain description. Recall (3.3.25) and (3.3.40).

The principal disadvantage of the frequency domain approach is that one performs two separate calculations; one, to assess the system stability, 'flutter', and a second, to determine the response to external loads such as gusts, etc. This will become clearer as we discuss the details of the procedures.

Let us now turn to the equations of motion, (3.5.1), and convert them to the frequency domain by taking the Fourier Transform of these equations. The result is

$$
-\omega^2 m\bar{h} - \omega^2 S_\alpha \bar{\alpha} + K_n \bar{h} = -\bar{L}
$$

$$
= \left\{ -H_{L\dot{h}}(k)\left[\frac{i\omega\bar{h}}{U} + \bar{\alpha}\right] - H_{L\dot{\alpha}}(k)\frac{i\omega b}{U}\bar{\alpha} \right.
$$

$$
\left. -H_{LG}(k)\frac{\bar{w}_G}{U} \right\} qb
$$

$$
-\omega^2 I_\alpha \bar{\alpha} - \omega^2 S_\alpha \bar{h} + K_\alpha \bar{\alpha} = \bar{M}_y \qquad (3.5.4)
$$

$$
= \left\{ H_{M\dot{h}}(k)\left[\frac{i\omega\bar{h}}{U} + \bar{\alpha}\right] + H_{M\dot{\alpha}}(k)\frac{i\omega b}{U}\bar{\alpha} \right.
$$

$$
\left. + H_{MG}\frac{\bar{w}_G}{U} \right\} qb^2
$$

where

$$
\bar{h} \equiv \int_{-\infty}^{\infty} h(t)e^{-i\omega}\,dt, \text{ etc.}
$$

Collecting terms and using matrix notation,

$$
\left[\begin{array}{cc} \left[-\omega^2 m + K_h + \left(H_{L\dot{h}}\frac{i\omega}{U}\right)qb\right] & \left[-\omega^2 S_\alpha + \left(H_{L\dot{h}} + H_{L\dot{\alpha}}\frac{i\omega b}{U}\right)qb\right] \\ \left[-\omega^2 S_\alpha - \left(H_{M\dot{h}}\frac{i\omega}{U}\right)qb^2\right] & \left[-\omega^2 I_\alpha + K_\alpha - \left(H_{M\dot{h}} + H_{M\dot{\alpha}}\frac{i\omega b}{U}\right)qb^2\right] \end{array} \right] \left\{ \begin{array}{c} \bar{h} \\ \bar{\alpha} \end{array} \right\}
$$

$$
= qb\frac{\bar{w}_G}{U}\left\{ \begin{array}{c} -H_{LG} \\ H_{MG}b \end{array} \right\} \quad (3.5.5)
$$

Formally, we may solve for $\bar{h}$ and $\bar{\alpha}$ by matrix inversion. The result will be:

$$
\frac{\dfrac{\bar{h}}{b}}{\dfrac{\bar{w}_G}{U}} \equiv H_{hG}
$$

3 Dynamic aeroelasticity

which is one of the aeroelastic transfer functions to a gust input and

$$\frac{\bar{\alpha}}{\frac{\bar{w}_G}{U}} \equiv H_{\alpha G} \tag{3.5.6}$$

It is left to the reader to evaluate these transfer functions explicitly from (3.5.5). Note these are aeroelastic transfer functions as opposed to the purely mechanical or structural transfer functions, H_{hF} and $H_{\alpha F}$, considered previously or the purely aerodynamic transfer functions, H_{Lh}, etc. That is, H_{hG} include not only the effects of structural inertia and stiffness, but also the aerodynamic forces due to structural motion.

With the aeroelastic transfer functions available one may now formally write the solutions in the frequency domain

$$\frac{h(t)}{b} = \frac{1}{2\pi} \int_{-\infty}^{\infty} H_{hG}(\omega) \mathscr{F}\left(\frac{w_G}{U}\right) e^{-i\omega t} \, d\omega \tag{3.5.7}$$

where the Fourier Transform of the gust velocity is written as

$$\mathscr{F}w_G \equiv \int_{-\infty}^{\infty} w_G(t) e^{i\omega t} \, dt \tag{3.5.8}$$

Compare (3.5.7) with (3.3.12).

Alternatively, one could write

$$\frac{h(t)}{b} = \int_{-\infty}^{\infty} I_{hG}(t-\tau) \frac{w_G(\tau)}{U} \, d\tau \tag{3.5.9}$$

where

$$I_{hG}(t) \equiv \frac{1}{2\pi} \int_{-\infty}^{\infty} H_{hG}(\omega) e^{i\omega t} \, d\omega \tag{3.5.10}$$

Compare (3.5.9) and (3.5.10) with (3.3.13) and (3.3.14). As mentioned in our discussion of time domain solutions, the integrals over frequency may be difficult to evaluate because of the oscillatory nature of the aerodynamic forces.

Finally, for random gust velocities one may write

$$\Phi_{(h/b)(h/b)} = |H_{hG}(\omega)|^2 \Phi_{(w_G/U)(w_G/U)} \tag{3.5.11}$$

where $\Phi_{(h/b)(h/b)}$, $\Phi_{(w_G/U)(w_G/U)}$, are the (auto) power spectra of $\dfrac{h}{b}$ and $\dfrac{w_G}{U}$,

96

respectively. Thus

$$\left(\frac{\bar{h}}{b}\right)^2 = \int_0^\infty |H_{hG}|^2 \Phi_{(w_G/U)(w_G/U)} \, d\omega \tag{3.5.12}$$

Compare (3.5.12) with (3.3.25). Since the transfer function is squared, the integral (3.5.12) may be somewhat easier to evaluate than (3.5.7) or (3.5.10). The gust velocity power spectra is generally a smoothly varying function. (3.5.12) is commonly used in applications.

To evaluate stability, 'flutter', of the system one need not evaluate any integrals over frequency. It suffices to consider the eigenvalues (or poles) of the transfer function. A pole of the transfer function, ω_p, will give rise to an aeroelastic impulse function of the form

$$I_{hG} \sim e^{i\omega_p t} = e^{i(\omega_p)_R t} e^{-(\omega_p)_I t}$$

see (3.5.10). Hence, the system will be stable if the imaginary part, $(\omega_p)_I$, of all poles is positive. If any one (or more) pole has a negative imaginary part, the system is unstable, i.e., it flutters. The frequency of oscillation is $(\omega_p)_R$, the real part of the pole. Note that the poles are also the eigenvalues of the determinant of coefficients of $\bar{h}$ and $\bar{\alpha}$ in (3.5.5).

Having developed the mathematical techniques for treating dynamic aeroelastic problems we will now turn to a discussion of results and some of the practical aspects of such calculations.

3.6   Representative results and computational considerations

We will confine ourselves to two important types of motion, 'flutter' and 'gust response'.

Time domain

If we give the typical section (or any aeroelastic system) an initial disturbance due to an impulsive force, the resultant motion may take one of two possible forms as shown in Figures 3.9 and 3.10. 'Flutter' is the more interesting of these two motions, since, if it is present, it will normally lead to catastrophic structural failure which will result in the loss of the flight vehicle. All flight vehicles are carefully analyzed for flutter and frequently the structure is stiffened to prevent flutter inside the flight envelope of the vehicle.

Even if flutter does not occur, however, other motions in response to continuous external forces may be of concern with respect to possible

3   Dynamic aeroelasticity

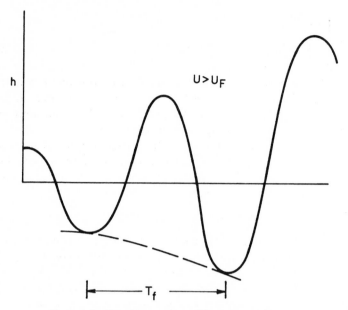

Figure 3.9   Time history of unstable motion or "flutter".

structural failure. An important example is the gust response of the flight vehicle. Consider a vertical gust velocity time history as shown in Figure 3.11. The resulting flight vehicle motion will have the form shown in Figure 3.12. Note that the time history of the response has a certain well defined average period or frequency with modulated, randomly varying amplitude. The more random input has been 'filtered' by the aeroelastic transfer function and only that portion of the gust velocity signal which has frequencies near the natural frequencies of the flight vehicle will be

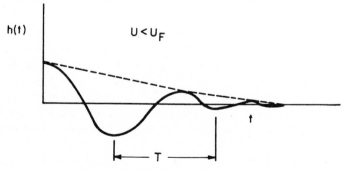

Figure 3.10   Time history of stable motion.

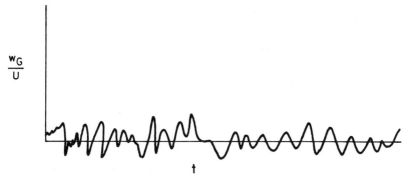

$\dfrac{w_G}{U}$

Figure 3.11   Time history of gust velocity.

identifiable in the response. This characteristic is perhaps more readily seen in the frequency domain than in the time domain.

Frequency domain

To examine flutter, we need only examine the poles of the transfer function. This is similar to a 'root locus' plot.* Typically, the real, ω_R, and imaginary, ω_I, parts of the complex frequency are plotted versus flight speed. For the typical section there will be two principal poles corresponding to two degrees of freedom and, at small flight speed or fluid velocity, these will approach the natural frequencies of the mechanical or structural system. See Figure 3.13. Flutter is identified by the lowest airspeed for which one of the ω_I becomes negative. Note the coming together or 'merging' of the ω_R of the two poles which is typical of some types of flutter. There are many variations on the above plot in practice but we shall defer a more complete discussion for the moment.

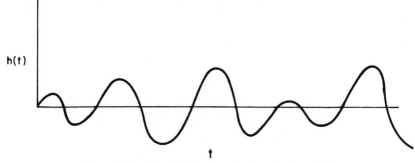

$h(t)$

Figure 3.12   Time history of motion due to gust velocity.

* Savant [18].

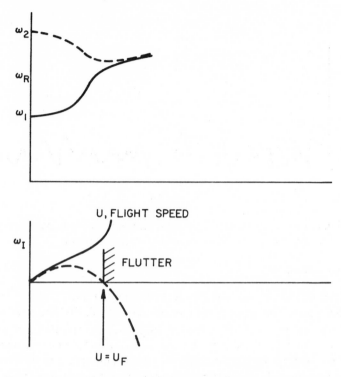

Figure 3.13   Real and imaginary components of frequency vs air speed.

Next, let us turn to the gust problem. A typical gust spectrum would be as in Figure 3.14. The transfer function (at some flight speed) would be as shown in Figure 3.15. Thus, the resultant response spectrum would appear as in Figure 3.16. As U approaches U_F, the resonant peaks of $|H_{hG}|^2$ and Φ_{hh} would approach each other for the system whose poles were sketched previously. For $U = U_F$ the two peaks would essentially collapse into one and the amplitude become infinite. For $U > U_F$ the

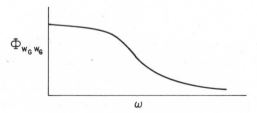

Figure 3.14   Gust power spectra.

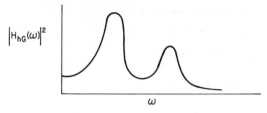

Figure 3.15   Transfer function.

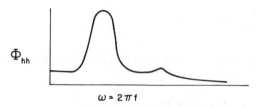

$$\omega = 2\pi f$$

Figure 3.16   Power spectra of motion.

amplitude predicted by the analytical model would become finite again for the power spectral approach and this physically unrealistic result is a possible disadvantage of the method.

Flutter ,and gust response classification including parameter trends

Here we shall study some of the important parameters which affect flutter and gust response of the typical section as well as more complex flight vehicles.

p. 57
(3.2.15)

Flutter.   If one nondimensionalizes the typical section equations of motion, one finds that the motion can be expressed formally as

$$\frac{h}{b} = F_1\left(\omega_\alpha t; \frac{S_\alpha}{mb}, \frac{I_\alpha}{mb^2}, \frac{m}{\rho(2b)^2}, \frac{e}{b}, \frac{\omega_h}{\omega_\alpha}, M, \frac{U}{b\omega_\alpha}\right)$$

$$\alpha = F_2(\omega_\alpha t; \dots) \qquad \alpha \text{ in rads} \tag{3.6.1}$$

where the functions F_1, F_2, symbolize the results of a calculated solution using one of the several methods discussed earlier.

The choice of nondimensional parameters is not unique but a matter of convenience. Some authors prefer a nondimensional dynamic pressure, or 'aeroelastic stiffness number'

$$\lambda \equiv \frac{1}{\mu k_\alpha^2} = \frac{4\rho U^2}{m\omega_\alpha^2},$$

to the use of a nondimensional velocity, $U/b\omega_\alpha$.

The following short-hand notation will be employed:

$\omega_\alpha t$ nondimensional time

$x_\alpha \equiv \dfrac{S_\alpha}{mb}$ static unbalance

$r_\alpha^2 \equiv \dfrac{I_\alpha}{mb^2}$ radius of gyration (squared)

$\mu \equiv \dfrac{m}{\rho(2b)^2}$ mass ratio

$a \equiv \dfrac{e}{b}$ location of elastic axis measured from aerodynamic center or mid-chord.

$\dfrac{\omega_h}{\omega_\alpha}$ frequency ratio

M Mach number

$k_\alpha = \dfrac{\omega_\alpha b}{U} \; = \; reduced \; freq.$

Time is an independent variable which we do not control; however, in some sense we can control the parameters, x_α, r_α, etc., by the design of our airfoil and choice of where and how we fly it. For some combination of parameters the airfoil will be dynamically unstable, i.e., it will 'flutter'.

An alternative parametric representation would be to assume sinusoidal motion

$h = \bar{h}e^{i\omega t}$

$\alpha = \bar{\alpha}e^{i\omega t}$

and determine the eigenvalues, ω. Formally, recalling $\omega = \omega_R + i\omega_I$,

$$\frac{\omega_R}{\omega_\alpha} = G_R\left(x_\alpha, r_\alpha, \mu, a, \frac{\omega_h}{\omega_\alpha}, M, \frac{U}{b\omega_\alpha}\right)$$

$$\frac{\omega_I}{\omega_\alpha} = G_I\left(x_\alpha, r_\alpha, \mu, a, \frac{\omega_h}{\omega_\alpha}, M, \frac{U}{b\omega_\alpha}\right)$$

(3.6.2)

If for some combination of parameters, $\omega_I < 0$, the system flutters.

Several types of flutter are possible. Perhaps these are most easily distinguished on the basis of the eigenvalues, ω_R/ω_α, ω_I/ω_α, and their variation with airspeed, $U/b\omega_\alpha$. Figures are shown below of the several possibilities with brief discussions of each.

Types of flutter
   'Coalescense' or 'merging frequency' flutter

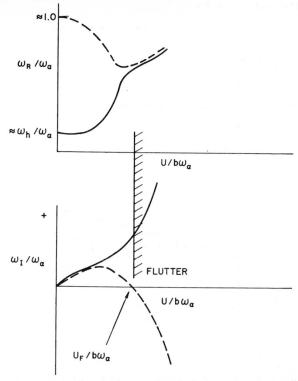

Figure 3.17   *Real and imaginary components of frequency vs air speed.*

   In this type of flutter (also called coupled mode or bending-torsion flutter) the distinguishing feature is the coming together of two (or more) frequencies, ω_R, near the flutter condition, $\omega_I \to 0$ and $U \to U_F$. For $U > U_F$ one of ω_I becomes large and positive (stable pole) and the other which gives rise to flutter becomes large and negative (unstable pole) while the corresponding ω_R remain nearly the same. Although one usually speaks of the torsion mode as being unstable and the bending mode stable, the airfoil normally is undergoing a flutter oscillation composed of important contributions of both h and α. For this type of flutter the out-of-phase or damping forces of the structure or fluid are not qualitatively important. Often one may neglect structural damping entirely in the model and use a quasi-steady or even a quasi-static aerodynamic assumption. This simplifies the analysis and, perhaps more

103

importantly, leads to generally accurate and reliable results based on theoretical calculations.

'Single-degree-of-freedom' flutter

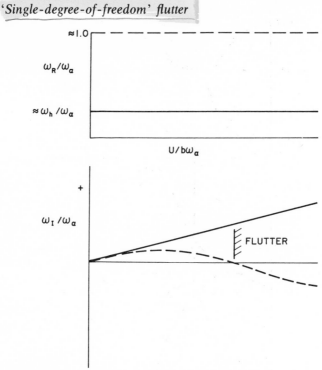

Figure 3.18   *Real and imaginary components of frequency vs air speed.*

In this type of flutter, the distinguishing feature is the virtual independence of the frequencies, ω_R, with respect to variations in airspeed, $U/b\omega_\alpha$. Moreover the change in the true damping, ω_I, with airspeed is also moderate. However, above some airspeed one of the modes (usually torsion) which has been slightly positively damped becomes slightly negatively damped leading to flutter. This type of flutter is very sensitive to structural and aerodynamic out-of-phase or damping forces. Since these forces are less well described by theory than the in-phase forces, the corresponding flutter analysis generally gives less reliable results. One simplification for this type of flutter is the fact that the flutter mode is virtually the same as one of the system natural modes at zero airspeed and thus the flutter mode and frequency (though not flutter speed!) are predicted rather accurately by theory. Airfoil blades in turbomachinery and bridges in a wind usually encounter this type of flutter.

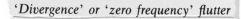

'Divergence' or 'zero frequency' flutter

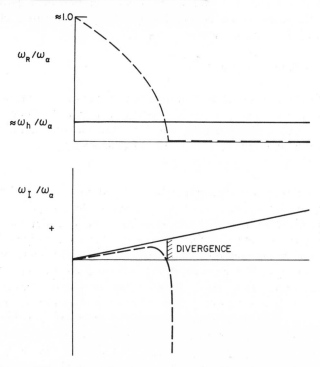

Figure 3.19   *Real and imaginary components of frequency vs air speed.*

This is also a one-degree-of-freedom type of flutter, but of a very special type. The flutter frequency is zero and hence this represents the static instability which we have previously analyzed in our discussion of static aeroelasticity under the name of 'divergence'. Because it is a static type of instability, out-of-phase forces are again unimportant and the theory is generally reliable.

We note that in all of the above we have considered only positive ω_R even though there are negative ω_R as well and these *are* physically meaningful. There are at least two reasons why this practice is usually followed. For those models where the aerodynamic transfer functions can be (approximately) expressed as a polynominal in $p \equiv i\omega$, the negative ω_R plane is (nearly) the mirror image of the positive ω_R plane and the ω_I are identical, i.e., all poles are complex conjugates in p. Secondly, some of the structural damping models employed in flutter analysis are only valid for $\omega_R > 0$; hence, the $\omega_R < 0$ in such cases cannot be interpreted in a

105

physically valid way. However, there are some types of traveling wave flutter in plates and shells for which a consideration of negative ω_R is essential. In such cases a change in sign of ω_R represents a change in direction of the traveling wave.

Flutter calculations in practice

At this point it should be emphasized that, in practice, one or another of several *indirect* methods is often used to compute the flutter velocity, e.g., the so called 'V–g method'. In this approach structural damping is introduced by multiplying the structural frequencies squared,

$$\omega_h^2, \, \omega_\alpha^2$$

by $1 + ig$ where g is a structural coefficient and pure sinusoidal motion is assumed, i.e., $\omega = \omega_R$ with $\omega_I \equiv 0$. For a given U the g required to sustain pure sinusoidal motion for each aeroelastic mode is determined. The computational advantage of this approach is that the aerodynamic forces only need be determined for real frequencies. The disadvantage is the loss of physical insight. For example, if a system (with no structural damping) is stable at a given airspeed, U, all the values of g so determined will be negative, but these values of g cannot be interpreted directly in terms of ω_I. Moreover, for a given system with some prescribed damping, only at one airspeed $U = U_F$ (where $\omega = \omega_R$ and $\omega_I \equiv 0$) will the mathematical solution be physically meaningful. The limitations of the 'V–g method' are fully appreciated by experienced practitioners and it is a measure of the difficulty of determining the aerodynamic forces for other than pure sinusoidal motion, that this method remains very popular. Here we digress from our main discussion to consider this and related methods in some detail.

For sinusoidal motion

$$h = \bar{h}e^{i\omega t}$$
$$\alpha = \bar{\alpha}e^{i\omega t}$$
$$L = \bar{L}e^{i\omega t}$$
$$M_y = \bar{M}_y e^{i\omega t}$$

The aerodynamic forces (due to motion *only*) can be expressed as

$$\bar{L} = 2\rho_\infty b^2 \omega^2 (2b) \left\{ [L_1 + iL_2] \frac{\bar{h}}{b} + [L_3 + iL_4]\bar{\alpha} \right\}$$

$$\bar{M}_y = -2\rho_\infty b^3 \omega^2 (2b) \left\{ [M_1 + iM_2] \frac{\bar{h}}{b} + [M_3 + iM_4]\bar{\alpha} \right\}$$

(3.6.3)

3.6 Representative results and computational considerations

This form of the aerodynamic forces is somewhat different from that previously used in this text and is only one of several alternative forms employed in the literature. Here L_1, L_2, L_3, L_4 are (nondimensional) real aerodynamic coefficients which are functions of reduced frequency and Mach number. L_1, L_2, L_3, L_4 are the forms in which the coefficients are generally tabulated for supersonic flow.*

$\dfrac{\omega b}{u}$

$(3.3.1+.2)$, $p58$

Using the above aerodynamic forms for $\bar{L}$ and $\bar{M}_y$ in (74) and setting the determinant of coefficients of $\bar{h}$ and $\bar{\alpha}$ to zero to determine nontrivial solutions, one obtains

$$\Delta(\omega) \equiv \left\{ \frac{m}{2\rho_\infty b(2b)} \left[1 + \left(\frac{\omega_\alpha}{i\omega}\right)^2 \left(\frac{\omega_h}{\omega_\alpha}\right)^2 \right] - [L_1 + iL_2] \right\}$$

$$\times \left\{ \frac{m}{2\rho_\infty b(2b)} r_\alpha^2 \left[1 + \left(\frac{\omega_\alpha}{i\omega}\right)^2 \right] - [M_3 + iM_4] \right\}$$

$$- \left\{ \frac{mx_\alpha}{2\rho_\infty b(2b)} - [L_3 + iL_4] \right\} \left\{ \frac{mx_\alpha}{2\rho_\infty b(2b)} - [M_1 + iM_2] \right\} = 0 \quad (3.6.4)$$

Because L_1, L_2, etc. are complicated, transcendental functions of k ** (and M) which are usually only known for real values of k (and hence real values of ω), often one does not attempt to determine from (3.6.4) the complex eigenvalue, $\omega = \omega_R + i\omega_I$. Instead one seeks to determine the conditions of neutral stability when ω is purely real. Several alternative procedures are possible; two [three] are described below.

↓   In the first the following parameters are chosen.

$\dfrac{\omega_h}{\omega_\alpha}$, r_α, x_α, M and (a real value of) k * *   ** reduced freq.

(3.6.4) is then a complex equation whose real and imaginary parts may be used independently to determine the two (real) unknowns

$$\left(\frac{\omega}{\omega_\alpha}\right)^2 \quad \text{and} \quad \frac{m}{2\rho_\infty bS}$$

From the imaginary part of (3.6.4), which is a linear equation in these two unknowns, one may solve for $(\omega/\omega_\alpha)^2$ in terms of $m/2\rho_\infty bS$. Substituting this result into the real part of (3.6.4) one obtains a quadratic equation in $m/2\rho_\infty bS$ which may be solved in the usual manner. Of course, only real positive values of $m/2\rho_\infty bS$ are meaningful and if negative or complex values are obtained these are rejected. By choosing various values of the parameters one may determine under what physically meaningful conditions flutter (neutrally stable oscillations) may occur. This procedure is

---
* Garrick [19].

3 Dynamic aeroelasticity

not easily extendable to more than two degrees of freedom and it is more readily applied for determining parameter trends than the flutter boundary of a specific structure. Hence, a different method which is described below is frequently used.

This method has the advantage of computational efficiency, though from a physical point of view it is somewhat artificial. Structural damping is introduced as an additional parameter by multiplying ω_α^2 and ω_h^2 by $1+ig$ where g is the structural damping coefficient. The following parameters are selected ω_h/ω_α, r_α, x_α, M, (a real value) of k, and $m/2\rho_\infty bS$. (3.6.4) is then identified as a complex polynomial in the complex unknown

(margin note: ② $V-g$)

(margin note: added "artificial" damping)

$$Z = \left(\frac{\omega_\alpha}{\omega}\right)^2 (1+ig)$$

Efficient numerical algorithms have been devised for determining the roots of such polynomials. A complex root determines

$$\frac{\omega_\alpha}{\omega} \quad \text{and} \quad g$$

From ω_α/ω and the previously selected value of $k \equiv \omega b/U_\infty$ one may compute

$$\frac{\omega_\alpha b}{U_\infty} = \frac{\omega_\alpha}{\omega} k$$

One may then plot g vs $U_\infty/b\omega_\alpha$*. A typical result is shown in Figure 3.20

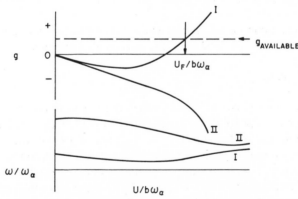

Figure 3.20. Structural damping and frequency required for neutrally stable motion vs air speed.

---

* (For each complex root of the polynomial.)

108

for two roots (two degrees of freedom). g is the value of structural damping *required* for neutral stability. If the *actual* structural damping is $g_{\text{AVAILABLE}}$ then flutter occurs when (see Figure 3.20)

$$g = g_{\text{AVAILABLE}}$$

It is normally assumed in this method that for $g < g_{\text{AVAILABLE}}$ and $U < U_F$, no flutter will occur. Sometimes more complicated velocity-damping or V–g curves are obtained, however. See Figure 3.21. Given the uncertainty as to what $g_{\text{AVAILABLE}}$ may be for a real physical system, it may then be prudent to define the flutter speed as the minimum value of $U_\infty/b\omega_\alpha$ for any $g > 0$. Here the physical interpretation of the result becomes more difficult, particularly when one recalls that the factor $1 + ig$ is only an approximate representation of damping in a structure. Despite this qualification, the V–g method remains a very popular approach to flutter analysis and is usually only abandoned or improved upon when the physical interpretation of the result becomes questionable.

One alternative to the V–g method is the so-called p–k method.[*] In this approach time dependence of the form h, $\alpha \sim e^{pt}$ is assumed where $p = \sigma + i\omega$. In the aerodynamic terms *only a* $k \equiv \omega b/U$ is *assumed*. The eigenvalues p are computed and the new ω used to compute a new k and the aerodynamic terms re-evaluated. The iteration continues until the process converges. For small σ, i.e., $|\sigma| \ll |\omega|$, the σ so computed may be interpreted as true damping of the system.

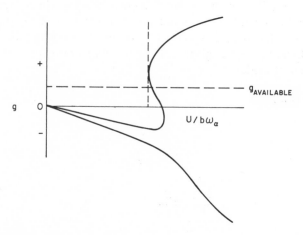

Figure 3.21 *Structural damping required for flutter vs air speed.*

[*] Hassig [20].

Nonlinear flutter behavior

There are two other types of flutter which are of importance, 'transonic buzz' and 'stall flutter'. Both of these involve significant aerodynamic nonlinearities and are, therefore, not describable by our previous models. Indeed, both are poorly understood theoretically and recourse to experiment and/or empirical rules-of-thumb is normally the only possibility.

'Transonic buzz'

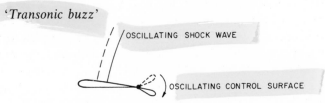

Figure 3.22   *Schematic of transonic buzz geometry.*

Typically an oscillating control surface gives rise to an oscillating shock which produces an oscillating pressure field which gives rise to an oscillating control surface which gives rise to an oscillating shock and so on and so forth.

The airfoil profile shape is known to be an important parameter and this fact plus the demonstrated importance of the shock means that any aerodynamic theory which hopes to successfully predict this type of flutter must accurately account for the nonuniform mean steady flow over the airfoil and its effect on the small dynamic motions which are superimposed due to control surface and shock oscillation. Perhaps the best theoretical model to date is that of Eckhaus; also see the discussion by Landahl. Lambourne has given a valuable summary of the experimental and theoretical evidence.*

'Stall' flutter

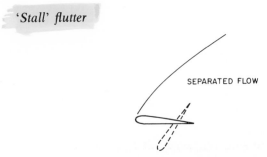

Figure 3.23   *Schematic of separated flow geometry.*

---

* Eckhaus [21], Landahl [22], Lambourne [23].

3.6 Representative results and computational considerations

An airfoil oscillating through large angles of attack will create a time lag in the aerodynamic moment which may give rise to *negative* aerodynamic damping in pitch and, hence, flutter, even though for small angles of attack the aerodynamic damping would be positive. Compressor, turbine and helicopter blades are particularly prone to this type of flutter, since they routinely operate through large ranges of angle of attack. A later chapter discusses this type of flutter in some detail.

Parameter trends

The coalescence flutter is perhaps most common for airfoils under conventional flow conditions (no shock oscillation and no stall). It is certainly the best understood. Hence, for this type of flutter, let us consider the variation of (nondimensional) flutter velocity with other important parameters.

Static unbalance. x_α:
If $x_\alpha < 0$ (i.e., c.g. is ahead of e.a.) frequently no flutter occurs. If $x_\alpha < 0$ the surface is said to be 'mass balanced'.

Frequency ratio. $\dfrac{\omega_h}{\omega_\alpha}$:

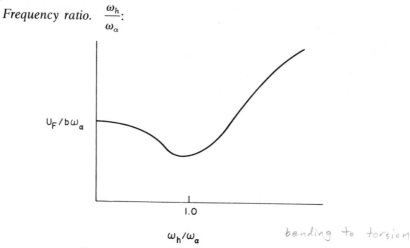

bending to torsion

Figure 3.24   *Flutter airspeed vs frequency ratio.*

Not unexpectedly, for coalescence flutter $U_F/b\omega_\alpha$ is a minimum when $\omega_h/\omega_\alpha \cong 1$.
Mach number. M:
The aerodynamic pressure on an airfoil is normally greatest near Mach number equal to one* and hence, the flutter speed tends to be a

* See Chapter 4.

111

minimum there. For $M \gg 1$ the aerodynamic piston theory predicts that the aerodynamic pressure is proportional to

$$p \sim \rho \frac{U^2}{M}$$

Hence, $U_F \sim M^{\frac{1}{2}}$ for $M \geqslant 1$ and constant μ. Also

$$\lambda_F \sim (\rho U^2)_F \sim M$$

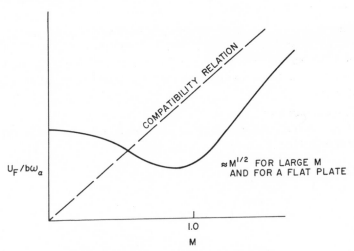

Figure 3.25   *Flutter air speed vs mach number.*

Note that for flight at constant altitude of a specific aircraft ρ (hence, μ) and a_∞ (speed of sound) are fixed. Since

$$U = M a_\infty$$

$U/b\omega_\alpha$ and M are not independent, but are related by

$$\left(\frac{U}{b\omega_\alpha}\right) = M\left(\frac{a_\infty}{b\omega_\alpha}\right)$$

Thus, a compatibility relation must also be satisfied as indicated by the dashed line in Figure 3.25. By repeating the flutter calculation for various altitudes (various ρ, a_∞ and hence various μ and $a_\infty/b\omega_\alpha$), one may obtain a plot of flutter Mach number versus altitude as given in Figure 3.26.

   Mass ratio. μ:
   For large μ the results are essentially those of a constant flutter dynamic pressure; for small μ they are often those of constant flutter

112

see p 102

$$\mu = \frac{m}{\rho(2v)^2}$$

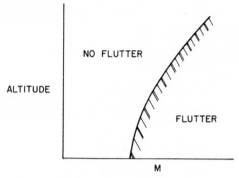

Figure 3.26 Altitude vs mach number.

velocity as indicated by the dashed line in Fig 3.27. However, for $M \equiv 0$ and two-dimensional airfoils theory predicts $\hat{U}_F \to \infty$ for some small but finite μ (solid line). This is contradicted by the experimental evidence and remains a source of some controversy in the literature.* Crisp† has recently suggested that the rigid airfoil chord assumption is untenable for small μ and that by including elastic chordwise bending the discrepancy between theory and experiment may be resolved.

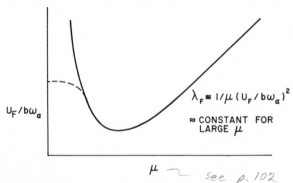

$$\lambda_F \equiv 1/\mu \, (U_F/b\omega_\alpha)^2$$

$\approx$ CONSTANT FOR LARGE μ

Figure 3.27 Flutter air speed vs mass ratio. see p. 102

Flutter prevention

After one has ascertained that there is a flutter problem then there is more than a casual curiosity as to how to fix it, i.e., increase U_F, without adding any weight, of course. There is no universal solution, but frequently one or more of the following are tried.

---

* Abramson [24]. Viscous fluid effects are cited as the source of the difficulty.
† Crisp [25].

(1) add mass or redistribute mass so that $x_\alpha < 0$, 'mass balance'

(2) increase torsional stiffness, i.e., increase ω_α

(3) increase (or decrease) $\dfrac{\omega_h}{\omega_\alpha}$ if it is near one (for fixed ω_α)

(4) add damping to the structure, particularly for single-degree-of-freedom flutter or stall flutter

(5) require the aircraft to be flown below its critical Mach number (normally used as a temporary expedient while one of the above items is studied).   Passive (slots at ¼ chord)

(6) Flutter suppression   Active (control system)

The above discussion was in the context of the typical section. For more complex aerospace vehicles, additional degrees of freedom, equations of motion and parameters will appear. Basically, these will have the form of additional frequency ratios (stiffness distribution) and inertial constants (mass distribution). Hence, *for example*, we might have

$$\frac{\omega_h}{\omega_\alpha} \quad \text{replaced by} \quad \frac{\omega_1}{\omega_\alpha}, \frac{\omega_2}{\omega_\alpha}, \frac{\omega_3}{\omega_\alpha}, \quad \text{etc.}$$

and x_α, r_α replaced by

$$\int \rho x \, dx, \int \rho x^2 \, dx, \int \rho x^3 \, dx, \text{ etc.}$$

$$\int \rho xy \, dx \, dy, \int \rho y \, dy, \int \rho y^2 \, dy, \text{ etc.}$$

Gust response.*   To the parameters for flutter we add

$$\frac{w_G}{U}$$

for gust response. Since w_G is a time history (deterministic or random) we actually add a function parameter rather than a constant. Hence, various gust responses will be obtained depending on the nature of the assumed gust time history.

The several approaches to gust response analysis can be categorized by the type of atmospheric turbulence model adopted. The simplest of these is the sharp edged gust; a somewhat more elaborate model is the 1-COSINE gust. Both of these are deterministic; in recent years a third gust model has been increasingly used where the gust velocity field is treated as a random process.

---

* Houbolt, Steiner and Pratt [6].

Discrete deterministic gust

An example of a gust time history is a *sharp edged gust,*

$$w_G = 50 \text{ ft/sec.} \quad \text{for} \quad x < Ut \left.\begin{matrix} \\ \text{or} \quad t > \dfrac{x}{U} \end{matrix}\right\} , \; x' < 0$$

$$= 0 \qquad \text{for} \quad x > Ut , \; x' > 0$$

x', t' fixed in atmosphere
x, t fixed with aircraft

(Galilean transformation) $x' = x - Ut$ (if $x' = x = 0$ at $t = t' = 0$)
$t' = t$

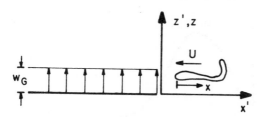

Figure 3.28   Sharp edged gust.

In this model w_G is constant with respect to space *and* time in the atmospheric fixed coordinate system for all $x' < 0$. We shall deal with the aerodynamic consequences of this property in the next chapter.

A somewhat more realistic gust model allows for the spatial scale of the gust field. In this model w_G is independent of time, t', but varies with distance, x', in the atmospheric fixed coordinate system, x', t'.

$$w_G = \frac{w_{G_{\max}}}{2} \left[1 - \cos \frac{2\pi x'}{x_G} \right]$$

$$\text{for} \quad t < \frac{x_G}{U'}, \qquad x' < 0$$

$$= 0 \quad \text{for} \quad t > \frac{x_G}{U'}, \qquad x' > 0$$

Recall

$$x' = x - U_\infty t$$

x_G is normally varied to obtain the most critical design condition and typically $w_{G_{\max}} \approx 50$ ft/sec. See sketch below.

115

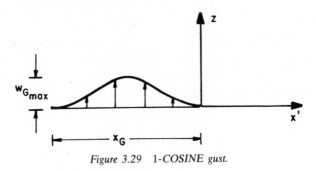

Figure 3.29 1-COSINE gust.

Schematic results for flight vehicle response to these deterministic gust models are shown below.

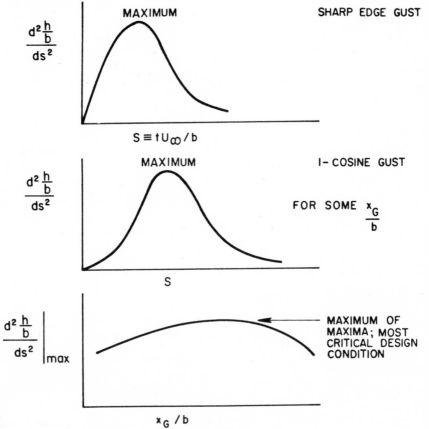

Figure 3.30 Response to deterministic gusts.

Random gust

In a random gust field, we still adopt the assumption that w_G, though now a random variable, varies only with x' and is independent of t'. In the theory of isotropic turbulence this is usually referred to as Taylor's hypothesis* or the 'frozen gust' assumption. Thus

$$w_G(x') = w_G(x - U_\infty t)$$

Since x and t only appear in the above combination, we may consider the alternative functional form

$$w_G = w_G\left(t - \frac{x}{U_\infty}\right)$$

The correlation function may then be defined as

$$\phi_{w_G w_G}(\tau) \equiv \lim_{T \to \infty} \frac{1}{2T} \int_{-\infty}^{\infty} w_G\left(t - \frac{x}{U_\infty}\right) w_G\left(t - \frac{x}{U_\infty} + \tau\right) dt$$

and the power spectral density as

$$\Phi_{w_G w_G}(\omega) \equiv \frac{1}{\pi} \int_{-\infty}^{\infty} \phi_{w_G w_G}(\tau) e^{-i\omega\tau} \, d\tau$$

The power spectral density is given in Figure 3.31. A useful approximate formula which is in reasonable agreement with measurements is†

$$\Phi_{w_G w_G} = \overline{w_G^2} \, \pi U \frac{1 + 3\left(\dfrac{\omega L_G}{U}\right)^2}{\left[1 + \left(\dfrac{\omega L_G}{U}\right)^2\right]^2}$$

Typically,

$\overline{w_G^2} \approx 33$ ft/sec.

$L_G \approx 50\text{–}500$ ft; gust scale length

We conclude this discussion with representative vehicle responses to random gust fields drawn from a variety of sources.‡ The analytical

---

* Houbolt, Steiner and Pratt [6]. The basis for the frozen gust assumption is that in the time interval for any part of the gust field to pass over the flight vehicle (the length/U_∞) the gust field does not significantly change its (random) spatial distribution. Clearly this becomes inaccurate as U_∞ becomes small.

† Houbolt, Steiner and Pratt [6].

‡ These particular examples were collected and discussed in Ashley, Dugundji and Rainey, [24].

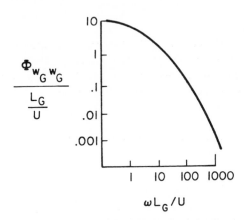

Figure 3.31   Gust power spectral density.

results are from mathematical models similar to those described above, but with more elaborate structural and aerodynamic ingredients as described in succeeding pages in this chapter and Chapter 4.

In the Figure 3.32, the measured and calculated power spectral densities for acceleration at the pilot station of the XB-70 aircraft are shown. The theoretical structural model allows for rigid body and elastic degrees of freedom using methods such as those described later in this chapter. The aerodynamic theory is similar to those described in Chapter

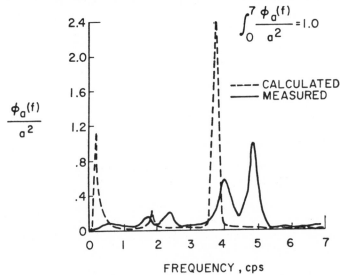

Figure 3.32   Acceleration power spectral density. From Stenton [26].

4. The dramatic conclusion drawn from this figure is that theory and experiment do not necessarily agree closely! If one assumes the peaks in the measured and calculated spectra are associated with resonances at natural frequencies of the (aeroelastic) system, then one concludes the theoretical model is not predicting these adequately. Since the resonances are determined primarily by mass and stiffness (springs), one concludes that for real vehicles even these characteristics may be difficult to model mathematically. This is quite aside from other complications such as structural damping and aerodynamic forces.

Usually when one is dealing with a real vehicle, physical small scale models are built and with these (as well as the actual vehicle when it is available) the resonant frequencies are measured (in the absence of any airflow). The results are then used to 'correct' the mathematical model, by one method or another, including a possible direct replacement of calculated resonant frequencies by their measured counterparts in the equations of motion. When this is done the peak frequencies in the measured and calculated spectra will then agree (necessarily so) and the question then becomes one of how well the peak levels agree.

A comparison for another aircraft, B-47, is shown in Figure 3.33. Here the measured and calculated resonant frequencies are in good agreement. Moreover the peak levels and indeed all levels are in good correspondence. The particular comparison shown is for the system transfer function which relates the acceleration at a point on the aircraft to the random gust input. The calculated transfer function has been obtained from an aeroelastic mathematical model. The measured transfer function (from flight test) is inferred from a measurement of gust power spectra and cross-spectra between the vehicle acceleration and gust velocity field using the relation (c.f. e.g. (3.3.31))

$$H_{\ddot{h}w_G} = \frac{\Phi_{\ddot{h}w_G}}{\Phi_{w_G w_G}}$$

Both the amplitude and phase of the transfer function are shown as a function of frequency for various positions along the wing span ($\bar{y} = 0$ is at the wing root and $\bar{y} = 1$ at the wing tip). Such good agreement between theory and experiment is certainly encouraging. However, clearly there is a major combined theoretical-experimental effort required to determine accurately the response of structures to gust loading. It should be noted that according to [6], Figure 3.33 is the bending strain transfer function. 'The dimensions of the ordinates ... are those for acceleration because the responses of the strain gages were calibrated in terms of the strain per unit normal acceleration experienced during a shallow pull-up maneuver.'

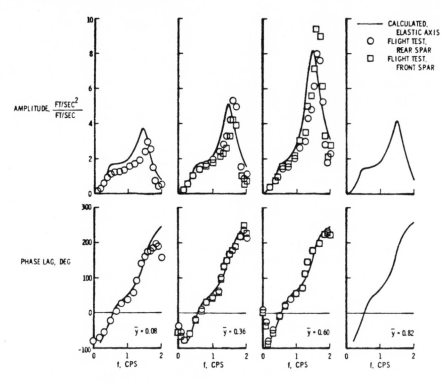

Figure 3.33 ȳ = *nondimensional distance along span. From Houbolt* [6].

3.7 Generalized equations of motion for complex structures

Lagrange's equations and modal methods (*Rayleigh-Ritz*)

The most effective method for deriving equations of motion for many complex dynamical systems is to use Lagrange's Equations.*

$$\frac{\mathrm{d}}{\mathrm{d}t}\frac{\partial L}{\partial \dot{q}_i} - \frac{\partial L}{\partial q_i} = Q_i$$

where

$L \equiv T - U$, Lagrangian
$T \equiv$ kinetic energy
$U \equiv$ potential energy
$Q_i \equiv$ generalized forces
$q_i \equiv$ generalized coordinates

---

* Recall Section 3.2.

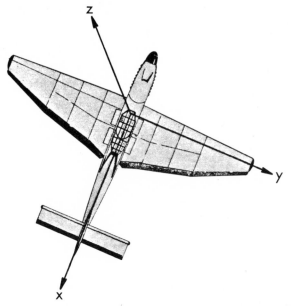

Figure 3.34   Two-dimensional (planar) representation of a flight vehicle.

The essential steps in the method are, first, a suitable choice of q_i and then an evaluation of T, U and Q_i in terms of q_i and $\dot{q}_i$.

Lagrange's equations have, as one of their principal advantages, the ability to obtain the equations of motion for complex systems with little or no more difficulty than that required for rather simple ones, such as the 'typical section'. Here we shall consider a two-dimensional (planar) representation of a flight vehicle. (See Figure 3.34).

We note that this formulation can include 'rigid' body as well as flexible body modes. For example, the following choices of modal functions, z_m, include rigid body vertical translation, pitching (rotation about y axis) and rolling (rotation about x axis), respectively

$z_1 = 1$ vertical translation
$z_2 = x$ pitching
$z_3 = y$ rolling

For such modes the potential elastic or strain energy is zero; however, in general, strain energy must be included for the flexible body modes.

The use of Lagrange's equations, while formally compact, does not reveal explicitly all of the complications which may arise in deriving equations of motion for an unrestrained vehicle or structure. These are

121

3 Dynamic aeroelasticity

seen more clearly in the discussion in a later section of integral equations of equilibrium.

Kinetic energy

The x–y plane is the plane of the (aircraft) structure. We consider deformations perpendicular to the x–y plane (in the z direction). The normal displacement with respect to a fixed inertial reference plane we call $z_a(x, y, t)$. We may then express the kinetic energy as

$$T = \tfrac{1}{2} \iint m \dot{z}_a^2 \, dx \, dy \qquad (3.7.1)$$

where m – mass/area and $\dot{z} \equiv \dfrac{\partial}{\partial t}$. If we expand the displacement in a modal series, say

$$z_a = \sum_m q_m(t) z_m(x, y) \qquad (3.7.2)$$

then the kinetic energy may be written as

$$T = \tfrac{1}{2} \sum_m \sum_n \dot{q}_m \dot{q}_n M_{mn} \qquad (3.7.3)$$

where the *generalized mass* is given by

$$M_{mn} \equiv \iint m z_m z_n \, dx \, dy$$

For small motions the above integral over the body may be taken as over the undeformed structure.

If the chosen modes, z_m, satisfy an orthogonality relation

$$M_{mn} = M_m \delta_{mn} \qquad \begin{aligned} \delta_{mn} &= 1 \quad \text{for} \quad m = m \\ &= 0 \quad \text{for} \quad m \neq n \end{aligned}$$

then (3.7.3) simplifies to

$$T = \tfrac{1}{2} \sum_m \dot{q}_m^2 M_m \qquad (3.7.4)$$

Strain (potential, elastic) energy

For the strain energy, we may write a similar relation to (3.7.3).

$$U = \tfrac{1}{2} \sum_m \sum q_m q_n K_{mn} \qquad (3.7.5)$$

122

where K_{mn} is a generalized spring constant which is determined from an appropriate structural theory.* Indeed if the z_m are the 'natural' or 'normal' modes of the structure, one may show that

$$K_{mn} = \omega_m^2 M_m \, \delta_{mn} \tag{3.7.6}$$

where ω_m is the mth 'natural frequency'.†

Equations (3.7.3)–(3.7.6) are the keys to the Lagrangian approach. Before continuing, we pause to consider K_{mn} in more detail.

Alternative determination of K_{mn}. A stiffness influence function, $K(x, y; \xi, \eta)$, may be defined which is the (static) force/area required at point x, y to give a unit deflection at point ξ, η. Hence

$$p(x, y) = \iint K(x, y; \xi, \eta) z_a(\xi, \eta) \, d\xi \, d\eta \tag{3.7.7}$$

$$F = K \, d \quad \text{simple spring analog}$$

The potential energy stored in the elastic body is thus

$$U = \tfrac{1}{2} \iint p(x, y) z_a(x, y) \, dx \, dy$$

$$U = \tfrac{1}{2} Fd \quad \text{simple spring analog}$$

Using (3.7.8) in (3.7.7),

$$U = \tfrac{1}{2} \iiiint K(x, y; \xi, \eta) z_a(\xi, \eta) z_a(x, y) \, d\xi \, d\eta \, dx \, dy \tag{3.7.9}$$

$$U = \tfrac{1}{2} Kd^2 \quad \text{simple spring analog}$$

Using our modal expansion

$$z_a(x, y, t) = \sum_m q_m(t) z_m(x, y)$$

in (3.7.9) we obtain

$$U = \tfrac{1}{2} \sum_m \sum_n K_{mn} q_m q_n$$

where

$$K_{mn} \equiv \iiiint K(x, y; \xi, \eta) z_m(\xi, \eta) z_n(x, y) \, d\xi \, d\eta \, dx \, dy$$

$$U = \tfrac{1}{2} Kd^2 \quad \text{simple spring analog} \tag{3.7.10}$$

---

* Recall Section 3.1.
† Meirovitch [4].

3   Dynamic aeroelasticity

From Maxwell's Reciprocity Theorem

$$K(x, y; \xi, \eta) = K(\xi, \eta; x, y)$$

and hence

$$K_{mn} = K_{nm} \tag{3.7.11}$$

$K(x, y; \xi, \eta)$ can be determined by a suitable theoretical analysis or it can be inferred from experiment. For the additional insight to be gained, let us consider the latter alternative. It is a difficult experiment to measure K directly since we must determine a distribution of force/area which gives unit deflection at one point and zero deflection elsewhere. Instead it is much easier to measure the inverse of K, a flexibility influence, *function* $C(x, y; \xi, \eta)$, which is the deflection at x, y due to a unit force/area at ξ, η. For $C(x, y; \xi, \eta)$ we have the following relation (recall Section 2.4)

$$z_a(x, y) = \iint C(x, y; \xi, \eta) p(\xi, \eta) \, d\xi \, d\eta \tag{3.7.12}$$

Using (3.7.6) and (3.7.1) it can be shown that

$$\iint C(x, y; \xi, \eta) K(\xi, \eta; r, s) \, d\xi \, d\eta = \delta(r - x, s - y) \tag{3.7.13}$$

where δ is a Dirac delta function. (3.7.13) is an integral equation for C or K given the other. However, it is rarely, if ever, used. Instead (3.7.6) and (3.7.1) are attacked directly by considering a finite number of loads and deflections over small (finite) areas of size $\Delta x \, \Delta y = \Delta \xi \, \Delta \eta$. Hence (3.7.7) and (3.7.12) are written

$$p(x_i, y_i) = \sum_j K(x_i, y_i; \xi_j, \eta_j) z_a(\xi_j, \eta_j) \, \Delta \xi \, \Delta \eta \tag{3.7.7}$$

$$z_a(x_j, y_j) = \sum_i C(x_j, y_j; \xi_i, \eta_i) p(\xi_i, \eta_i) \, \Delta \xi \, \Delta \eta \tag{3.7.12}$$

In matrix notation

$$\{p\} = [K]\{z_a\}\Delta \xi \, \Delta \eta \tag{3.7.7}$$

$$\{z_a\} = [C]\{p\}\Delta \xi \, \Delta \eta \tag{3.7.12}$$

Substitution of (3.7.12) into (3.7.7) and solving, gives

$$[K] = [C]^{-1}/(\Delta \xi)^2 (\Delta \eta)^2 \tag{3.7.14}$$

(3.7.14) is essentially a finite difference solution to (3.7.13). Hence, *in practice*, if (3.7.10) is used to compute K_{mn}, one measures C, computes K from (3.7.14) and then evaluates K_{mn} by numerical integration of

124

(3.7.10). For a fuller discussion of influence functions, the reader may wish to consult Bisplinghoff, Mar and Pian [2].

There is one further subtlety which we have not discussed as yet. If rigid body motions of the structure are possible, then one may wish to use a C measured with respect to a fixed point. For example, it may be convenient to measure C with the center of mass fixed with respect to translation and rotation. This matter is discussed more fully later in this chapter when integral equations of equilibrium are reviewed.

We now continue the general discussion from which we digressed to consider K_{mn}. Two examples will be considered next.

Examples

(a) *Torsional vibrations of a rod.* To illustrate the key relations (3.7.3)–(3.7.6) in a more familiar situation, consider the torsional vibrations of a rod.

Here

$$z_a = -x\alpha(y, t) \qquad \text{cf. (3.7.2)}$$

and thus (3.7.1) becomes

$$T = \tfrac{1}{2}\int I_\alpha \dot\alpha^2 \, dy \qquad (3.7.15)$$

where

$$I_\alpha \equiv \int mx^2 \, dx$$

$$\alpha \equiv \text{angle of twist}$$

From structural theory [2],

$$U = \tfrac{1}{2}\int GJ\left(\frac{d\alpha}{dy}\right)^2 dy \qquad (3.7.16)$$

Let

$$\alpha = \sum_{m=1}^{M} q_m^\alpha \alpha_m(y) \qquad (3.7.17)$$

then

$$T = \tfrac{1}{2}\sum_m \sum_n \dot q_m^\alpha \dot q_n^\alpha M_{mn} \qquad (3.7.18)$$

where

$$M_{mn} \equiv \int I_\alpha \alpha_m \alpha_n \, dy \qquad \text{cf. (3.7.3)}$$

125

and

$$U = \tfrac{1}{2} \sum_m \sum_n q_m^\alpha q_n^\alpha K_{mn} \qquad (3.7.19)$$

where

$$K_{mn} = \iint GJ \frac{\mathrm{d}\alpha_m}{\mathrm{d}y} \frac{\mathrm{d}\alpha_n}{\mathrm{d}y} \, \mathrm{d}y \qquad \text{cf. } (3.7.5)$$

The specific structural model chosen determines the accuracy with which the generalized masses and stiffnesses are determined, but they always exist.

(b) *Bending-torsional motion of a beam-rod.* The above is readily generalized to include bending as well as torsional vibrations of a beam-rod.

Let

$$z_a(x, y, t) = -x\alpha(y, t) - h(y, t) \qquad \text{cf. } (3.7.2)$$

$\alpha \equiv$ twist about elastic axis

$h \equiv$ bending deflection of elastic axis

and thus (3.7.1) becomes

$$T = \tfrac{1}{2} \left\{ \int M\dot{h}^2 \, \mathrm{d}y + 2 \int S_\alpha \dot{h}\dot{\alpha} \, \mathrm{d}y + \int I_\alpha \dot{\alpha}^2 \, \mathrm{d}y \right\} \qquad (3.7.15)$$

where

$$M \equiv \int m \, \mathrm{d}x, \qquad S_\alpha \equiv \int mx \, \mathrm{d}x, \qquad I_\alpha \equiv \int mx^2 \, \mathrm{d}x$$

Also from structural theory [2],

$$U = \frac{1}{2} \left\{ \int GJ \left(\frac{\partial \alpha}{\partial y} \right)^2 \mathrm{d}y + \int EI \left(\frac{\partial^2 h}{\partial y^2} \right)^2 \right\} \mathrm{d}y \qquad (3.7.16)$$

Let

$$h = \sum_{r=1}^{R} q_r^h h_r(y)$$

$$\alpha = \sum_{m=1}^{M} q_m^\alpha \alpha_m(y) \qquad (3.7.17)$$

Then

$$T = \tfrac{1}{2} \sum_m \sum_n \dot{q}_m^\alpha \dot{q}_n^\alpha M_{mn}^{\alpha\alpha}$$

$$+ 2 \sum_m \sum_r \dot{q}_m^\alpha \dot{q}_r^h M_{mr}^{\alpha h} + \sum_r \sum_s \dot{q}_r^h \dot{q}_s^h M_{rs}^{hh} \qquad \text{cf. } (3.7.3)$$

where

$$M_{mn}^{\alpha\alpha} \equiv \int I_\alpha \alpha_m \alpha_n \, dy, \qquad M_{mr}^{\alpha h} \equiv \int S_\alpha \alpha_m h_r \, dy, \qquad M_{rs}^{hh} \equiv \int m h_r h_s \, dy$$

$$(3.7.18)$$

and

$$U = \tfrac{1}{2} \left\{ \sum_m \sum_n q_m^\alpha q_n^\alpha K_{mn}^{\alpha\alpha} + \sum_r \sum_s q_r^h q_s^h K_{rs}^{hh} \right\} \qquad \text{cf. (3.7.5)}$$

where

$$K_{mn}^{\alpha\alpha} \equiv \int GJ \frac{d\alpha_m}{dy} \frac{d\alpha_n}{dy} \, dy, \qquad K_{rs}^{hh} \equiv \int EI \frac{d^2 h_r}{dy^2} \frac{d^2 h_s}{dy^2} \, dy \qquad (3.7.19)$$

Of all possible choices of modes, the 'free vibration, natural modes' are often the best choice. These are discussed in more detail in the next section.

Natural frequencies and modes-eigenvalues and eigenvectors

Continuing with our general discussion, consider Lagrange's equations with the generalized forces set to zero,

$$\frac{d}{dt} \left(\frac{\partial(T-U)}{\partial \dot{q}_i} \right) + \frac{\partial U}{\partial q_i} = 0 \qquad i = 1, 2, \ldots, M$$

and thus obtain, using (3.7.3) and (3.7.5) in the above,

$$\sum M_{mi} \ddot{q}_m + K_{mi} q_m = 0 \qquad i = 1, \ldots, M \qquad (3.7.20)$$

Consider sinusoidal motion

$$q_m = \bar{q}_m e^{i\omega t} \qquad (3.7.21)$$

then, in matrix notation, (3.7.20) becomes

$$-\omega^2 [M]\{q\} + [K]\{q\} = \{0\} \qquad (3.7.22)$$

This is an eigenvalue problem, for the eigenvalues, ω_j, $j = 1, \ldots, M$ and corresponding eigenvectors, $(q)_j$. If the function originally chosen, z_m or α_m and h_r, were 'natural modes' of the system then the M and K matrices will be diagonal and the eigenvalue problem simplifies.

$$-\omega^2 \left[\begin{smallmatrix} \ddots \\ & M \\ & & \ddots \end{smallmatrix} \right] \{q\} + \left[\begin{smallmatrix} \ddots \\ & M\omega_j^2 \\ & & \ddots \end{smallmatrix} \right] \{q\} = \{0\} \qquad (3.7.23)$$

127

and

$$\omega_1^2, \begin{Bmatrix} q_1 \\ 0 \\ 0 \\ 0 \\ 0 \end{Bmatrix}_1$$

$$\omega_2^2, \begin{Bmatrix} 0 \\ q_2 \\ 0 \\ 0 \\ 0 \end{Bmatrix}_2$$

etc.

$$\omega_M^2, \begin{Bmatrix} 0 \\ 0 \\ 0 \\ q_M \end{Bmatrix}_M$$

If this is not so then the eigenvalues may be determined from (3.7.22) and a linear transformation may be made to diagonalize the M and K matrices. The reader may wish to determine the eigenvalues and eigenvectors of the typical section as an exercise.

For our purposes, the key point is that expressions like (3.7.3)–(3.7.6) exist. For a more extensive discussion of these matters, the reader may consult Meirovitch [4].

Evaluation of generalized aerodynamic forces

The generalized forces in Lagrange's equations are evaluated from their definition in terms of virtual work.

$$\delta W_{NC} = \sum_m Q_m \delta q_m \tag{3.7.24}$$

Now the virtual work may be evaluated independently from

$$\delta W_{NC} = \int\int p \delta z_\alpha \, dx \, dy \tag{3.7.25}$$

where p is the net aerodynamic pressure on an element of the structure with (differential) area $dx \, dy$. Using (3.7.2) in (3.7.25)

$$\delta W_{NC} = \sum_m \delta q_m \int\int p z_m \, dx \, dy \tag{3.7.26}$$

and we may identify from (3.7.24) and (3.7.26)

$$Q_m \equiv \iint p z_m \, \mathrm{d}x \, \mathrm{d}y \qquad (3.7.27)$$

From aerodynamic theory,* one can establish a relation of the form

$$p(x, y, t) = \iiint\limits_0^t A(x - \xi, y - \eta, t - \tau)$$

$$\times \underbrace{\left[\frac{\partial z_a}{\partial \tau} (\xi, \eta, \tau) + U \frac{\partial z_a}{\partial \xi} (\xi, \eta, \tau) \right]}_{\text{'downwash'}} \mathrm{d}\xi \, \mathrm{d}\eta \, \mathrm{d}\tau \quad (3.7.28)$$

A may be physically interpreted as the pressure at point x, y at time t due to a unit impulse of downwash at point ξ, η at time τ. Using (3.7.2) and (3.7.28) in (3.7.27) we may evaluate Q_m in more detail,

$$Q_m = \sum_n \int_0^t \left[\dot{q}_n(\tau) I_{nm\dot{q}}(t - \tau) + q_n(\tau) I_{nmq}(t - \tau) \right] \mathrm{d}\tau \qquad (3.7.29)$$

where

$$I_{nm\dot{q}}(t - \tau) \equiv \iiiint A(x - \xi, y - \eta, t - \tau) z_n(\xi, \eta) z_m(x, y) \, \mathrm{d}x \, \mathrm{d}y \, \mathrm{d}\xi \, \mathrm{d}\eta$$

$$I_{nmq}(t - \tau) \equiv \iiiint A(x - \xi, y - \eta, t - \tau)$$

$$\times U \frac{\partial z_n}{\partial \xi} (\xi, \eta) z_m(x, y) \, \mathrm{d}x \, \mathrm{d}y \, \mathrm{d}\xi \, \mathrm{d}\eta$$

$I_{nm\dot{q}}$, I_{nmq} may be thought of as generalized aerodynamic impulse functions.

Equations of motion and solution methods

Finally applying Lagrange's equations, using 'normal mode' coordinates for simplicity,

$$M_m[\ddot{q}_m + \omega_m^2 q_m] = \sum_{n=1}^M \int_0^t \left[\dot{q}_n(\tau) I_{nm\dot{q}}(t - \tau) + q_n(\tau) I_{nmq}(t - \tau) \right] \mathrm{d}\tau$$

$$m = 1, \ldots, M \quad (3.7.30)$$

---

* See Chapter 4, and earlier discussion in Section 3.4.

Note the form of (3.7.30). It is identical, mathematically speaking, to the earlier results for the typical section.* Hence similar mathematical solution techniques may be applied.

Time domain solutions. Taylor Series expansion

$$q_n(t + \Delta t) = q_n(t) + \dot{q}_n \Big|_t \Delta t + \frac{\ddot{q}_n}{2} \Big|_t (\Delta t)^2$$

One may solve for $\ddot{q}_n$ from (3.7.30) and hence $q_n(t + \Delta t)$ is determined. $q_n(t)$, $\dot{q}_n(t)$ are known from initial conditions and

$$\dot{q}_n(t + \Delta t) = \dot{q}_n(t) + \ddot{q}_n(t)\Delta t + \cdots \qquad (3.7.31)$$

Frequency domain solutions. Taking a Fourier Transform of (3.7.30)

$$M_m[-\omega^2 + \omega_m^2]\bar{q}_m = \sum_n^M [i\omega H_{nm\dot{q}} + H_{nmq}]\bar{q}_n$$

where

$$\bar{q}_m \equiv \int_{-\infty}^{\infty} q_m e^{-i\omega t}\, dt$$

In matrix notation

$$\left[\left[\diagdown M_m(-\omega^2 + \omega_m^2) \diagdown \right] - [i\omega H_{nm\dot{q}} + H_{nmq}] \right] \{\bar{q}_n\} = \{0\} \qquad (3.7.32)$$

By examining the condition for nontrivial solutions

$$\|[\cdots]\| = 0$$

we may find the 'poles' of the aeroelastic transfer functions and assess the stability of the system.

Response to gust excitation. If we wish to examine the gust response problem then we must return to (3.7.28) and add the aerodynamic pressure due to the gust loading

$$p_G(x, y, t) = \iiint A(x - \xi, y - \eta, t - \tau) w_G(\xi, \eta, \tau)\, d\xi\, d\eta\, d\tau$$

---

* Provided $S_\alpha \equiv 0$ so that h, α are normal mode coordinates for the typical section.

The resulting generalized forces are

$$Q_{mG}(t) = \iiint\!\!\iint A(x - \xi, y - \eta, t - \tau)$$

$$\times w_G(\xi, \eta, \tau) z_m(x, y) \, d\xi \, d\eta \, dx \, dy \, d\tau \quad (3.7.33)$$

Adding (3.7.33) to (3.7.30) does not change the mathematical technique for the time domain solution. In the frequency domain, the right hand column of (3.7.32) is now $(\bar{Q}_{mG})$

$$\bar{Q}_{mG} = \int_{-\infty}^{\infty} Q_{mG} e^{-i\omega t} \, dt$$

Hence by solving (3.7.32) we may obtain generalized aeroelastic transfer functions

$$\frac{\bar{q}_n}{\bar{Q}_{mG}} \equiv H_{q_n Q_{mG}}(\omega; \cdots) \tag{3.7.34}$$

and employ the usual techniques of the frequency domain calculus including power spectral methods.

Integral equations of equilibrium

As an alternative approach to Lagrange's Equations, we consider an integral equation formulation using the concept of a structural influence (Green's) function. We shall treat a flat (two-dimensional) structure which deforms under (aerodynamic) loading in an arbitrary way. We shall assume a symmetrical vehicle and take the origin of our coordinate system at the vehicle center of mass with the two axes in the plane of the vehicle as principal axes, x, y. See Figure 3.34. Note the motion is assumed sufficiently small so that no distinction is made between the deformed and undeformed axes of the body. For example the inertia and elastic integral properties are evaluated using the (undeformed) axes x, y. The axes x, y are inertial axes, i.e., fixed in space. If we consider small deflections normal to the x, y plane, the x, y axes are approximately the principal axes of the deformed vehicle.

It will be useful to make several definitions.

z_a absolute vertical displacement of a point from x, y plane, positive up

m mass/area

p_E external applied force/area, e.g., aerodynamic forces due to gust, p_G

p_M force/area due to motion, e.g., aerodynamic forces (but not including inertial forces)

$$p_z = p_E + p_M - m\frac{\partial^2 z_a}{\partial t^2}$$

total force/area, including inertial forces. Let us first consider equilibrium of rigid body motions.

Translation: $\qquad \iint p_z \, dx \, dy = 0$ $\qquad\qquad\qquad$ (3.7.35)

Pitch: $\qquad\quad \iint x p_z \, dx \, dy = 0$ $\qquad\qquad\qquad$ (3.7.36)

Roll: $\qquad\quad\; \iint y p_z \, dx \, dy = 0$ $\qquad\qquad\qquad$ (3.7.37)

Now consider equilibrium of deformable or elastic motion.

$$z_a^{\text{elastic}} \equiv z_a(x, y, t) - z_a(0, 0, t) - x\frac{\partial z_a}{\partial x}(0, 0, t) - y\frac{\partial z_a}{\partial y}(0, 0, t)$$

$$= \iint C(x, y; \xi, \eta) p_z(\xi, \eta, t) \, d\xi \, d\eta \qquad\qquad (3.7.38)$$

where

$z_a^{\text{elastic}} \equiv$ deformation (elastic) of a point on vehicle

$\quad C \equiv$ structural influence or Green's function; the (static) elastic deformation at x, y due to unit force/area at ξ, η for a vehicle fixed* at origin, $x = y = 0$.

Since the method of obtaining the subsequent equations of motion involves some rather extensive algebra, we outline the method here.

(1) Set $p_E = p_M = 0$.

(2) Obtain 'natural frequencies and modes'; prove orthogonality of modes.

(3) Expand deformation, z_a, for nonzero p_E and p_M in terms of normal modes or natural modes and obtain a set of equations for the (time dependent) coefficients of the expansion. The final result will again be (3.7.30).

---

* By fixed we mean 'clamped' in the sense of the structural engineer, i.e., zero displacement and slope. It is sufficient to use a static influence function, since invoking D'Alambert's Principle the inertial contributions are treated as equivalent forces.

Natural frequencies and modes

Set $p_E = p_M = 0$. Assume sinusoidal motion, i.e.,

$$z_a(x, y, t) = \bar{z}_a(x, y)e^{i\omega t} \qquad (3.7.39)$$

then (3.7.38) becomes

$$\bar{z}_a(x, y) - \bar{z}_a(0, 0) - x\frac{\partial \bar{z}_a}{\partial x}(0, 0) - y\frac{\partial \bar{z}_a}{\partial y}(0, 0)$$

$$= \omega^2 \iint C(x, y; \xi, \eta)m(\xi, \eta)\bar{z}_a(\xi, \eta)\, d\xi\, d\eta \qquad (3.7.40)$$

The frequency, ω, has the character of an eigenvalue. (3.7.40) can be put into the form of a standard eigenvalue problem by solving for $\bar{z}_a(0, 0)$, $\frac{\partial \bar{z}_a}{\partial x}(0, 0)$, $\frac{\partial \bar{z}_a}{\partial y}(0, 0)$ and substituting into (3.7.40). For example, consider the determination of $\bar{z}_a(0, 0)$. Multiply (3.7.40) by m and integrate over the flight vehicle area. The result is:

$$\iint m\bar{z}_a\, dx\, dy - \bar{z}_a(0, 0)\iint m\, dx\, dy$$

$$-\frac{\partial \bar{z}_a}{\partial x}(0, 0)\iint mx\, dx\, dy - \frac{\partial \bar{z}_a}{\partial y}(0, 0)\iint my\, dx\, dy$$

$$= \omega^2 \iint m(x, y)\left[\iint C(x, y; \xi, \eta)\bar{z}_a(\xi, \eta)m(\xi, \eta)\, d\xi\, d\eta\right]$$

$$\cdot dx\, dy \qquad (3.7.41)$$

Examining the left-hand side of (3.7.41), the first integral is zero from (3.7.35), the third and fourth integrals are zero because of our use of the center-of-mass as our origin of coordinates. The second integral is identifiable as the total mass of the vehicle.

$$M \equiv \iint m\, dx\, dy$$

Hence,

$$\bar{z}_a(0, 0) = -\frac{\omega^2}{M}\iint m(x, y)\left[\iint Cm\bar{z}_a\, d\xi\, d\eta\right]dx\, dy$$

$$= -\frac{\omega^2}{M}\iint m(\xi, \eta)\bar{z}_a(\xi, \eta)$$

$$\times \left[\iint C(x, y; \xi, \eta)m(x, y)\, dx\, dy\right]d\xi\, d\eta \qquad (3.7.42)$$

133

where the second line follows by change of order of integration. In a similar fashion $\dfrac{\partial \bar{z}_a}{\partial x}(0,0)$, $\dfrac{\partial \bar{z}_a}{\partial y}(0,0)$ may be determined by multiplying (3.7.40) by mx and my respectively with integration over the flight vehicle. The results are

$$\frac{\partial \bar{z}_a}{\partial x}(0,0) = -\frac{\omega^2}{I_y}\iint m(\xi,\eta)\bar{z}_a(\xi,\eta)\left[\iint C(x,y;\xi,\eta)xm(x,y)\,dx\,dy\right]$$
$$\cdot d\xi\,d\eta \tag{3.7.43}$$

etc. where

$$I_y \equiv \iint x^2 m(x,y)\,dx\,dy$$

$$I_x \equiv \iint y^2 m(x,y)\,dx\,dy$$

In (3.7.42), and (3.7.43) note that x,y are now dummy integration variables, not to be confused with the x, y which appear in (3.7.40). Using (3.7.43) and (3.7.44) in (3.7.40) we have

$$\bar{z}_a(x,y) = \omega^2 \iint G(x,y;\xi,\eta)m(\xi,\eta)\bar{z}_a(\xi,\eta)\,d\xi\,d\eta \tag{3.7.44}$$

where

$$G(x,y;\xi,\eta) \equiv C(x,y;\xi,\eta)$$
$$- \iint C(r,s;\xi,\eta)\left[\frac{1}{M}+\frac{xr}{I_y}+\frac{ys}{I_x}\right]m(r,s)\,dr\,ds$$

(3.7.44) has the form of a standard eigenvalue problem. In general, there are infinite number of nontrivial solutions (eigenfunctions), ϕ_m, with corresponding eigenvalues, ω_m, such that

$$\phi_m(x,y) = \omega_m^2 \iint G(x,y;\xi,\eta)m(\xi,\eta)\phi_m(\xi,\eta)\,d\xi\,d\eta \tag{3.7.45}$$

These eigenfunctions could be determined in a number of ways; perhaps the most efficient method being the replacement of (3.7.45) by system of linear algebraic equations through approximation of the integral in (3.7.45) by a sum.

$$\phi_m(x_i,y_i) = \omega_m^2 \sum_j G(x_i,y_i;\xi_j,\eta_j)m(\xi_j,\eta_j)\phi_m(\xi_j,\eta_j)\Delta\xi\,\Delta\eta \tag{3.7.46}$$

In matrix notation,

$$\{\phi\} = \omega^2 [G_{ij} \, \Delta\xi \, \Delta\eta] \left[\diagdown m \diagdown \right] \{\phi\}$$

or

$$\left[\left[\diagdown 1 \diagdown \right] - \omega^2 [G_{ij} \, \Delta\xi \, \Delta\eta] \left[\diagdown m \diagdown \right] \right] \{\phi\} = \{0\} \tag{3.7.47}$$

Setting the determinant of coefficients equal to zero, we obtain a polynomial in ω^2 which gives us (approximate) eigenvalues as roots. The related eigenvector of (3.7.47) is an approximate description of the eigenfunctions of (3.7.44) or (3.7.45).

An important and useful property of eigenfunctions is their orthogonality, i.e.,

$$\iint \phi_m(x, y)\phi_n(x, y)m(x, y) \, dx \, dy = 0 \quad \text{for} \quad m \neq n \tag{3.7.48}$$

We shall digress briefly to prove (3.7.48).

Proof of orthogonality.   Consider two different eigenvalues and eigenfunctions.

$$\phi_m(x, y) = \omega_m^2 \iint Gm\phi_m \, d\xi \, d\eta \tag{3.7.49a}$$

$$\phi_n(x, y) = \omega_n^2 \iint Gm\phi_n \, d\xi \, d\eta \tag{3.7.49b}$$

Multiply (3.7.49a) and (3.7.49b) by $m\phi_n(x, y)$ and $m\phi_m(x, y)$ respectively and $\iint \cdots dx \, dy$.

$$\frac{1}{\omega_m^2} \iint \phi_n \phi_m m \, dx \, dy = \iint \phi_n m \left[\iint G\phi_m m \, d\xi \, d\eta \right] \cdot dx \, dy \tag{3.7.49c}$$

$$\frac{1}{\omega_n^2} \iint \phi_m \phi_n m \, dx \, dy = \iint \phi_m m \left[\iint G\phi_n m \, d\xi \, d\eta \right] \cdot dx \, dy \tag{3.7.49d}$$

Interchanging the order of integration in (3.7.49c) and transferring x, y to ξ, η, and vice versa on the right-hand side gives:

$$\frac{1}{\omega_m^2} \iint \phi_m \phi_n m \, dx \, dy = \iint \phi_m m \left[\iint G(\xi, \eta; x, y) \right.$$

$$\left. \cdot \phi_n(\xi, \eta)m(\xi, \eta) \, d\xi \, d\eta \right] dx \, dy \tag{3.7.50}$$

135

If G were symmetric, i.e.,

$$G(\xi, \eta; x, y) = G(x, y; \xi, \eta) \tag{3.7.51}$$

then the right-hand side of (3.7.49d) and (3.7.50) would be equal and hence one could conclude that

$$\left[\frac{1}{\omega_m^2} - \frac{1}{\omega_n^2}\right] \int\int \phi_m \phi_n m \, dx \, dy = 0$$

or

$$\int\int \phi_m \phi_n m \, dx \, dy = 0 \quad \text{for} \quad m \neq n \tag{3.7.52}$$

Unfortunately, the situation is more complicated since G is *not* symmetric. However, from (3.7.44), et. seq., one can write

$$G(\xi, \eta; x, y) - G(x, y; \xi, \eta)$$

$$= \int\int C(r, s; \xi, \eta)\left[\frac{1}{M} + \frac{ys}{I_x} + \frac{xr}{I_y}\right] m(r, s) \, dr \, ds$$

$$- \int\int C(r, s; x, y)\left[\frac{1}{M} + \frac{\eta s}{I_x} + \frac{\xi r}{I_y}\right] m(r, s) \, dr \, ds \tag{3.7.53}$$

Using the above to substitute for $G(\xi, \eta; x, y)$ in (3.7.50) and using (3.7.35)–(3.7.37) to simplify the result, one sees that the terms on the right-hand side of (3.7.53) contribute nothing. Hence, the right-hand sides of (3.7.49d) and (3.7.50) are indeed equal.

The orthogonality result follows. Note that the rigid body modes

$$
\begin{aligned}
\omega_1 &= 0 & \phi_1 &= 1 \\
\omega_2 &= 0 & \phi_2 &= x \\
\omega_3 &= 0 & \phi_3 &= y
\end{aligned}
\tag{3.7.54}
$$

are orthogonal as well. One can verify readily that the above satisfy the equations of motion, (3.7.35)–(3.7.38), and that the orthogonality conditions follow from (3.7.35)–(3.7.37).

Forced motion including aerodynamic forces

We will simplify the equations of motion to a system of ordinary integral-differential equations in time by expanding the deformation in terms of normal modes.

$$z_a(x, y, t) = \sum_{m=1}^{\infty} q_m(t)\phi_m(x, y) \tag{3.7.55}$$

Recall the natural modes, ϕ_m, must satisfy the equations of motion with $p_E = p_M = 0$ and

$$z_a \sim e^{i\omega_m t}$$

Substituting (3.7.55) in (3.7.35)–(3.7.37) and using orthogonality, (3.7.52), and (3.7.54),

$$\ddot{q}_1 \iint m \, dx \, dy = \iint [p_E + p_M] \, dx \, dy \tag{3.7.56}$$

$$\ddot{q}_2 \iint x^2 m \, dx \, dy = \iint x[p_E + p_M] \, dx \, dy \tag{3.7.57}$$

$$\ddot{q}_3 \iint y^2 m \, dx \, dy = \iint y[p_E + p_M] \, dx \, dy \tag{3.7.58}$$

The reader should be able to identify readily the physical significance of the several integrals in the above equations. Substituting (3.7.55) into (3.7.38) gives

$$\sum_{m=1}^{\infty} q_m \left[\phi_m(x, y) - \phi_m(0, 0) - x \frac{\partial \phi_m}{\partial x}(0, 0) - y \frac{\partial \phi_m}{\partial y}(0, 0) \right]$$

$$= \iint C(x, y; \xi, \eta) \left[p_E + p_M - m \sum_{m=1}^{\infty} \ddot{q}_m \phi_m(\xi, \eta) \right] d\xi \, d\eta \tag{3.7.59}$$

Now the normal modes, ϕ_m, satisfy

$$\phi_m(x, y) - \phi_m(0, 0) - x \frac{\partial \phi_m}{\partial x}(0, 0) - y \frac{\partial \phi_m}{\partial y}(0, 0)$$

$$= \omega_m^2 \iint C(x, y; \xi, \eta) m(\xi, \eta) \phi_m(\xi, \eta) \, d\xi \, d\eta \quad m = 1, \ldots, \infty \tag{3.7.60}$$

Also the left-hand side of (3.7.59) is identically zero for the rigid body modes, $m = 1, 2, 3$. Further using (3.7.60) in the right-hand side of (3.7.59) for $m = 4, 5, \ldots$, gives finally

$$\sum_{m=4}^{\infty} \left(q_m + \frac{\ddot{q}_m}{\omega_m^2} \right) \left[\phi_m(x, y) - \phi_m(0, 0) - x \frac{\partial \phi_m}{\partial x}(0, 0) - y \frac{\partial \phi_m}{\partial y}(0, 0) \right]$$

$$= \iint C(x, y; \xi, \eta)[p_E + p_M - m\ddot{q}_1 - m\xi\ddot{q}_2 - m\eta\ddot{q}_3] \, d\xi \, d\eta \tag{3.7.61}$$

Multiplying (3.7.61) by $m(x, y)\phi_n(x, y)$ and $\iint \cdots dx\, dy$, invoking orthogonality, gives

$$M_n\left(q_n + \frac{\ddot{q}_n}{\omega_n^2}\right) = \iint \phi_n m\left\{\iint C[p_E + p_M - m\ddot{q}_1 - m\xi\ddot{q}_2\right.$$
$$\left. - m\eta\ddot{q}_3]\,d\xi\,d\eta\right\}dx\,dy \quad (3.7.62)$$

where the 'generalized mass', M_n, is defined as

$$M_n \equiv \iint \phi_n^2 m\, dx\, dy$$

Now the structural influence function, C, is symmetric, i.e.,

$$C(x, y; \xi, \eta) = C(\xi, \eta; x, y) \quad (3.7.63)$$

This follows from Maxwell's reciprocity theorem* which states that the deflection at x, y due to a unit load at ξ, η is equal to the deflection at ξ, η due to a unit load at x, y.

Using (3.7.63) and interchanging the order of integration in (3.7.62), one obtains

$$M_n\left(q_n + \frac{\ddot{q}_n}{\omega_n^2}\right) = \iint [p_E + p_M - m\ddot{q}_1 - m\xi\ddot{q}_2 - m\eta\ddot{q}_3]$$
$$\cdot\left\{\iint C(\xi, \eta; x, y)\phi_n(x, y)m(x, y)\,dx\,dy\right\}\cdot d\xi\,d\eta$$
$$(3.7.64)$$

Using (3.7.60) in (3.7.64),

$$M_n\left(q_n + \frac{\ddot{q}_n}{\omega_n^2}\right) = \frac{1}{\omega_n^2}\iint [p_E + p_M - m\ddot{q}_1 - m\xi\ddot{q}_2 - m\eta\ddot{q}_3]$$
$$\cdot\left[\phi_n(\xi, \eta) - \phi_n(0, 0) - \xi\frac{\partial\phi_n}{\partial\xi}(0, 0) - \eta\frac{\partial\phi_n}{\partial\eta}(0, 0)\right]$$
$$\cdot d\xi\,d\eta \quad (3.7.65)$$

By using orthogonality, (3.7.52), and the equations of rigid body equilibrium, (3.7.56)–(3.7.58), one may show that the right-hand side of (3.7.65) can be simplified as follows:

$$M_n\left(q_n + \frac{\ddot{q}_n}{\omega_n^2}\right) = \frac{1}{\omega_n^2}\iint [p_E + p_M]\phi_n\,d\xi\,d\eta \quad (3.7.66)$$

---

* Bisplinghoff, Mar and Pian [2].

Defining the generalized force,

$$Q_n \equiv \iint [p_E + p_M] \phi_n \, d\xi \, d\eta$$

one has

$$M_n[\ddot{q}_n + \omega_n^2 q_n] = Q_n \qquad n = 1, 2, 3, 4, \ldots \tag{3.7.67}$$

Note that there is no inertial or structural coupling in the equations
(3.7.67). However, p_M generally depends upon $q_1, q_2, \ldots$ and hence the
equations are aerodynamically coupled.* The lack of inertial and structural coupling is due to our use of natural or normal modes. Finally, note
that the rigid body equation of motions, (3.7.56)–(3.7.58), also have the
form of (3.7.67). Hence n may run over all integer values.

Examples

(a) *Rigid wing undergoing translation responding to a gust.* One mode
only, $\phi_1 = 1$, q_1 ($\equiv -h$ was notation used previously in typical section
model) and thus

$$M_1 \ddot{q}_1 = Q_1^M + Q_1^E \tag{3.7.68}$$

$$Q_1^M = \iint p_M \phi_1 \, dx \, dy = \int L_M \, dy \tag{3.7.69}$$

$$Q_1^E = \iint p_E \phi_1 \, dx \, dy = \int L_G \, dy \tag{3.7.70}$$

where

$$L_M \equiv \int p_M \, dx \qquad \text{lift/span} \tag{3.7.71}$$

$$L_G \equiv \int p_E \, dx \qquad \text{lift/span} \tag{3.7.72}$$

Introducing nondimensional time, $s \equiv tU/b$, (3.7.68) may be written

$$\frac{U^2}{b^2} M_1 q'' = \int_0^l L_M \, dy + \int_0^l L_G \, dy$$

$$\text{where} \qquad ' \equiv \frac{d}{ds} \tag{3.7.73}$$

---

* cf. (3.7.29).

3 Dynamic aeroelasticity

Assuming strip-theory aerodynamics, two dimensional, incompressible flow, one has (recall Section 3.4 and see Chapter 4)

$$L_M(s) = -\pi\rho U_\infty^2 \left[q''(s) + 2 \int_0^s q''(\sigma)\phi(s-\sigma)\,d\sigma \right] \tag{3.7.74}$$

Note we have assumed $q_1'(0) = 0$ in the above. Similarly

$$L_G = 2\pi\rho U_\infty b \left[w_G(0)\psi(s) + \int_0^s \frac{dw_G(\sigma)}{d\sigma}\,\psi(s-\sigma)\,d\sigma \right]$$

$$= 2\pi\rho U^2 b \left[\int_0^s \frac{w_G(\sigma)}{U}\,\psi'(s-\sigma)\,d\sigma \right] \tag{3.7.75}$$

$$\psi'(s) \equiv \frac{d\psi}{ds}$$

Here we have assumed w_G is independent of y for simplicity. Substituting (3.7.74) and (3.7.75) into (3.7.73) we have

$$\frac{U_\infty^2}{b^2} M q_1''(s) = \pi\rho U_\infty^2 (2bl) \left[-\frac{q_1''}{2b} - \frac{1}{b}\int_0^s q_1''(\sigma)\phi(s-\sigma)\,d\sigma \right.$$

$$\left. + \int_0^s \frac{w_G(\sigma)}{U_\infty}\,\psi'(s-\sigma)\,d\sigma \right] \tag{3.7.76}$$

$$M \equiv \iint m\phi_1\,dx\,dy, \qquad \text{total mass of wing}$$

Note $\int L\,dy = lL$ since we have assumed b a constant and $l \equiv$ half-span of wing. (3.7.76) may be solved in several ways which have previously been discussed in the context of the typical section. Here we shall pursue the method of Laplace Transforms. Transforming (3.7.76) (p is the Laplace Transform variable) gives

$$\frac{U^2}{b^2} M p^2 \bar{q}_1(p) = \pi\rho U^2 (2bl) \left[\frac{\bar{w}_G}{U} p\bar{\psi} - \frac{p^2 \bar{q}_1}{2b} - \frac{p^2 \bar{q}_1}{b}\bar{\phi} \right] \tag{3.7.77}$$

We have taken $q(0) = q'(0) = 0$ while using the convolution theorem, i.e.,

$$\overline{\left\{ \int_0^s w_G(\sigma)\psi'(s-\phi)\,d\sigma \right\}} = \bar{w}_G p\bar{\psi}$$

$$\overline{\left\{ \int_0^s q_1''(\sigma)\phi(s-\sigma)\,d\sigma \right\}} = p^2 \bar{q}_1 \bar{\phi}$$

140

and a bar $(\bar{\ })$ denotes Laplace Transform. Solving (3.7.77) for $\bar{q}_1$ gives

$$\bar{q}_1(p) = \frac{\dfrac{b}{2}\dfrac{\bar{w}_G}{U}\bar{\psi}}{p\left(\dfrac{\mu}{2}+\dfrac{1}{4}+\dfrac{1}{2}\bar{\phi}\right)} \tag{3.7.78}$$

where

$$\mu \equiv \frac{M}{\pi(2bl)b\rho}, \quad \text{mass ratio.}$$

To complete the solution we must invert (3.7.78). To make this inversion tractable, ϕ and ψ are approximated by

$$\psi(s) = 1 - 0.5e^{-0.13s} - 0.5e^{-s}$$
$$\phi(s) = 1 - 0.165e^{-0.0455s} - 0.335e^{-0.3s} \tag{3.7.79}$$

Thus

$$\bar{\psi} = (0.565p + 0.13)/p(p + 0.0455)(p + 0.3)$$
$$\bar{\phi} = \frac{0.5p^2 + 0.2805p + 0.01365}{p^3 + 0.3455p^2 + 0.01365p} \tag{3.7.80}$$

and

$$\bar{q}_1 = \frac{b\dfrac{\bar{w}_G}{U}0.565(p^3 + 0.575p^2 + 0.093p + 0.003)}{(\mu + 0.5)p(p + 0.13)(p + 1)(p^3 + a_1p^2 + a_2p + a_3)}$$

$$\tag{3.7.81}$$

where

$$a_1 \equiv \frac{0.3455\mu + 0.67}{\mu + 0.5}$$

$$a_2 \equiv \frac{0.01365\mu + 0.28}{\mu + 0.5}$$

$$a_3 \equiv \frac{0.01365}{\mu + 0.5}$$

Often one is interested in the acceleration,

$$\ddot{q}_1 = \frac{U^2}{b^2}q_1''.$$

$$\ddot{q}_1 = \frac{U^2}{b^2}\mathscr{L}^{-1}\{p^2\bar{q}_1\}^*$$

---

* For $q_1(0) = \dot{q}_1(0) = 0$. $\mathscr{L}^{-1} \equiv$ inverse Laplace Transform.

$$= \frac{0.565}{\mu + 0.5} \int_0^s \frac{U_\infty}{b} w_G(\sigma) \{ A_1 e^{-0.13(s-\sigma)}$$
$$+ A_2 e^{-(s-\sigma)} + B_1 e^{\gamma_1(s-\sigma)}$$
$$+ B_2 e^{\gamma_2(s-\sigma)} + B_3 e^{\gamma_3(s-\sigma)} \} \, d\sigma \qquad (3.7.82)$$

where

$$A_1 = \frac{N(-0.13)}{D'(-0.13)}$$

$$A_2 = \frac{N(-1)}{D'(-1)}$$

$$B_{k=1\,2\,3} = \frac{N(\gamma_k)}{D'(\gamma_k)}$$

and

$$N(p) \equiv p(p^3 + 0.5756p^2 + 0.09315p + 0.003141)$$
$$D(p) \equiv (p + 0.13)(p + 1)(p^3 + a_1 p^2 + a_1 p + a_3)$$
$$\gamma_k \quad \text{roots of} \quad p^3 + a_1 p^2 + a_1 p + a_3 = 0$$

Note that bracketed term in (3.7.82) must be a real quantity though components thereof may be complex (conjugates). What does it mean physically if the real part of γ_1, γ_2, or γ_3 is positive?

An even simpler theory of gust response is available if one further approximates the aerodynamic forces. For example, using a quasi-static aerodynamic theory (recall Section 3.4), one has

$$\psi = 1 \quad \text{and thus} \quad L_G = 2\pi\rho U_\infty^2 b \frac{w_G}{U_\infty}$$

and

$$\phi = 0, \quad \text{and thus} \quad L_M = 0 \text{ (ignoring virtual inertia term)}$$

Hence

$$M_1 \ddot{q}_1 = \int L^G \, dy = 2\pi\rho U^2 bl \frac{w_G}{U}$$

$$\qquad (3.7.83)$$

$$\ddot{q}_{1_s} = \pi \frac{\rho U^2}{M} (2bl) \frac{w_G}{U_\infty} = \frac{U_\infty}{b} \frac{w_G}{\mu} \qquad (3.7.83)$$

The subscripted quantity, $\ddot{q}_{1_s}$, is called the *static approximation to the gust response*. Figure 3.35 is a schematic of the result from the full theory, (3.7.82), referenced to the static result, (3.7.83). Here we have further assumed a sharp-edge gust, i.e., $w_G = $ constant. After Figure 10.22 BAH.

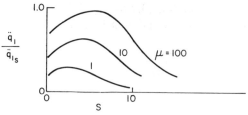

Figure 3.35   *Acceleration time history.*

The maxima of the above curves are presented in Figure 3.36. As can be seen the static approximation is a good approximation for large mass ratio, μ. For smaller μ the acceleration is less than the static result. Hence the quantity,

$$\frac{\ddot{q}_{1_{max}}}{\ddot{q}_{1_s}}$$

is sometimes referred to as a 'gust alleviation' factor.

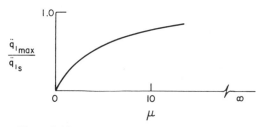

Figure 3.36   *Maximum acceleration vs mass ratio.*

A somewhat more sophisticated aerodynamic approximation is to let (again recall Section 3.4)

$$\psi = 1 \quad \text{and thus} \quad L_G = 2\pi\rho U^2 b \frac{w_G}{U}$$

$$\phi = 1 \quad \text{and thus} \quad L_M = -\pi\rho U^2 [q''(s) + 2q'(s)]$$

(3.7.84)

assuming $q'(0) = 0$. In the motion derived lift, the first term is a virtual inertial term which is generally negligible compared to the inertia of the flight vehicle. However, the second term is an aerodynamic damping term which provides the only damping in the system and hence may be important. It is this aerodynamic damping, even in the guise of the full (linear) aerodynamic theory, which gives results substantially different from the static approximation. (3.7.84) is termed a quasi-steady aerodynamic approximation.

143

3 Dynamic aeroelasticity

Using the approximation (3.7.84), (3.7.68) becomes for a constant chord wing (span: l) and in nondimensional form

$$(\mu + 0.5)q_1''(s) + q_1'(s) = \frac{bw_G(s)}{U_\infty} \tag{3.7.85}$$

where

$$\mu \equiv \frac{M_1}{\pi\rho(2bl) \cdot b} \quad \text{mass ratio}$$

Taking the Laplace transform of (3.7.85) with initial conditions

$$q_1'(0) = q(0) = 0, \qquad w_G(0) = 0,$$

we have

$$(\mu + 0.5)p^2\bar{q}_1(p) + p\bar{q}_1(s) = \frac{b\bar{w}_G(p)}{U_\infty}$$

Solving

$$\bar{q}_1(p) = \frac{\dfrac{b}{U_\infty}\bar{w}_G(p)}{p\{(\mu + 0.5)p + 1\}}$$

and thus

$$q_1''(s) = \mathcal{L}^{-1}p^2\bar{q}_1(p)$$

$$= \frac{1}{(\mu + 0.5)}\mathcal{L}^{-1}\frac{b}{U_\infty}\bar{w}_G(p) \cdot \left[1 - \frac{\dfrac{1}{\mu + 0.5}}{p + \dfrac{1}{\mu + 0.5}}\right]$$

$$= \frac{1}{\mu + 0.5}\int_0^s \frac{b}{U_\infty}w_G(\sigma) \cdot \left\{\delta(s - \sigma) - \frac{1}{\mu + 0.5}\exp\left(-\frac{s - \sigma}{\mu + 0.5}\right)\right\}d\sigma \tag{3.7.86}$$

or

$$\ddot{q}_1 = \frac{U_\infty^2}{b^2}q_1''(s) = \frac{1}{\mu + 0.5}\int_0^s \frac{U_\infty}{b}w_G(\sigma)$$

$$\times \left\{\delta(s - \sigma) - \frac{1}{\mu + 0.5}\exp\left(-\frac{s - \sigma}{\mu + 0.5}\right)\right\}d\sigma$$

Since

$$\ddot{q}_{1s} = \frac{U_\infty}{b} \frac{w_G(s)}{\mu} \quad \text{(static result)},$$

$$\frac{\ddot{q}_1}{\ddot{q}_{1s}} = \frac{\mu}{\mu + 0.5} \frac{1}{w_G(s)} \int_0^s w_G(\sigma)$$

$$\times \left\{ \delta(s - \sigma) - \frac{1}{\mu + 0.5} \exp\left(-\frac{s-\sigma}{\mu + 0.5}\right) \right\} d\sigma \quad (3.7.87)$$

For a sharp edge gust,

$$w_G(s) = w_0 : \text{const} \qquad (s > 0),$$
$$= 0 \qquad\qquad (s < 0)$$

(3.7.87) becomes

$$\frac{\ddot{q}_1}{\ddot{q}_{1s}} = \frac{\mu}{\mu + 0.5} \exp\left(-\frac{s}{\mu + 0.5}\right) \qquad\qquad (3.7.88)$$

(3.7.88) is presented graphically in the Figure 3.37. From (3.7.88) one may plot the maxima (which occur at $s = 0$ for the quasi-steady aerodynamic theory) vs. μ. These are shown in Figure 3.38 where the results are compared with those using the full unsteady aerodynamic theory and the static aerodynamic theory. What conclusion do you draw concerning the adequacy of the various aerodynamic theories?

(b) *Wing undergoing translation and spanwise bending*

$$M_n \ddot{q}_n + M_n \omega_n^2 q_n = Q_n^M + Q_n^G \qquad n = 1, 2, 3, \dots \qquad (3.7.89)$$

q_1 rigid body mode of translation

$q_2, q_3, \dots$ beam bending modal amplitude of wing

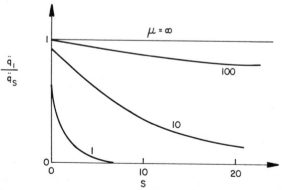

Figure 3.37 *Acceleration time history: Quasi-steady aerodynamics.*

3 Dynamic aeroelasticity

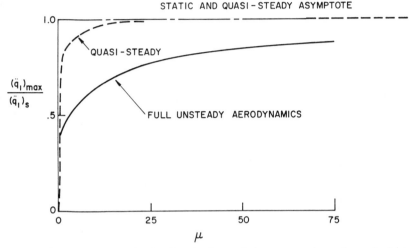

Figure 3.38   Maximum acceleration for wing in translation encountering a sharp edged (step function) gust as given by various aerodynamic models vs mass ratio.

The mode shapes are denoted by $\phi_n(y)$ and are normalized such that the generalized masses are given by

$$M_n \equiv \iint \phi_n^2 m \, dx \, dy = \int \left[\int m \, dx \right] \phi_n^2 \, dy = M \tag{3.7.90}$$

The generalized forces are given by

$$Q_n^M = \iint P_M \phi_n \, dx \, dy = \int L_M \phi_n \, dy$$

$$Q_n^G = \iint P_G \phi_n \, dx \, dy = \int L_G \phi_n \, dy \tag{3.7.91}$$

Introduce $s \equiv \dfrac{Ut}{b_r}$ where b_r is reference half chord. Also let chord vary spanwise, i.e.,

$$b(y) = b_r g(y) \tag{3.7.92}$$

where g is given from wing geometry. (3.7.89) may be written

$$\frac{U^2}{b_r^2} M q_n'' + M \omega_n^2 q_n = Q_n^M + Q_n^G \tag{3.7.93}$$

Using two-dimensional aerodynamics in a 'strip theory' approximation and assuming the gust velocity is uniform spanwise, the aerodynamic lift

146

forces are

$$L_M(y, s) = -\pi\rho(b_r g)^2 \frac{U^2}{b_r^2} \sum_m \phi_m q_m''$$

$$-2\pi\rho U \left(\frac{U}{b_r}\right)(b_r g) \int_0^s \left(\sum_m \phi_m q_m''(\sigma)\right)\phi(s-\sigma)\,\mathrm{d}\sigma$$

and

$$L_G(y, s) = 2\pi\rho U(b_r g)\int_0^s w_G(\sigma)\psi'(s-\sigma)\,\mathrm{d}\sigma \qquad (3.7.94)$$

Substituting (3.7.94) into (3.7.91) and the result into (3.7.89) gives (when nondimensionalized)

$$\mu[q_n'' + \Omega_n^2 q_n] + \sum_{m=1}^\infty A_{nm}q_m'' + 2\sum_m B_{nm}\int_0^s q_m''(\sigma)\phi(s-\sigma)\,\mathrm{d}\sigma$$

$$= 2b_r B_{1n}\int_0^s \frac{w_G(\sigma)}{U}\psi'(s-\sigma)\,\mathrm{d}\sigma \qquad n = 1, 2, 3, \ldots \quad (3.7.95)$$

where

$$\mu \equiv \frac{M}{\pi\rho S b_r}, \qquad \Omega_n \equiv \frac{\omega_n b_r}{U}$$

$$A_{nm} \equiv \frac{b_r}{S}\int_{-l/2}^{l/2} g^2 \phi_n \phi_m\,\mathrm{d}y$$

$$B_{nm} \equiv \frac{b_r}{S}\int_{-l/2}^{l/2} g\phi_n \phi_m\,\mathrm{d}y$$

$$(3.7.96)$$

$$S \equiv \int_{-l/2}^{l/2} 2b\,\mathrm{d}y = 2b_r\int_{-l/2}^{l/2} g\,\mathrm{d}y, \quad \text{wing area}$$

(3.7.95) is a set of integral-differential equations in one variable, time. They are mathematically similar to the typical section equations. If we further restrict ourselves to consideration of translation plus the first wing bending mode, we have two equations in two unknowns. These may be solved as in Example (a) by Laplace Transformation. Alternatively, Examples (a) and (b) could be handled by numerical integration in the time domain. Yet another option is to work the problem in the frequency domain.

(c) *Random gusts—solution in the frequency domain.* Pursuing the latter option, we only need replace the Laplace transform variable, p, by $i\omega$ where ω is the Fourier frequency. For simplicity, consider again Example

(a). (3.7.81) may be written

$$\frac{\bar{q}_1}{b} = H_{qG}(\omega)\frac{w_{G(\omega)}}{U}$$

where

$$H_{qG}(\omega) \equiv \frac{0.565[(i\omega)^3 + 0.5756(i\omega)^2 + 0.093i\omega + 0.003]}{(\mu + 0.5)(i\omega)[i\omega + 0.13][i\omega + 1][(i\omega)^3 + a_1(i\omega)^2 + a_2(i\omega)a_3]}$$

is a transfer function relating sinusoidal rigid body response to sinusoidal gust velocity. The poles of the transfer function can be examined for stability. The mean square response to a random gust velocity can be written as (cf. equation (3.7.40) in Section 3.3)

$$\left(\frac{\overline{q_1}}{b}\right)^2 = \int_0^\infty |H_{qG}(\omega)|^2 \Phi_{(w_G/U)(w_G/U)}\, d\omega \tag{3.7.98}$$

Similar expressions can be obtained for two or more degrees of freedom.

3.8 Nonairfoil physical problems

Fluid flow through a flexible pipe

This problem has received a good deal of attention in the research literature. It has a number of interesting features, including some analogies to the flutter of plates. Possible technological applications include oil pipelines, hydraulic lines, rocket propellant fuel lines, and human lung airways.* The equation of motion is given by

$$EI\frac{\partial^4 w}{\partial x^4} + m\frac{\partial^2 w}{\partial t^2} + \rho A\left[\frac{\partial^2 w}{\partial t^2} + 2U\frac{\partial^2 w}{\partial x\,\partial t} + U^2\frac{\partial^2 w}{\partial x^2}\right] = 0 \tag{3.8.1}$$

EI	bending stiffness of pipe	A	open area of pipe
m	mass/length of pipe	w	transverse deflection of pipe
ρ	fluid density	a	pipe length
U	fluid velocity		

Figure 3.39 Geometry.

---

* Weaver and Paidoussis [27].

We consider a cantilevered pipe clamped at one end and free at the other. Previously we had considered a pipe pinned at both ends and discovered that a static instability occurred.* The present boundary conditions allow a dynamic instability, flutter. We shall consider a classical eigenvalue analysis of this differential equation. Let

$$w = \bar{w}(x)e^{i\omega t} \qquad (3.8.2)$$

where the ω are to be determined by the requirement that nontrivial solutions, $\bar{w}(x) \neq 0$, are sought. Substituting (3.8.2) into (3.8.1) we have

$$\left\{ EI\frac{d^4\bar{w}}{dx^4} - m\omega^2\bar{w} + \rho A\left[-\omega^2\bar{w} + 2Ui\omega\frac{d\bar{w}}{dx} + U^2\frac{d^2\bar{w}}{dx^2} \right] \right\}e^{i\omega t} = 0 \qquad (3.8.3)$$

This ordinary differential equation may be solved by standard methods. The solution has the form

$$\bar{w}(x) = \sum_{i=1}^{4} C_i e^{p_i x}$$

where $p_1, \ldots, p_4$ are the four roots of

$$EIp^4 - m\omega^2 + \rho A[-\omega^2 + 2Ui\omega p + U^2 p^2] = 0 \qquad (3.8.4)$$

The four boundary conditions give four equations for $C_1, \ldots, C_4$. They are

$$w(x = 0) = 0 \Rightarrow C_1 + C_2 + C_3 + C_4 = 0$$

$$\frac{\partial w}{\partial x}(x = 0) = 0 \Rightarrow C_1 p_1 + C_2 p_2 + C_3 p_3 + C_4 p_4 = 0$$

$$EI\frac{\partial^2 w}{\partial x^2}(x = a) = 0 \Rightarrow C_1 p_1^2 e^{p_1 a} + C_2 p_2^2 e^{p_2 a} + C_3 p_3^2 e^{p_3 a} + C_4 p_4^2 e^{p_4 a} = 0$$

$$EI\frac{\partial^3 w}{\partial x^3}(x = a) = 0 \Rightarrow C_1 p_1^3 e^{p_1 a} + C_2 p_2^3 e^{p_2 a} + C_3 p_3^3 e^{p_3 a} + C_4 p_4^3 e^{p_4 a} = 0$$

$$(3.8.5)$$

Setting the determinant of coefficients of (3.8.5) equal to zero gives

$$D \equiv \begin{vmatrix} 1 & 1 & 1 & 1 \\ p_1 & p_2 & p_3 & p_4 \\ p_1^2 e^{p_1 a} & p_2^2 e^{p_2 a} & p_3^2 e^{p_3 a} & p_4^2 e^{p_4 a} \\ p_1^3 e^{p_1 a} & p_2^3 e^{p_2 a} & p_3^3 e^{p_3 a} & p_4^3 e^{p_4 a} \end{vmatrix} = 0 \qquad (3.8.6)$$

---

* Section 2.5.

3   Dynamic aeroelasticity

(3.8.6) is a transcendental equation for ω which has no known analytical solution. Numerical solutions are obtained as follows. For a given pipe at a given U one makes a guess for ω (in general a complex number with real and imaginary parts). The $p_1, \ldots, p_4$ are then evaluated from (3.8.4). D is evaluated from (3.8.6); in general it is not zero and one must improve upon the original guess for ω (iterate) until D is zero. A new U is selected and the process repeated. For $U = 0$, the ω will be purely real and correspond to the natural frequencies of the pipe including the virtual mass of the fluid. Hence, it is convenient to first set $U = 0$ and then systematically increase it. A sketch of ω vs U is shown below in nondimensional form. These results are taken from a paper by Paidoussis who has worked extensively on this problem. When the imaginary part of ω_I becomes negative, flutter occurs. The nondimensional variables used in presenting these results are (we have changed the notation from Paidoussis with respect to frequency)

$$\beta \equiv \rho A/(\rho A + m)$$
$$u \equiv \left(\rho \frac{A U^2}{EI} \right)^{\frac{1}{2}} a$$
$$\Omega \equiv [(m + \rho A)/EI]^{\frac{1}{2}} \omega a^2$$

Also shown are results obtained by a Galerkin procedure using the natural modes of a cantilevered beam.

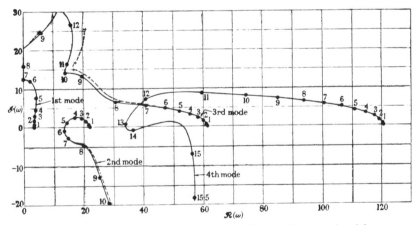

Figure 3.40a   *The dimensionless complex frequency of the four lowest modes of the system as a function of the dimensionless flow velocity for* $\beta = 0.200$. ——, *Exact analysis* ---- *four-mode approximation (Galerkin). Numbers on graph are values of u.*

150

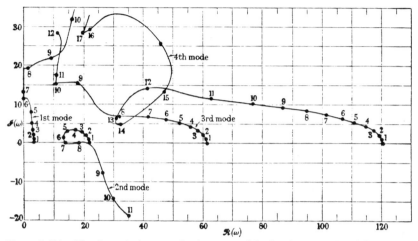

Figure 3.40b   The dimensionless complex frequency of the four lowest modes of the system as a function of the dimensionless flow velocity for $\beta = 0.295$.

The stability boundary for this system may be presented in terms of u and β as given in Figure 3.41. Also shown is the frequency, Ω_F, of the flutter oscillation. These results have been verified experimentally by Gregory and Paidoussis.* For a very readable historical and technical review of this problem, see the paper by Paidoussis and Issid.† A similar physical problem arises in nuclear reactor fuel bundles where one has a pipe in an external flow. The work of Chen is particularly noteworthy.‡

(High speed) fluid flow over a flexible wall—a simple prototype for plate flutter

One type of flutter which became of considerable technological interest with the advent of supersonic flight is called 'panel flutter'. Here the concern is with a thin elastic panel supported at its edges. For simplicity consider two dimensional motion. The physical situation is sketched below. Over the top of the elastic plate, which is mounted flush in an otherwise rigid wall, there is an airflow. The elastic bending of the plate in the direction of the airflow (streamwise) is the essential difference between this type of flutter and classical flutter of an airfoil as exemplified by the typical section. It is not our purpose to probe deeply into this

---

* Gregory and Paidoussis [28].
† Paidoussis and Issid [29].
‡ Chen [30].

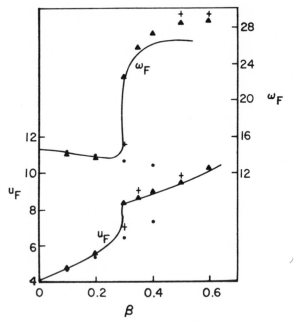

Figure 3.41   *Flutter boundary.*

problem here; for a thorough treatment the reader is referred to Dowell.*
We shall instead be content to consider a highly simplified model (somewhat analogous to the typical section model for airfoil flutter) which will
bring out some of the important features of this type of problem. Thus we
consider the alternative physical model shown below.† Here our model
consists of three rigid plates each hinged at both ends. The hinges
between the first and second plates and also the second and third plates
are supported by springs. The plates have mass per unit length, m, and
are of length, l. At high supersonic Mach number, $M \gg 1$, the
aerodynamic pressure change (perturbation) p, due to plate motion is

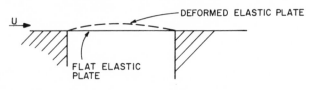

Figure 3.42

---

* Dowell [31]. Also see Bolotin [32].
† This was suggested by Dr. H. M. Voss.

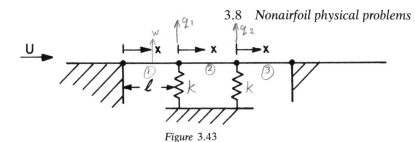

Figure 3.43

given by (see Chapter 4)

$$p = \frac{\rho_\infty U_\infty^2}{M_\infty} \frac{\partial w}{\partial x} \tag{3.8.7}$$

where $w(x, t)$, $\dfrac{\partial w}{\partial x}$ are deflection and slope of any one of the rigid plates.

To write the equations of motion for this physical model we must recognize that there are two degrees of freedom. It is convenient to choose as generalized coordinates, q_1, q_2, the vertical deflections of the springs.

The potential energy of the model is then

$$U = \tfrac{1}{2}kq_1^2 + \tfrac{1}{2}kq_2^2 \tag{3.8.8}$$

The kinetic energy requires expressions for w in terms of q_1 and q_2 since the mass is distributed. For each plate we have, in turn,

Plate 1: $w = q_1 \dfrac{x}{l}$, $\dfrac{\partial w}{\partial x} = q_1/l$

Plate 2: $w = q_1\left[1 - \dfrac{x}{l}\right] + q_2 x/l$, $\dfrac{\partial w}{\partial x} = \dfrac{q_2 - q_1}{l}$

Plate 3: $w = q_2\left[1 - \dfrac{x}{l}\right]$, $\dfrac{\partial w}{\partial x} = \dfrac{-q_2}{l}$ $\tag{3.8.9}$

Because the plates are rigid, the slopes are constant within each plate. x is measured from the front (leading) edge of each plate. The kinetic energy is

$$T = \frac{1}{2} \int m\left(\frac{\partial w}{\partial t}\right)^2 dx \tag{3.8.10}$$

Using (3.8.9) in (3.8.10), we obtain after integration

$$T = \tfrac{1}{2}ml\left[(\tfrac{2}{3})\dot{q}_1^2 + (\tfrac{2}{3})\dot{q}_2^2 + \tfrac{2}{6}\dot{q}_1\dot{q}_2\right] \tag{3.8.11}$$

The virtual work done by the aerodynamic pressure is given by

$$\delta W = \int (-p)\delta w \, dx \tag{3.8.12}$$

153

3 Dynamic aeroelasticity

and using (3.8.9) in (3.8.12) we obtain

$$\delta W = Q_1 \delta q_1 + Q_2 \delta q_2 \tag{3.8.13}$$

where

$$Q_1 \equiv -\frac{\rho_\infty U_\infty^2}{M_\infty} q_2/2$$

$$Q_2 \equiv \frac{\rho_\infty U_\infty^2}{M_\infty} q_1/2$$

Using Lagrange's Equations and (3.8.8), (3.8.11), (3.8.13) the equations of motion are

Verify

$$\begin{cases} \dfrac{2}{3} ml\ddot{q}_1 + \dfrac{ml}{6}\ddot{q}_2 + kq_1 + \dfrac{\rho_\infty U_\infty^2}{2M_\infty} q_2 = 0 \\[2mm] \dfrac{ml}{6}\ddot{q}_1 + \dfrac{2}{3} ml\ddot{q}_2 + kq_2 - \dfrac{\rho_\infty U_\infty^2}{2M_\infty} q_1 = 0 \end{cases} \tag{3.8.14}$$

In the usual way we seek an eigenvalue solution to assess the stability of the system, i.e., let

$$q_1 = \bar{q}_1 e^{i\omega t}$$

$$q_2 = \bar{q}_2 e^{i\omega t}$$

then (3.8.14) becomes (in matrix notation)

$$\left[-\omega^2 ml \begin{bmatrix} \frac{2}{3} & \frac{1}{6} \\ \frac{1}{6} & \frac{2}{3} \end{bmatrix} + \begin{bmatrix} k & 0 \\ 0 & k \end{bmatrix} + \frac{\rho_\infty U_\infty^2}{2M_\infty} \begin{bmatrix} 0 & 1 \\ -1 & 0 \end{bmatrix} \right] \left\{ \begin{matrix} \bar{q}_1 e^{i\omega t} \\ \bar{q}_2 e^{i\omega t} \end{matrix} \right\} = \left\{ \begin{matrix} 0 \\ 0 \end{matrix} \right\} \tag{3.8.15}$$

We seek nontrivial solutions by requiring the determinant of coefficients to vanish which gives the following (nondimensional) equation after some algebraic manipulation

verify

$$\Omega^4 \left(\tfrac{15}{36}\right) - \Omega^2 \left(\tfrac{4}{3}\right) + 1 + \lambda^2 = 0 \tag{3.8.16}$$

where

$$\Omega^2 \equiv \frac{\omega^2 ml}{k}, \qquad \lambda \equiv \frac{\rho_\infty U_\infty^2}{2M_\infty k}$$

Solving (3.8.16) for Ω^2 we obtain

$$\Omega^2 = \tfrac{8}{5} \pm \tfrac{2}{5}[1 - 15\lambda^2]^{\frac{1}{2}} \tag{3.8.17}$$

When the argument of the square root becomes negative, Ω^2 becomes complex conjugate and hence one solution for Ω will have a negative imaginary part corresponding to unstable motion. Hence, flutter will

154

occur for

$$\lambda^2 > \lambda_F^2 \equiv \tfrac{1}{15} \tag{3.8.18}$$

The frequency at this λ_F is given by (3.8.17).

$$\Omega_F = \left[\tfrac{8}{5}\right]^{\frac{1}{2}}$$

For reference the natural frequencies ($\lambda \equiv 0$) are from (3.8.17)

$$\Omega_1 = \left(\tfrac{6}{5}\right)^{\frac{1}{2}} \quad \text{and} \quad \Omega_2 = (2)^{\frac{1}{2}} \qquad \Omega_1 - \text{1st natural mod}$$

From (3.8.15) (say the first of the equations) the eigenvector ratio may be determined

1st Natural Mode: $\dfrac{\bar{q}_1}{\bar{q}_2} = +1$ for $\Omega = \Omega_1$ at $\lambda = 0$

2nd Natural Mode: $\dfrac{\bar{q}_1}{\bar{q}_2} = -1$ for $\Omega = \Omega_2$ at $\lambda = 0$

and at flutter

Flutter Mode: $\dfrac{\bar{q}_1}{\bar{q}_2} = -4 + 15^{\frac{1}{2}}$ for $\Omega = \Omega_F$, $\lambda = \lambda_F$

Sketches of the corresponding plate shapes are given below. The important features of this hinged rigid plate model which carry over to an elastic plate are:

(1) The flutter mechanism is a convergence of natural frequencies with increasing flow velocity. The flutter frequency is between the first and second natural frequencies. In this respect it is similar to classical bending-torsion flutter of an airfoil.

(2) The flutter mode shape shows a maximum nearer the rear edge of the plate (rather than the front edge).

There are some oversimplifications in the rigid plate model. For example,

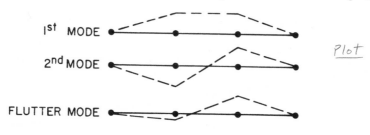

Figure 3.44

the plate length does not affect the flow velocity at which flutter occurs. For an elastic plate, it would. Also in subsonic flow the curvature of the plate has a strong influence on the aerodynamic pressure. In the rigid plate model, the curvature is identically zero, of course. Nevertheless the model serves a useful purpose in introducing this type of flutter problem.

References for Chapter 3

[1] Meirovitch, L., *Methods of Analytical Dynamics*, McGraw-Hill Book Co., New York, 1970.

[2] Bisplinghoff, R. L., Mar, J. W. and Pian, T. H. H., *Statics of Deformable Solids*, Addison-Wesley, 1965.

[3] Timoshenko, S. P. and Goodier, J. N., *Theory of Elasticity*, McGraw-Hill, 1951.

[4] Meirovitch, L., *Elements of Vibration Analysis*, McGraw-Hill, 1975.

[5] Crandall, S. and Mark, W. D., *Random Vibration in Mechanical Systems*, Academic Press, 1963.

[6] Houbolt, J. C., Steiner, R. and Pratt, K. G., 'Dynamic Response of Airplanes to Atmospheric Turbulence Including Flight Data on Input and Response', *NASA TR R-199* (June 1964).

[7] Acum, W. E. A., *The Comparison of Theory and Experiment of Oscillating Wings*, Vol. II, Chapter 10, AGARD Manual on Aeroelasticity.

[8] Pines, S., 'An Elementary Explanation of the Flutter Mechanism', *Proceedings Nat. Specialists Meeting on Dynamics and Aeroelasticity*, Institute of the Aeronautical Sciences, Ft. Worth, Texas (November 1958) pp. 52–58.

[9] Ashley, H. and Zartarian, G., 'Piston Theory—A New Aerodynamic Tool for the Aeroelastician', *J. Aero. Sci.* Vol. 23, No. 12 (December 1956) pp. 1109–1118.

[10] Sears, W. R., 'Operational Methods in the Theory of Airfoils in Non-uniform Motion', *J. of the Franklin Institute*, Vol. 230, pp. 95–111, 1940.

[11] Jones, R. T., 'Properties of Low Aspect-Ratio Pointed Wings at Speeds Below and Above the Speed of Sound', *NACA Report 835*, 1946.

[12] Dowell, E. H. and Widnall, S. E., 'Generalized Aerodynamic Forces on an Oscillating Cylindrical Shell: Subsonic and Supersonic Flow', *AIAA Journal*, Vol. 4, No. 4 (April 1966) pp. 607–610.

[13] Widnall, S. E. and Dowell, E. H., 'Aerodynamic Forces on an Oscillating Cylindrical Duct with an Internal Flow', *J. Sound Vibration*, Vol. 1, No. 6 (1967) pp. 113–127.

[14] Dowell, E. H., 'Generalized Aerodynamic Forces on a Flexible Cylindrical Shell Undergoing Transient Motion', *Quarterly of Applied Mathematics*, Vol. 26, No. 3 (October 1968) pp. 343–353.

[15] Hamming, R. W., *Numerical Methods for Scientists and Engineers*, McGraw-Hill, 1973.

[16] Houbolt, J. C., 'A Recurrence Matrix Solution for the Dynamic Response of Elastic Aircraft', *J. Aero. Sc.*, Vol. 17, No. 9 (September 1950) pp. 540–550.

[17] Hausner, A., *Analog and Analog/Hybrid Computer Programming*, Prentice-Hall, Inc. 1971.

[18] Savant, C. J., *Basic Feedback Control System Design*, McGraw-Hill, 1958.

[19] Garrick, I. E. and Rubinow, S. L., 'Flutter and Oscillating Air Force Calculations for an Airfoil in a Two-Dimensional Supersonic Flow', *NACA TR 846*, 1946.

[20] Hassig, H. J., 'An Approximate True Damping Solution of the Flutter Equation by Iteration', *J. Aircraft*, Vol. 8, No. 11 (November 1971) pp. 885–889.

[21] Eckhaus, W., 'Theory of Transonic Aileron Buzz, Neglecting Viscous Effects', *J. Aerospace Sciences*, Vol. 29 (June 1962) pp. 712–718.

[22] Landahl, M., *Unsteady Transonic Flow*, Pergamon Press, 1961.

[23] Lambourne, N. C., *Flutter in One Degree of Freedom*, Vol. V., Chapter 5, AGARD Manual on Aeroelasticity.

[24] Abramson, H. N., 'Hydroelasticity: A Review of Hydrofoil Flutter', *Applied Mechanics Reviews*, Vol. 22, No. 2, p. 115, 1969.

[25] Crisp, J. D. C., 'On the Hydrodynamic Flutter Anomaly' *Noise, Shock, and Vibration Conference*, Monash University, Melbourne, Australia, 1974.

[26] Stenton, T. E., 'Turbulence Response Calculations for the XB-70 Airplane and Preliminary Comparison with Flight Data', *presented at the Meeting on Aircraft Response to Turbulence*, NASA Langley Research Center, Sept. 24–25, 1968.

[27] Weaver, D. S. and Paidoussis, M. P., 'On Collapse and Flutter Phenomena in Thin Tubes Conveying Fluid', *J. of Sound Vibration*, 50 (Jan. 8, 1977) pp. 117–132.

[28] Gregory, R. W. and Paidoussis, M. P., 'Unstable Oscillation of Tubular Cantilevers Conveying Fluid. I. Theory. II. Experiments'. *Proc. of the Royal Society A* 293, pp. 512–527, 528–542, 1966.

[29] Paidoussis, M. P. and Issid, N. T., 'Dynamic Instability of Pipes Conveying Fluid', *J. Sound and Vibration*, Vol. 33, No. 3, pp. 267–294, 1974.

[30] Chen, S. S., *Vibration of Nuclear Fuel Bundles*, Nuclear Engineering Design, Vol. 35, pp. 399–422, 1975.

[31] Dowell, E. H., *Aeroelasiticty of Plates and Shells*, Noordhoff International Publishing, Leyden, The Netherlands, 1974.

[32] Bolotin, V. V., *Non-conservative Problems of the Elastic Theory of Stability*, Pergamon Press, 1963.

4

Nonsteady aerodynamics of lifting and non-lifting surfaces

4.1 Basic fluid dynamic equations

Nonsteady aerodynamics is concerned with the *time dependent* fluid motion generated by (solid) bodies moving in a fluid. Normally (and as distinct from classical acoustics) the body motion is composed of a (large) steady motion plus a (small) time dependent motion. In classical acoustics no (large) steady motions are examined. On the other hand, it should be said, in most of classical aerodynamic theory small time dependent motions are ignored, i.e., only small *steady* perturbations from the original steady motion are usually examined. However, in a number of problems arising in aeroelasticity, such as flutter and gust analysis, and also in fluid generated noise, such as turbulent boundary layers and jet wakes, the more general problem must be attacked. It shall be our concern here.[*] The basic assumptions concerning the nature of fluid are that it be inviscid and its thermodynamic processes be isentropic. We shall first direct our attention to a derivation of the equations of motion, using the apparatus of vector calculus and, of course, allowing for a large mean flow velocity.

First, let us recall some purely mathematical relations developed in the vector calculus. These are all variations of what is usually termed Gauss' theorem.[†]

---

[*] References: Chapter 7, Liepmann [1].

Chapter 4, BA pp. 70–81, Brief Review of Fundamentals; pp. 82–152, Catalog of available results with some historical perspective (1962).

Chapters 5, 6, 7, BAH, Detailed discussion of the state-of-the art (1955) now largely of interest to aficionados. Read pp. 188–200 and compare with Chapter 4, BA.

AGARD, Vol. II.

[†] Hildebrand [2].

$$\text{I} \quad \iint c\vec{n}\, \mathrm{d}A = \iiint \nabla c\, \mathrm{d}V$$

$$\text{II} \quad \iint \vec{b}\cdot\vec{n}\, \mathrm{d}A = \iiint \nabla\cdot b\, \mathrm{d}V$$

$$\text{III} \quad \iint \vec{a}(\vec{b}\cdot\vec{n})\, \mathrm{d}A = \iiint [\vec{a}(\nabla\cdot\vec{b})+(\vec{b}\cdot\nabla)\vec{a}]\, \mathrm{d}V$$

Also

$$\text{IV} \quad \nabla(\vec{a}\cdot\vec{a}) = 2(\vec{a}\cdot\nabla)\vec{a} + 2\vec{a}\times(\nabla\times\vec{a})$$

In the above, V is an arbitrary closed volume, A its surface area and $\vec{n}$, the unit normal to the surface, positive outward. $\vec{a}$ and $\vec{b}$ are arbitrary vectors and c an arbitrary scalar.

Conservation of mass

Consider an arbitrary but fixed volume of fluid, V, enclosed by a surface, A. $\vec{q}$ is the (vector) fluid velocity, $\mathrm{d}A$ is the surface elemental area, $\vec{n}$ is surface normal, $\vec{q}\cdot\vec{n}$ is the (scalar) velocity component normal to surface, $\iint \rho\vec{q}\cdot\vec{n}\, \mathrm{d}A$ is the rate of mass flow (mass flux) through surface, positive outward, $\partial/\partial t \iiint \rho\, \mathrm{d}V$ is the rate of mass increase inside volume and $= \iiint (\partial\rho/\partial t)\, \mathrm{d}V$ since V, though arbitrary, is fixed.

The physical principle of continuity of mass states that the fluid increase inside volume = rate of mass flow *into* volume through surface.

$$\iiint \frac{\partial\rho}{\partial t}\, \mathrm{d}V = -\iint \rho\vec{q}\cdot\vec{n}\, \mathrm{d}A \tag{4.1.1}$$

Using II, the area integral may be transformed to a volume integral. (4.1.1) then reads:

$$\iiint \frac{\partial\rho}{\partial t}\, \mathrm{d}V = -\iiint \nabla\cdot(\rho\vec{q})\, \mathrm{d}V$$

or

$$\tag{4.1.2}$$

$$\iiint \left[\frac{\partial\rho}{\partial t}+\nabla\cdot(\rho\vec{q})\right]\mathrm{d}V = 0$$

Since V is arbitrary, (4.1.2) can only be true for any and all V, if the integrand is zero.

$$\frac{\partial\rho}{\partial t}+\nabla\cdot(\rho\vec{q}) = 0 \tag{4.1.3}$$

This is the conservation of mass, differential equation in three dimensions. Alternative forms are:

$$\frac{\partial \rho}{\partial t} + \rho \nabla \cdot \vec{q} + (\vec{q} \cdot \nabla)\rho = 0$$

$$\frac{D\rho}{Dt} + \rho(\nabla \cdot \vec{q}) = 0 \qquad (4.1.4)$$

where

$$\frac{D}{Dt} \equiv \frac{\partial}{\partial t} + (\vec{q} \cdot \nabla)$$

Conservation of momentum

The conservation or balance of momentum equation may be derived in a similar way.

$$\iiint \frac{\partial}{\partial t}(\rho\vec{q})\, dV$$

is the rate of momentum increase inside the volume

$$\iint \rho\vec{q}(\vec{q} \cdot \vec{n})\, dA$$

is the rate of momentum flow (momentum flux) through surface, positive outward

$$\iint -p\vec{n}\, dA$$

is the force acting on volume (recall $\vec{n}$ is positive outward)

The physical principle is that the total rate of change of momentum = force acting on V.

$$\iiint \frac{\partial(\rho\vec{q})}{\partial t}\, dV + \iint \rho\vec{q}(\vec{q} \cdot \vec{n})\, dA = \iint -p\vec{n}\, dA \qquad (4.1.5)$$

Using I and III to transform the area integrals and rearranging terms,

$$\iiint \left\{ \frac{\partial}{\partial t}(\rho\vec{q}) + \rho\vec{q}(\nabla \cdot \vec{q}) + (\vec{q} \cdot \nabla)\rho\vec{q} + \nabla p \right\} dV = 0 \qquad (4.1.6)$$

4 Nonsteady aerodynamics

Again because V is arbitrary,

$$\frac{\partial}{\partial t}(\rho\vec{q}) + \rho\vec{q}(\nabla \cdot \vec{q}) + (\vec{q} \cdot \nabla)\rho\vec{q} = -\nabla p \qquad (4.1.7)$$

Alternative forms are

$$\frac{D}{Dt}(\rho\vec{q}) + \rho\vec{q}(\nabla \cdot \vec{q}) = -\nabla p$$

or

$$\rho\frac{D\vec{q}}{Dt} + \vec{q}\left[\rho\nabla \cdot \vec{q} + \frac{D\rho}{Dt}\right] = -\nabla p \qquad (4.1.8)$$

where the bracketed term vanishes from (4.1.8).

Finally to complete our system of equations we have the isentropic relation,

$$p/\rho^\gamma = \text{constant} \qquad (4.1.9)$$

(4.1.3), (4.1.8) and (4.1.9) are five scalar equations (or two scalar plus one vector equations) in five scalar unknowns: p, ρ, and three scalar components of (vector) velocity, $\vec{q}$.

Irrotational flow, Kelvin's theorem and Bernoulli's equation

To solve these nonlinear, partial differential equations we must integrate them. Generally, this is an impossible task except by numerical procedures. However, there is one integration which may be performed which is both interesting theoretically and useful for applications.

Consider the momentum equation which may be written

$$\frac{D\vec{q}}{Dt} = \frac{-\nabla p}{\rho} \qquad (4.1.10)$$

On the right-hand side, using Leibnitz' Rule,* we may write

$$\frac{\nabla p}{\rho} = \nabla \int_{P_{\text{ref}}}^{p} \frac{dp_1}{\rho_1(p_1)} \qquad (4.1.11)$$

---

* Hildebrand [2], pp. 348–353.

162

where ρ_1, p_1 are dummy integration variables, and p_{ref} some constant reference pressure On the left-hand side

$$\frac{D\vec{q}}{Dt} \equiv \frac{\partial \vec{q}}{\partial t} + (\vec{q} \cdot \nabla)\vec{q}$$

In the above the second term may be written

$$(\vec{q} \cdot \nabla)\vec{q} = \nabla \frac{(\vec{q} \cdot \vec{q})}{2} \quad \text{from IV}$$

and, if we assume the flow is irrotational,

$$\vec{q} = \nabla \phi \tag{4.1.12}$$

where ϕ is the scalar velocity potential. (4.1.12) implies and is implied by

$$\nabla \times \vec{q} = 0 \tag{4.1.13}$$

The vanishing of the curl of velocity is a consequence of Kelvin's Theorem which states that a flow which is initially irrotational, $\nabla \times \vec{q} = 0$, remains so at all subsequent time in the absence of dissipation, e.g., viscosity or shock waves. It can be proven using (4.1.3), (4.1.8) and (4.1.9). No additional assumptions are needed.

Let us pause to prove this result. We shall begin with the momentum equation.

$$\frac{D\vec{q}}{Dt} = -\frac{\nabla p}{\rho}$$

First form $\nabla \times$ and then dot the result into $\vec{n}_A \, dA$ and integrate over A. $\vec{n}_A$ is a unit normal to A and A itself is an arbitrary area of the fluid. The result is

$$\frac{D}{Dt} \iint (\nabla \times \vec{q}) \cdot \vec{n}_A \, dA = -\iint \frac{(\nabla \times \nabla p)}{\rho} \cdot \vec{n}_A \, dA$$

From Stokes Theorem,*

$$-\iint \frac{(\nabla \times \nabla p)}{\rho} \cdot \vec{n}_A \, dA = -\iint \frac{\nabla p}{\rho} \cdot d\vec{r}$$

$$= -\oint \left(\frac{\partial p}{\partial r} \Big/ \rho\right) dr$$

$$= -\oint \frac{dp}{\rho}$$

---

* Hildebrand [2], p. 318.

$\mathrm{d}\vec{r} \equiv$ arc length along contour of bounding arc of A. Since the bounding contour is closed, and ρ is solely a function of p,

$$\oint \frac{\mathrm{d}p}{\rho} = 0$$

Hence

$$\frac{D}{Dt} \iint (\nabla \times \vec{q}) \cdot \vec{n}_A \, \mathrm{d}A = 0$$

Since A is arbitrary

$$\nabla \times \vec{q} = \text{constant}$$

and if $\nabla \times \vec{q} = 0$ initially, it remains so thereafter.

Now let us return to the integration of the momentum equation, (4.1.10). Collecting the several terms from (4.1.10)–(4.1.12), we have

$$\frac{\partial}{\partial t}(\nabla \phi) + \nabla \frac{(\nabla \phi \cdot \nabla \phi)}{2} + \nabla \int_{p_{\mathrm{ref}}}^{p} \frac{\mathrm{d}p_1}{\rho_1} = 0 \tag{4.1.14}$$

or

$$\nabla \left[\frac{\partial \phi}{\partial t} + \frac{\nabla \phi \cdot \nabla \phi}{2} + \int_{p_{\mathrm{ref}}}^{p} \frac{\mathrm{d}p_1}{\rho_1} \right] = 0$$

or

$$\frac{\partial \phi}{\partial t} + \frac{\nabla \phi \cdot \nabla \phi}{2} + \int_{p_{\mathrm{ref}}}^{p} \frac{\mathrm{d}p_1}{\rho_1} = F(t) \tag{4.1.15}$$

We may evaluate $F(t)$ by examining the fluid at some point where we know its state. For example, if we are considering an aircraft or missile flying at constant velocity through the atmosphere we know that far away from the body

$$\vec{q} = U_\infty \vec{i}$$

$$\phi = U_\infty x$$

$$p = p_\infty$$

If we choose as the lower limit, $p_{\mathrm{ref}} = p_\infty$, then (4.1.15) becomes

$$0 + \frac{U_\infty^2}{2} + 0 = F(t)$$

and we find that F is a constant independent of space *and* time. Hence finally

$$\frac{\partial \phi}{\partial t} + \frac{\nabla \phi \cdot \nabla \phi}{2} + \int_{p_\infty}^{p} \frac{\mathrm{d} p_1}{\rho_1} = \frac{U_\infty^2}{2} \tag{4.1.16}$$

(4.1.16) is usually referred to as Bernoulli's equation although the derivation for nonsteady flow is due to Kelvin.

The practical value of Bernoulli's equation is that it allows one to relate p to ϕ. Using

$$\frac{p}{p_\infty} = \left(\frac{\rho}{\rho_\infty}\right)^\gamma$$

one may compute from (4.1.16) (the reader may do the computation)

$$C_p \equiv \frac{p - p_\infty}{\frac{\gamma}{2} p_\infty M^2}$$

$$= \frac{2}{\gamma M^2} \left\{ \left[1 + \frac{\gamma - 1}{2} M^2 \left(1 - \frac{\left(\vec{q} \cdot \vec{q} + 2 \frac{\partial \phi}{\partial t}\right)}{U_\infty^2} \right) \right]^{\gamma/(\gamma - 1)} - 1 \right\} \tag{4.1.17}$$

where the Mach number is

$$M^2 \equiv \frac{U_\infty^2}{a_\infty^2}$$

and

$$a^2 \equiv \frac{\mathrm{d} p}{\mathrm{d} \bar{\rho}} = \frac{\gamma p}{\rho}$$

is the speed of sound.

Derivation of single equation for velocity potential

Most solutions are obtained by solving this equation.

We shall begin with the conservation of mass equation (4.1.4).

$$\frac{1}{\rho} \frac{\partial \rho}{\partial t} + \frac{\vec{q} \cdot \nabla \rho}{\rho} + \nabla \cdot \vec{q} = 0 \tag{4.1.4}$$

Consider the first term. Using Leibnitz' rule we may write

$$\frac{\partial}{\partial t} \int_{p_\infty}^{p} \frac{dp_1}{\rho_1} = \frac{\partial \rho}{\partial t} \frac{dp}{d\rho} \frac{d}{dp} \int_{p_\infty}^{p} \frac{dp_1}{\rho_1}$$

$$= \frac{\partial \rho}{\partial t} a^2 \frac{1}{\rho}$$

Thus

$$\frac{1}{\rho} \frac{\partial \rho}{\partial t} = \frac{1}{a^2} \frac{\partial}{\partial t} \int_{p_\infty}^{p} \frac{dp_1}{\rho_1}$$

$$= -\frac{1}{a^2} \frac{\partial}{\partial t} \left[\frac{\partial \phi}{\partial t} + \frac{\nabla \phi \cdot \nabla \phi}{2} \right] \tag{4.1.18}$$

from Bernouilli's equation (4.1.16).

In a similar fashion, the second term may be written

$$\vec{q} \cdot \frac{\nabla \rho}{\rho} = \frac{-\vec{q} \cdot \nabla}{a^2} \left[\frac{\partial \phi}{\partial t} + \frac{\nabla \phi \cdot \nabla \phi}{2} \right] \tag{4.1.19}$$

Finally, the third term

$$\nabla \cdot \vec{q} = \nabla \cdot \nabla \phi = \nabla^2 \phi \tag{4.1.20}$$

Collecting terms, and rearranging

$$-\frac{1}{a^2} \left\{ \frac{\partial^2 \phi}{\partial t^2} + \frac{\partial}{\partial t} \left(\frac{\nabla \phi \cdot \nabla \phi}{2} \right) + \nabla \phi \cdot \frac{\partial}{\partial t} \nabla \phi + \nabla \phi \cdot \nabla \left(\frac{\nabla \phi \cdot \nabla \phi}{2} \right) \right\} + \nabla^2 \phi = 0$$

$$\nabla^2 \phi - \frac{1}{a^2} \left[\frac{\partial}{\partial t} (\nabla \phi \cdot \nabla \phi) + \frac{\partial^2 \phi}{\partial t^2} + \nabla \phi \cdot \nabla \left(\frac{\nabla \phi \cdot \nabla \phi}{2} \right) \right] = 0 \tag{4.1.21}$$

Note we have not yet accomplished what we set out to do, since (4.1.21) is a single equation with *two* unknowns, ϕ and a. A second independent relation between ϕ and a is needed.

The simplest method of obtaining this is to use

$$a^2 \equiv \frac{dp}{d\rho}$$

and

$$\frac{p}{\rho^\gamma} = \text{constant}$$

in Bernoulli's equation. The reader may verify that

$$\frac{a^2 - a_\infty^2}{\gamma - 1} = \frac{U_\infty^2}{2} - \left(\frac{\partial \phi}{\partial t} + \frac{\nabla \phi \cdot \nabla \phi}{2}\right) \tag{4.1.22}$$

Small perturbation theory

(4.1.21) and (4.1.22) are often too difficult to solve. Hence a simpler approximate theory is sought.

As in acoustics we shall linearize about a uniform equilibrium state. Assume

$$a = a_\infty + \hat{a}$$
$$p = p + \hat{p}$$
$$\rho = \rho_\infty + \hat{\rho}$$
$$\vec{q} = U_\infty \vec{i} + \vec{q} \qquad \nabla \phi = U_\infty \vec{i} + \nabla \hat{\phi} \tag{4.1.23}$$
$$\phi = U_\infty x + \hat{\phi}$$

Note in the present case we linearize about a uniform flow with velocity, U_∞. Using (4.1.23) in (4.1.21) and retaining lowest order terms:
First term:

$$\nabla^2 \phi \rightarrow \nabla^2 \hat{\phi}$$

Second term:

$$\frac{\partial}{\partial t}(\nabla \phi \cdot \nabla \phi) + \frac{\partial^2 \phi}{\partial t^2} + \nabla \phi \cdot \nabla \left(\frac{\nabla \phi \cdot \nabla \phi}{2}\right)$$

$$= 2[U_\infty \vec{i} + \nabla \hat{\phi}] \cdot \frac{\partial}{\partial t}[U_\infty \vec{i} + \nabla \hat{\phi}] + \frac{\partial^2 \hat{\phi}}{\partial t^2}$$

$$+ [U_\infty \vec{i} + \nabla \hat{\phi}] \cdot \nabla \left[\frac{U_\infty^2}{2} + U_\infty \vec{i} \cdot \nabla \hat{\phi} + \frac{1}{2} \nabla \hat{\phi} \cdot \nabla \hat{\phi}\right]$$

$$= 2U_\infty \frac{\partial^2 \hat{\phi}}{\partial x \, \partial t} + \frac{\partial^2 \hat{\phi}}{\partial t^2} + U_\infty^2 \frac{\partial^2 \hat{\phi}}{\partial x^2} + 0(\hat{\phi}^2)$$

Thus the linear or small perturbation equation becomes

$$\nabla^2 \hat{\phi} - \frac{1}{a_\infty^2}\left[\frac{\partial^2 \hat{\phi}}{\partial t^2} + 2U_\infty \frac{\partial^2 \hat{\phi}}{\partial x \, \partial t} + U_\infty^2 \frac{\partial^2 \hat{\phi}}{\partial x^2}\right] = 0 \tag{4.1.24}$$

167

Note that we have replaced a by a_∞ which is correct to lowest order. By examining (4.1.22) one may show that

$$\hat{a} = -\frac{\gamma - 1}{2} \frac{\left[\dfrac{\partial \hat{\phi}}{\partial t} + U_\infty \dfrac{\partial \hat{\phi}}{\partial x}\right]}{a_\infty} \tag{4.1.25}$$

Hence it is indeed consistent to replace a by a_∞ as long as M is not too large where $M \equiv U_\infty / a_\infty$.

In a similar fashion the relationship between pressure and velocity potential, (4.1.17), may be linearized

$$C_p \simeq \frac{\hat{p}}{\dfrac{\gamma}{2} p_\infty M^2} = -\frac{2}{U_\infty} \frac{\partial \hat{\phi}}{\partial x} - \frac{2}{U_\infty^2} \frac{\partial \hat{\phi}}{\partial t}$$

or

$$\hat{p} = -\rho_\infty \left[\frac{\partial \hat{\phi}}{\partial t} + U_\infty \frac{\partial \hat{\phi}}{\partial x}\right] \tag{4.1.26}$$

Reduction to acoustics.   By making a transformation of coordinates to a system at rest with respect to the fluid, we may formally reduce the problem to that of classical acoustics.

Define

$$x' \equiv x - U_\infty t$$
$$y' \equiv y$$
$$z' \equiv z$$
$$t' \equiv t$$

then

$$\frac{\partial}{\partial x} = \frac{\partial}{\partial x'}$$
$$\frac{\partial}{\partial t} = \frac{\partial x'}{\partial t} \frac{\partial}{\partial x'} + \frac{\partial t'}{\partial t} \frac{\partial}{\partial t'}$$
$$= -U_\infty \frac{\partial}{\partial x'} + \frac{\partial}{\partial t'}$$

and (4.1.24) becomes the classical wave equation

$$\nabla'^2 \hat{\phi} - \frac{1}{a_\infty^2} \frac{\partial^2 \hat{\phi}}{\partial t'^2} = 0 \tag{4.1.27}$$

as well as (4.1.26) becomes

$$\hat{p} = -\rho_\infty \frac{\partial \hat{\phi}}{\partial t'}$$

The general solution to (4.1.27) is

$$\hat{\phi} = f(\alpha x' + \beta y' + \varepsilon z' + a_\infty t')$$
$$+ g(\alpha x' + \beta y' + \varepsilon z' - a_\infty t')$$

where

$$\alpha^2 + \beta^2 + \varepsilon^2 = 1$$

Unfortunately the above solution is not very useful, nor is the primed coordinate system, as it is difficult to satisfy the boundary conditions on the moving body in a coordinate system at rest with respect to the air (and hence moving with respect to the body). That is, obtaining solutions of (4.1.24) or (4.1.27) is not especially difficult per se. It is obtaining solutions subject to the boundary conditions of interest which is challenging.

Boundary conditions

We shall need to consider boundary conditions of various types and also certain continuity conditions as well. In general we shall see that, at least in the small perturbation theory, it is the boundary conditions, rather than the equations of motion per se, which offer the principal difficulty.

The BODY BOUNDARY CONDITION states the normal velocity of the fluid at the body surface equals the normal velocity of the body.

Consider a body whose surface is described by $F(x, y, z, t) = 0$ at some time, t, and at some later time, $t + \Delta t$, by $F(x + \Delta x, y + \Delta y, z + \Delta z, t + \Delta t) = 0$.

Now

$$\Delta F \equiv F(\vec{r} + \Delta \vec{r}, t + \Delta t) - F(\vec{r}, t) = 0$$

also

$$\Delta F = \frac{\partial F}{\partial x} \Delta x + \frac{\partial F}{\partial y} \Delta y + \frac{\partial F}{\partial z} \Delta z + \frac{\partial F}{\partial t} \Delta t$$

$$= \nabla F \cdot \Delta \vec{r} + \frac{\partial F}{\partial t} \Delta t$$

169

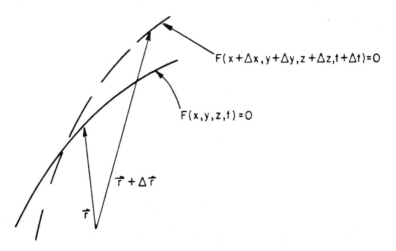

Figure 4.1   Body geometry

Thus

$$\nabla F \cdot \Delta \vec{r} + \frac{\partial F}{\partial t} \Delta t = 0 \qquad (4.1.28)$$

Now

$$\vec{n} = \frac{\nabla F}{|\nabla F|} \quad \text{unit normal} \qquad (4.1.29)$$

also

$$\vec{V} \equiv \lim_{\Delta t \to 0} \frac{\Delta \vec{r}}{\Delta t} \equiv \text{body velocity}$$

Thus the body *normal* velocity is

$$\vec{V} \cdot \vec{n} = \frac{\Delta \vec{r}}{\Delta t} \cdot \frac{\nabla F}{|\nabla F|}$$
$$= -\frac{\partial F}{\partial t} \frac{1}{|\nabla F|} \quad \text{from (4.1.28)} \qquad (4.1.30)$$

The boundary condition on the body is, as stated before, the normal fluid velocity equals the normal body velocity on the body. Thus, using (4.1.28) and (4.1.29) one has

$$\vec{q} \cdot \vec{n} = \vec{q} \cdot \frac{\nabla F}{|\nabla F|} = -\frac{\partial F}{\partial t} \frac{1}{|\nabla F|} \qquad (4.1.31)$$

or

$$\frac{\partial F}{\partial t} + \vec{q} \cdot \nabla F = 0 \qquad (4.1.32)$$

on the body surface

$$F = 0$$

Example. Planar (airfoil) surface

$$F(x, y, z, t) \equiv z - f(x, y, t)$$

where f is height above the plane, $z = 0$, of the airfoil surface. (4.1.32) may be written:

$$-\frac{\partial f}{\partial t} + [(U_\infty + u)\vec{i} + v\vec{j} + w\vec{k}] \cdot \left[-\frac{\partial f}{\partial x}\vec{i} - \frac{\partial f}{\partial y}\vec{j} + \vec{k} \right] = 0$$

or

$$\frac{\partial f}{\partial t} + (U_\infty + u)\frac{\partial f}{\partial x} + v\frac{\partial f}{\partial y} = w \qquad (4.1.33)$$

on

$$z = f(x, y, t) \qquad (4.1.34)$$

One may approximate (4.1.33) and (4.1.34) using the concept of a Taylor series about $z = 0$ and noting that $u \ll U_\infty$.

$$\frac{\partial f}{\partial t} + U_\infty \frac{\partial f}{\partial x} = w \quad \text{on} \quad z = 0 \qquad (4.1.35)$$

Note

$$w_{\text{on} z = f} = w_{z=0} + \frac{\partial w}{\partial z}\bigg|_{z=0} f + \text{H.O.T.}$$

$$\simeq w_{z=0}$$

to a consistent approximation within the context of small perturbation theory.

Symmetry and anti-symmetry

One of the several advantages of linearization is the ability to divide the aerodynamic problem into two distinct cases, symmetrical (thickness) and

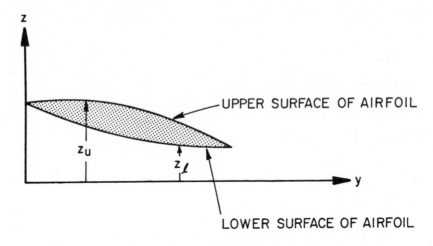

Figure 4.2 Airfoil geometry.

anti-symmetrical (lifting) flow. If one denotes the upper surface by

$$f_{\text{upper}} = z_u(x, y, t)$$

and the lower surface by

$$f_{\text{lower}} = z_l(x, y, t)$$

then it is useful to write

$$z_u \equiv z_t + z_L$$

$$z_l \equiv -z_t + z_L$$

(4.1.36)

where (4.1.36) defines z_t, thickness, and z_L, lifting, contributions to z_u and z_l.

One may treat the thickness and lifting cases separately (due to linearity) and superimpose their results for any z_u and z_l. The thickness case is much simpler than the lifting case as we shall see.

Recall (4.1.35), (we henceforward drop the $\hat{}$ on ϕ, p)

$$\frac{\partial f}{\partial t} + U_\infty \frac{\partial f}{\partial x} = \frac{\partial \phi}{\partial z} \bigg|_{z=0^+ \text{ or } 0^-}$$

(4.1.35)

where + denotes upper surfaces and − denotes lower. From (4.1.35) and (4.1.36), one sees that

Thickness case

$\dfrac{\partial \phi}{\partial z}$ is anti-symmetric with respect to z
(continuous across airfoil)

hence ϕ is symmetric (and also p).

Lifting case

$\dfrac{\partial \phi}{\partial z}$ is symmetric with respect to z
(discontinuous across airfoil)

hence ϕ is anti-symmetric (and also p).

Consider now the pressure difference across the airfoil.

$$\Delta p \equiv p_l - p_u = -\rho \left[\frac{\partial \Delta \phi}{\partial t} + U_\infty \frac{\partial \Delta \phi}{\partial x} \right]$$

Thus $\Delta p = 0$ for the thickness case, i.e., there is no lift on the airfoil.

The OFF-BODY BOUNDARY CONDITIONS (these are really continuity conditions), state that p and $\vec{q} \cdot \vec{n}$ are continuous across any fluid surface. In particular, for $z = 0$,

$$p_u = p_l \quad \text{and} \quad \left. \frac{\partial \phi}{\partial z} \right|_u = \left. \frac{\partial \phi}{\partial z} \right|_l \tag{4.1.37}$$

(4.1.37) may be used to prove some interesting results.

Thickness case

$$\frac{\partial \phi}{\partial z} = 0 \quad \text{off wing}$$

This follows from the fact that since $\partial \phi / \partial z$ is anti-symmetric, one has

$$\left. \frac{\partial \phi}{\partial z} \right|_{0^+} = - \left. \frac{\partial \phi}{\partial z} \right|_{0^-}$$

But from the second of (4.1.37), this can only be true if

$$\left. \frac{\partial \phi}{\partial z} \right|_{0^+} = \left. \frac{\partial \phi}{\partial z} \right|_{0^-} = 0$$

173

4 Nonsteady aerodynamics

Lifting case

$p = 0$ off wing

This follows in a similar way using the anti-symmetry of p and the first of (4.1.37).

The BOUNDARY CONDITIONS AT INFINITY are conditions of finiteness or outwardly propagating waves (Sommerfeld radiation condition) which will be imposed at infinity, $z \to \pm\infty$.

4.2 Supersonic flow

It is convenient to distinguish between various flow regimes on the basis of geometry (two or three dimensions) and Mach number (subsonic or supersonic). We shall not give a historical development, but shall instead begin with the simplest regime and proceed to the more difficult problems. Our main focus will be the determination of pressure distributions on airfoils.

*Two-dimensional flow**

This flow regime is the simplest as the fluid ahead of the body remains undisturbed and that behind the body does not influence the pressure distribution on the body. These results follow from the mathematics, but they also can be seen from reasonably simple physical considerations. Take a body moving with velocity, U_∞, through a fluid whose undisturbed speed of sound is a_∞, where $M \equiv U_\infty/a_\infty > 1$. At any point in the fluid disturbed by the passage of the body, disturbances will propagate to the right with velocity, $+a_\infty$, and to the left, $-a_\infty$, *with respect to the fluid*. That is, as viewed in the prime coordinate system. The corresponding propagation velocities as seen with respect to the body or airfoil will be:

$$U_\infty - a_\infty \quad \text{and} \quad U_\infty + a_\infty$$

Note these are both positive, hence the fluid ahead of the airfoil is never disturbed; also disturbances behind the airfoil never reach the body. For subsonic flow, $M < 1$, life is more complicated. Even for *three-dimensional, supersonic* flow one must consider possible effects of disturbances off side edges in the third dimension. Hence the two-dimensional, supersonic problem offers considerable simplification.

---

* See van der Vooren [3].

One of the consequences of the simplicity, as we will see, is that no distinction between thickness and lifting cases need be made as far as the mathematics is concerned. Hence the body boundary condition is (considering $z > 0$)

$$\frac{\partial \phi}{\partial z}\bigg|_{z=0} = \frac{\partial z_a}{\partial t} + U_\infty \frac{\partial z_a}{\partial x} \equiv w_a \tag{4.2.1}$$

where one may use $z_a \equiv f$ interchangably and the equation of motion is

$$\nabla^2_{x,z}\,\phi - \frac{1}{a^2_\infty}\left[\frac{\partial}{\partial t} + U_\infty \frac{\partial}{\partial x}\right]^2 \phi = 0 \tag{4.2.2}$$

Simple harmonic motion of the airfoil. Almost all of the available literature is for simple harmonic motion, that is:

$$\begin{aligned}
z_a &= \bar{z}_a(x)\mathrm{e}^{i\omega t}\\
w_a &= \bar{w}_a(x)\mathrm{e}^{i\omega t}\\
\phi &= \bar{\phi}(x, z)\mathrm{e}^{i\omega t}\\
p &= \bar{p}(x, z)\mathrm{e}^{i\omega t}
\end{aligned} \tag{4.2.3}$$

Hence we shall consider this case first. Thus (4.2.1) becomes:

$$\frac{\partial \bar{\phi}}{\partial z}\mathrm{e}^{i\omega t} = \bar{w}_a\mathrm{e}^{i\omega t} \tag{4.2.4}$$

and (4.2.2)

$$\bar{\phi}_{xx} + \bar{\phi}_{zz} - \frac{1}{a^2_\infty}\left[-\omega^2\bar{\phi} + 2i\omega U_\infty \frac{\partial \bar{\phi}}{\partial x} + U^2_\infty \frac{\partial^2 \bar{\phi}}{\partial x^2}\right] = 0 \tag{4.2.5}$$

Since $\bar{\phi}$, $\partial\bar{\phi}/\partial x$, etc., are zero for $x < 0$, this suggests the possibility of using a Laplace Transform with respect to x, i.e.,

$$\Phi(p, z) \equiv \mathscr{L}\{\bar{\phi}\} = \int_0^\infty \bar{\phi}\mathrm{e}^{-px}\,\mathrm{d}x \tag{4.2.6}$$

$$W(p) \equiv \mathscr{L}\{\bar{w}_a\} = \int_0^\infty \bar{w}_a\mathrm{e}^{-px}\,\mathrm{d}x \tag{4.2.7}$$

Taking a transform of (4.2.4) and (4.2.5) gives:

$$\frac{\mathrm{d}\Phi}{\mathrm{d}z}\bigg|_{z=0} = W \tag{4.2.8}$$

$$\frac{\mathrm{d}^2\Phi}{\mathrm{d}z^2} = \mu^2\Phi \tag{4.2.9}$$

175

where

$$\mu^2 \equiv (M^2 - 1)p^2 + 2Mi\frac{\omega p}{a_\infty} - \frac{\omega^2}{a_\infty^2}$$

$$= (M^2 - 1)\left\{\left[p + \frac{iM\omega}{a_\infty(M^2 - 1)}\right]^2 + \frac{\omega^2}{a_\infty^2(M^2 - 1)^2}\right\}$$

Note $M \equiv U_\infty/a_\infty$. (4.2.8) and (4.2.9) are now equations we can solve.

$$\Phi = Ae^{\mu z} + Be^{-\mu z} \qquad (4.2.10)$$

Select $A \equiv 0$ to keep Φ finite as $z \to +\infty$. Hence

$$\Phi = Be^{-\mu z}$$

where B can be determined using (4.2.8). From the above,

$$\frac{d\Phi}{dz}\bigg|_{z=0} = -\mu B$$

Using this result and (4.2.8), one has

$$-\mu B = W$$

or

$$B = -W/\mu$$

and hence

$$\Phi = -(W/\mu)e^{-\mu z} \qquad (4.2.11)$$

Inverting (4.2.11), using the convolution theorem,

$$\bar{\phi} = -\int_0^x \bar{w}_a(\xi)\mathscr{L}^{-1}\left\{\frac{e^{-\mu z}}{\mu}\right\}d\xi \qquad (4.2.12)$$

and, in particular,

$$\bar{\phi}(x, z = 0) = -\int_0^x \bar{w}_a(\xi)\mathscr{L}^{-1}\left\{\frac{1}{\mu}\right\}d\xi$$

From H. Bateman, 'Table of Integral Transforms', McGraw-Hill, 1954,

$$\mathscr{L}^{-1}\left\{\frac{1}{\sqrt{p^2 + \alpha^2}}\right\} = J_0(\alpha x)$$

$$\mathscr{L}^{-1}\{F(p + a)\} = e^{-ax}f(x)$$

where $\mathscr{L}^{-1}\{F(p)\} \equiv f(x)$. Thus

$$\mathscr{L}^{-1}\left\{\frac{1}{\mu}\right\} = \frac{\exp\left[-\dfrac{iM\omega}{a_\infty(M^2-1)}(x-\xi)\right]}{(M^2-1)^{\frac{1}{2}}} J_0\left[\frac{\omega}{a_\infty(M^2-1)}(x-\xi)\right] \quad (4.2.13)$$

$\mathscr{L}^{-1}\{e^{-\mu z}/\mu\}$ may be computed by similar methods. In nondimensional terms,

$$\bar{\phi}(x^*, 0) = -\frac{2b}{(M^2-1)^{\frac{1}{2}}} \int_0^{x^*} \bar{w}(\xi^*)\exp[-i\bar{\omega}(x^*-\xi^*)]J_0\left[\frac{\bar{\omega}}{M}(x^*-\xi^*)\right]d\xi^* \tag{4.2.14}$$

where

$$\bar{\omega} \equiv \frac{kM^2}{M^2-1}, \qquad k \equiv \frac{2b\omega}{U_\infty} \text{ is a reduced frequency and}$$

$$x^* \equiv x/2b, \qquad \xi^* \equiv \xi/2b$$

One can now use Bernoulli's equation to compute p.

$$p = -\rho_\infty\left[\frac{\partial\phi}{\partial t} + U_\infty\frac{\partial\phi}{\partial x}\right]$$

or

$$\bar{p} = -\rho_\infty\left[i\omega\bar{\phi} + U_\infty\frac{\partial\bar{\phi}}{\partial x}\right]$$

$$= -\frac{\rho_\infty U_\infty}{2b}\left[\frac{\partial\bar{\phi}}{\partial x^*} + ik\bar{\phi}\right]$$

Using Leibnitz' rule,

$$\bar{p} = -\frac{\rho_\infty U_\infty^2}{(M^2-1)^{\frac{1}{2}}}\left\{\int_0^{x^*}\left[ik\frac{\bar{w}_a}{U_\infty} + \frac{1}{U_\infty}\frac{d\bar{w}}{d\xi^*}\right]e^{-\cdots}J_0[\cdots]d\xi^*\right.$$

$$\left. + \frac{\bar{w}(0)}{U_\infty}e^{-i\omega x^*}J_0\left[\frac{\bar{\omega}}{M}x^*\right]\right\} \quad (4.2.15)$$

Discussion of inversion. The above inversion was something less than rigorous and, what is more important, in at least one substantial aspect it was misleading. Let us reconsider it, therefore, now that the general outline of the analysis is clear.

Formally the inversion formula reads:

$$\bar{\phi}(x, z) = \frac{1}{2\pi i}\int_{-i\infty}^{i\infty}\Phi(p, z)e^{px}\,dp \tag{4.2.16}$$

Define $\alpha \equiv ip$, (α can be thought of as a Fourier Transform variable), then

$$\bar{\phi}(x, z) = \frac{1}{2\pi} \int_{-\infty}^{\infty} \Phi(-i\alpha, z) e^{-i\alpha x} \, d\alpha \qquad (4.2.17)$$

and

$$\mu = \sqrt{M^2 - 1} \sqrt{-\left[-\alpha + \frac{M\omega}{a_\infty(M^2 - 1)}\right]^2 + \frac{\omega^2}{a_\infty^2(M^2 - 1)^2}}$$

where

$$\Phi = \pm \frac{W}{\mu} e^{\pm \mu z} \qquad (4.2.18)$$

Consider μ as $\alpha = -\infty \rightarrow +\infty$.

The quantity under the radical changes sign at

$$\alpha = \alpha_1, \; \alpha_2 = \frac{\omega}{a_\infty} \frac{1}{M \pm 1}$$

where $\mu = 0$. Thus

$$\mu = \pm i \, |\mu| \quad \text{for} \quad \alpha < \alpha_1 \quad \text{or} \quad \alpha > \alpha_2$$
$$= \pm |\mu| \quad \text{for} \quad \alpha_1 < \alpha < \alpha_2$$

where

$$|\mu| = (M^2 - 1) \left| -\left[-\alpha + \frac{M\omega}{a_\infty(M^2 - 1)}\right]^2 + \frac{\omega^2}{a_\infty^2(M^2 - 1)^2} \right|^{\frac{1}{2}}$$

In the interval, $\alpha_1 < \alpha < \alpha_2$, we have seen we must select the minus sign so that Φ is finite at infinity. What about elsewhere? In particular, when $\alpha < \alpha_1$ and/or $\alpha > \alpha_2$?

The solution for $\phi = \bar{\phi} e^{i\omega t}$ has the form

$$\phi = -\frac{1}{2\pi} \int_{-\infty}^{\infty} \pm \frac{W}{\mu} \exp\left(\pm \mu z - iax + i\omega t\right) d\alpha \qquad (4.2.19)$$

In the intervals $\alpha < \alpha_1$ and/or $\alpha > \alpha_2$, (4.2.19) reads:

$$\phi = -\frac{1}{2\pi} \int_{-\infty}^{\infty} \pm i \frac{W}{|\mu|} \exp\left(\pm i \, |\mu| \, z - iax + i\omega t\right) d\alpha \qquad (4.2.20)$$

To determine the proper sign, we require that solution represent an outgoing wave in the fluid fixed coordinate system, i.e., in the prime system. In the prime system $x' = x - U_\infty t$, $z' = z$, $t' = t$ and thus

$$\phi = -\frac{1}{2\pi} \int_{-\infty}^{\infty} \pm i \frac{W}{|\mu|} \exp\left[\pm i |\mu| z' - iax' + i(\omega - U_\infty \alpha)t'\right] d\alpha \quad (4.2.21)$$

Consider a z', t' wave for fixed x'. For a wave to be outgoing, if $\omega - U_\infty \alpha > 0$ then one must choose $-$ sign while if $\omega - U_\infty \alpha < 0$ then choose $+$ sign. Note that

$$\omega - U_\infty \alpha = 0$$

when

$$\alpha = \alpha_3 \equiv \frac{\omega}{U_\infty} = \frac{\omega}{a_\infty M}$$

also note that

$$\frac{\omega}{a_\infty(M+1)} \equiv \alpha_1 < \alpha_3 < \alpha_2 \equiv \frac{\omega}{a_\infty(M-1)}$$

Thus the signs are chosen as sketched below.

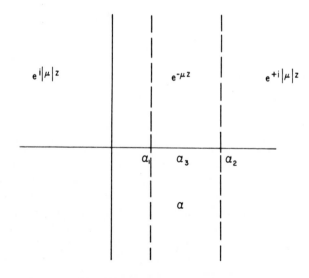

Here again

$$\alpha_1 \equiv \frac{\omega}{a_\infty} \frac{1}{M+1} \qquad \alpha_2 \equiv \frac{\omega}{a_\infty} \frac{1}{M-1} \qquad \alpha_3 \equiv \frac{\omega}{a_\infty} \frac{1}{M}$$

179

The reader may find it of interest to consider the subsonic case, $M < 1$, using similar reasoning.

Knowing the appropriate choice for μ in the several intervals, (4.2.19)–(4.2.21) may be integrated numerically, or by contour integration. The inversion formulae used previously were obtained by contour integration.

Discussion of physical significance of results.   Because of the complicated mathematical form of our solution, it is difficult to understand its physical significance. Perhaps it is most helpful for this purpose to consider the limits of low and high frequency.

One may show that (from (4.2.11) et. seq. or (4.2.15))*
$\omega \to 0$: *steady flow*

$$p(x) \to \frac{\rho_\infty U_\infty^2}{\sqrt{M^2-1}} \frac{w}{U}(x), \; p(x, z) = \frac{\rho_\infty U_\infty^2}{\beta} \frac{w(x-\beta z)}{U_\infty}$$

$$\frac{w}{U_\infty} \to \frac{\partial f}{\partial x} \qquad \beta \equiv \sqrt{M^2-1}$$

$\omega \to \infty$: *highly unsteady flow*

$$p(x, t) \to \frac{\rho_\infty U_\infty^2}{M} \frac{w(x, t)}{U_\infty}, \; p(x, z, t) = \frac{\rho_\infty U_\infty^2}{M} \frac{w}{U_\infty}(x - Mz, t)$$

$$\frac{w}{U_\infty} \to \frac{1}{U_\infty}\frac{\partial f}{\partial t} + \frac{\partial f}{\partial x}$$

The latter result may be written as

$$p = \rho_\infty a_\infty w$$

which is the pressure on a piston in a long, narrow (one-dimensional) tube with w the velocity of the piston. It is, therefore, termed 'piston theory' for obvious reasons. Note that in the limits of low and high frequency the pressure at point x depends only upon the downwash at that same point. For arbitrary ω, the pressure at one point depends in general upon the downwash at all other points. See (4.2.15). Hence the flow has a simpler behavior in the limits of small and large ω than for intermediate ω. Also recall that low and high frequency may be interpreted in the time domain for transient motion as long and short time respectively. This follows from the initial and final value Laplace Transform theorems.† For example, if

---

* See the appropriate example problem in Appendix II for details.
† Hildebrand [2].

we consider a motion which corresponds to a step change in angle of attack, α, we have

$$f = -x\alpha \quad \text{for} \quad t > 0$$
$$\quad = 0 \quad \quad \text{for} \quad t < 0$$
$$w = -\alpha \quad \text{for} \quad t > 0$$
$$w/U_\infty = 0 \quad \text{for} \quad t < 0$$

Hence for short time, (large ω)

$$p = \frac{-\rho_\infty U_\infty^2}{M}\alpha$$

and long time, (small ω)

$$p = \frac{-\rho_\infty U_\infty^2 \alpha}{\sqrt{M^2 - 1}}$$

The result for short time may also be deduced by applying a Laplace Transform with respect to time and taking the limit $t \to 0$ of the formal inversion.

General comments.   A few general comments should be made about the solution. First of all, the solution has been obtained for simple harmonic motion. In principle, the solution for arbitrary time dependent motion may be obtained via Fourier superposition of the simple harmonic motion result. Actually it is more efficient to use a Laplace Transform with respect to time and invert the time variable *prior* to inverting the spatial variable, x. Secondly, with regard to the distinction between the lifting and thickness cases, one can easily show by direct calculation and using the method applied previously that

thickness	$z = 0^+$	$w = w_a$	$p = p^+$
	$z = 0^-$	$w = -w_a$	$p = p^+$
lifting	$z = 0^+$	$w = w_a$	$p = p^+$
	$z = 0^-$	$w = w_a$	$p = -p^+$

where p^+ is the solution previously obtained. Of course these results also follow from our earlier general discussion of boundary conditions.

Gusts.   Finally it is of interest to consider how aerodynamic pressures develop on a body moving through a nonuniform flow, i.e., a 'gust'. If the body is motionless, the body boundary condition is that the total fluid velocity be zero on the body.

$$\left.\frac{\partial\phi}{\partial z}\right|_{z=0} + w_G = 0$$

where w_G is the specified vertical 'gust' velocity and $\partial\phi/\partial z$ is the perturbation fluid velocity resulting from the body passing through the gust field. Hence in our previous development we may replace w by $-w_G$ and the same analysis then applies. Frequently one assumes that the gust field is 'frozen', i.e., fixed with respect to the fluid fixed coordinates, x', y', z', t'. Hence

$$w_G = w_G(x', y')$$
$$= w_G(x - U_\infty t, y)$$

Further a special case is a 'sharp edge' gust for which one simply has

$$w_G = w_0 \quad \text{for} \quad x' < 0$$
$$= 0 \quad \text{for} \quad x' > 0$$

or

$$w_G = w_0 \quad \text{for} \quad t > x/U_\infty$$
$$= 0 \quad \text{for} \quad t < x/U_\infty$$

These special assumptions are frequently used in applications.

Solutions for the sharp edge gust can be obtained through superposition of (simple harmonic motion) sinusoidal gusts. However, it is more efficient to use methods developed for transient motion. Hence before turning to three-dimensional supersonic flow, we consider transient motion. Transient solutions can be obtained directly (in contrast to Fourier superposition of simple harmonic motion results) for a two-dimensional, supersonic flow.

Transient motion. Taking a Laplace transform with respect to *time* and a Fourier transform with respect to the *streamwise coordinate*, x, the analog of (4.2.11) is

$$\mathscr{L}F\{\phi\}_{\text{at } z=0} = -\frac{\mathscr{L}Fw}{\mu} \tag{4.2.22}$$

$i\omega \equiv s$ is the Laplace Transform variable (where ω was the frequency in the simple harmonic motion result), α is the Fourier Transform variable (where $i\alpha \equiv p$ was the Laplace transform variable used in the previous simple harmonic motion result), $\mathcal{L} \equiv$ Laplace transform, $F \equiv$ Fourier transform, and

$$\mu^2 \equiv -(M^2-1)\alpha^2 + 2\frac{Msi}{a_\infty}\alpha + \frac{s^2}{a_\infty^2}$$

Inverting the Laplace Transform, and using $*$ to denote a Fourier transform

$$\phi^*|_{\text{at } z=0} = -\int_0^t w^*(\tau)\mathcal{L}^{-1}\left\{\frac{1}{\mu}\right\}\Bigg|_{t-\tau} \mathrm{d}\tau$$

$$= -a_\infty \int_0^t w^*(\tau) \exp\left[-i\alpha Ma_\infty(t-\tau)\right]J_0[a_\infty\alpha(t-\tau)]\,\mathrm{d}\tau \quad (4.2.23)$$

Now from (4.1.26),

$$p^* = -\rho_\infty\left[\frac{\partial\phi^*}{\partial t} + U_\infty i\alpha\phi^*\right]$$

Thus using (4.2.23) and the above,

$$p^* = \rho_\infty\left\{a_\infty w^*(t) - a_\infty^2 \int_0^t w^*(\tau)\alpha \exp\left[-i\alpha Ma_\infty(t-\tau)\right]J_1[\alpha a_\infty(t-\tau)]\,\mathrm{d}\tau\right\}$$

$$\equiv p_0^* + p_1^* \quad (4.2.24)$$

Finally, a formal solution is obtained using

$$p = \frac{1}{2\pi}\int_{-\infty}^{\infty} p^* e^{i\alpha x}\,\mathrm{d}\alpha \quad (4.2.25)$$

The lift is obtained by using (4.2.24) and (4.2.25) in its definition below.

$$L \equiv -2\int_0^{2b} p\,\mathrm{d}x = -2\rho_\infty a_\infty \int_0^{2b} w\,\mathrm{d}x - \frac{1}{\pi}\int_{-\infty}^{\infty} p_1^*\left[\frac{e^{i\alpha 2b}-1}{i\alpha}\right]\mathrm{d}\alpha \quad (4.2.26)$$

In the second term the integration over x has been carried out explicitly.

Lift, due to airfoil motion. Considering a translating airfoil, $w = -\mathrm{d}h/\mathrm{d}t$, for example, we have

$$w^* = -\frac{\mathrm{d}h}{\mathrm{d}t}\frac{[e^{-i\alpha 2b}-1]}{-i\alpha}$$

and

$$L = 2\rho_\infty a_\infty \frac{dh}{dt} (2b)$$

$$+ \rho_\infty a_\infty^2 \int_0^t \frac{dh}{dt}(\tau) K(t-\tau) \, d\tau \qquad (4.2.27)$$

where

$$K(t-\tau) \equiv -\frac{1}{\pi} \int_{-\infty}^\infty \frac{e^{-()}}{\alpha} J_1[e^{i\alpha 2b} - 1][e^{-i\alpha 2b} - 1] \, d\alpha$$

K may be simplified to

$$K(t-\tau) = -\frac{4}{\pi} \int_0^\infty \frac{J_1[a_\infty \alpha(t-\tau)] \cos[\alpha M a_\infty(t-\tau)]}{\alpha} \cdot [1 - \cos \alpha 2b] \, d\alpha$$

One can similarly work out aerodynamic lift (and moment) for pitching and other motions.

Lift, due to atmospheric gusts.   For a 'frozen gust',

$$w_G(x - U_\infty t) = w_G(x')$$

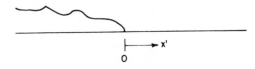

Figure 4.3   *Frozen gust geometry in fluid fixed coordinate system.*

x, t are coordinates fixed with respect to airfoil and x', t' are coordinates fixed with respect to atmosphere. At $t = t' = 0$ the airfoil enters the gust; the boundary condition is $w_a + w_G = 0$; $w_a = -w_G$ on airfoil. Short and long time correspond to high and low frequency; hence it is of interest to use our previously developed approximate theories for these limits. Subsequently we treat the full transient case.

(i) Piston Theory (short t) on the upper and lower airfoil surfaces

$$p_u = -\rho a_\infty w_G$$

and

$$p_l = +\rho a_\infty w_G$$

Thus

$$L(t) = \int (p_l - p_u)\, dx$$

$$= 2\rho_\infty a_\infty \int_0^{2b} w_G(x - U_\infty t)\, dx$$

For simplicity, we first consider a sharp edge gust.
Let

$$w_G = w_0 \quad \text{for} \quad x' < 0 \quad \text{or} \quad x < Ut, t > x/U_\infty$$
$$= 0 \quad \text{for} \quad x' > 0 \quad \text{or} \quad x > Ut, t < x/U_\infty$$

Thus

$$L(t) = 2\rho_\infty a_\infty w_0 \int_0^{U_\infty t} dx$$

$$= 2\rho_\infty a_\infty w_0 U_\infty t \quad \text{for} \quad t < \frac{2b}{U_\infty}$$

$$= 2\rho_\infty a_\infty w_0 2b \quad \text{for} \quad t > \frac{2b}{U_\infty} \qquad (4.2.28)$$

(ii) Static theory (large t)

$$L(t) = \frac{2\rho_\infty U_\infty^2}{\sqrt{M^2 - 1}} \frac{w_0}{U_\infty} \int_0^{2b} dx = 4b \frac{\rho_\infty a_\infty w_0 M}{\sqrt{M^2 - 1}} \qquad (4.2.29)$$

(iii) Full Transient Theory from (4.2.24),

$$p = \rho_\infty a_\infty \left[w_a(x, t) - a_\infty \int_0^t \alpha w_a^*(\alpha, \tau) e^{-(\cdots)} J_1(\cdots)\, d\tau \right] \qquad (4.2.24)$$

Special case.   Sharp Edge Gust

$$w_a = -w_G(x - U_\infty t) = -w_0 \quad \text{for} \quad x < U_\infty t$$
$$= 0 \quad \text{for} \quad x > U_\infty t$$

Thus

$$w_a^*(\alpha, \tau) = -\int_{-\infty}^{\infty} e^{-i\alpha x} w_G(x - U_\infty \tau)\, dx$$

$$= -w_0 \int_{-\infty}^{U_\infty \tau} e^{-i\alpha x}\, dx$$

$$= \frac{-w_0}{-i\alpha} e^{-i\alpha x} \Big|_{-\infty}^{U_\infty \tau}$$

$$= \frac{w_0}{i\alpha} e^{i\alpha U_\infty \tau}$$

185

4 Nonsteady aerodynamics

Using the above and (4.2.24),

$$p = \rho_\infty a_\infty \Big[-w_G(x - U_\infty t) - \frac{a_\infty w_0}{2\pi}$$

$$\times \int_{-\infty}^{\infty} \int_0^t \alpha \frac{e^{-i\alpha U_\infty \tau}}{i\alpha} e^{-(\cdot)} J_1(\,) \, d\tau e^{i\alpha x} \, d\alpha \Big] \quad (4.2.30)$$

Again one may proceed further by computing the lift.

$$L = 2\rho_\infty a_\infty w_0 \quad \begin{matrix} U_\infty t, & \text{for} & U_\infty t < 2b \\ 2b, & \text{for} & U_\infty t > 2b \end{matrix}$$

$$+ 2\rho_\infty \frac{a_\infty^2 w_0}{2\pi} \int_0^{2b} \int_{-\infty}^{\infty} \int_0^t \cdots d\tau \, d\alpha \, dx$$

Integrating over x first, and introducing nondimensional notation

$$s \equiv \frac{tU_\infty}{2b} \qquad \alpha^* \equiv \alpha 2b$$

$$\sigma \equiv \frac{\tau U_\infty}{2b}$$

one obtains

$$\frac{L}{2\rho_\infty U_\infty^2 2b} = \Big[\frac{w_0}{U_\infty} \frac{s}{M} - \frac{1}{M^2} \int_0^s F(s, \sigma) \, d\sigma \Big] \quad (4.2.31)$$

where

$$F(s, \sigma) \equiv \frac{1}{\pi} \int_0^\infty \frac{[-\cos \alpha^* s + \cos \alpha^*(1-s)] J_1 \Big[\alpha^* \dfrac{(s-\sigma)}{M} \Big]}{\alpha^*} \, d\alpha^*$$

General case.   Arbitrary Frozen Gust

$$w_a^*(\alpha, \tau) = -\int_{-\infty}^{\infty} e^{-i\alpha x} w_G(x - U_\infty \tau) \, dx$$

$$= -\int_{-\infty}^{U\tau} e^{-i\alpha x} w_G(x - U_\infty \tau) \, dx$$

Let $x' = x - U_\infty t$, $dx' = dx$, then

$$w_a^* = -e^{-i\alpha U_\infty \tau} \int_{-\infty}^0 e^{-i\alpha x'} w_G(x') \, dx$$

$$= -e^{-i\alpha U_\infty \tau} w_G^*(\alpha)$$

186

Using above in (4.2.24), the pressure is

$$p = \rho_\infty a_\infty \left[- w_G(x - U_\infty t) \right.$$

$$\left. + \frac{a_\infty}{2\pi} \int_{-\infty}^{\infty} \int_0^t w_G^* \alpha e^{-i\alpha U_\infty \tau} e^{-(\ )} J_1(\ ) \, d\tau e^{i\alpha x} \, d\alpha \right]$$

and the lift,

$$L = 2\rho_\infty a_\infty \int_0^{2b} w_G(x - U_\infty t) \, dx$$

$$- \frac{2\rho_\infty a_\infty^2}{2\pi} \int_0^{2b} \int_{-\infty}^{\infty} \int_0^t \cdots d\tau \, d\alpha \, dx$$

Integrating over x first,

$$\frac{L}{2\rho_\infty U_\infty^2 2b} = \int_0^{2b} \frac{w_G/U_\infty}{M} \frac{dx}{2b} - \frac{1}{M^2} \int_0^s F(s, \sigma) \, d\sigma \qquad (4.2.32)$$

where now

$$F(s, \sigma) \equiv \frac{1}{\pi} \int_0^{\infty} \left\{ W_I^* \{\cos [\alpha^*(1-s)] - \cos \alpha^* s\} J_1 \left[\frac{\alpha^*(s-\sigma)}{M} \right] \right.$$

$$\left. + W_R^* \{\sin [\alpha^*(1-s)] + \sin \alpha^* s\} J_1 \left[\frac{\alpha^*(s-\sigma)}{M} \right] \right\} \, d\alpha^*$$

and

$$W^* \equiv \frac{w^*}{U_\infty 2b}$$

For an alternative approach to transient motion which makes use of an analogy between two-dimensional time dependent motion and three-dimensional steady motion, the reader may consult Lomax [4].

This completes our development for two-dimensional, supersonic flow. We now have the capability for determining the aerodynamic pressures necessary for flutter, gust and even, in principle, acoustic analyses for this type of flow. For the latter the pressure in the 'far field' (large z) is usually of interest. Now let us consider similar analyses for three-dimensional, supersonic flow.

Three dimensional flow †

We shall now add the third dimension to our analysis. As we shall see there is no essential complication with respect to solving the governing

---

† References: BA, pp. 134–139; Landahl and Stark [5], Watkins [6].

differential equation; the principal difficulty arises with respect to satisfying all of the relevant boundary conditions.

The convected wave equation reads in three spatial dimensions and time

$$\nabla^2 \phi - \frac{1}{a_\infty^2} \left[\frac{\partial^2 \phi}{\partial t^2} + 2U_\infty \frac{\partial^2 \phi}{\partial x \, \partial t} + U_\infty^2 \frac{\partial^2 \phi}{\partial x^2} \right] = 0 \tag{4.2.33}$$

As before we assume simple harmonic time dependence.

$$\phi = \bar{\phi}(x, y, z)e^{i\omega t}$$

Further taking a Laplace transform with respect to x, gives

$$\frac{\partial^2 \Phi}{\partial z^2} + \frac{\partial^2 \Phi}{\partial y^2} = \mu^2 \Phi \tag{4.2.34}$$

where

$$\Phi \equiv \mathscr{L}\bar{\phi} = \int_0^\infty \bar{\phi}e^{-px} \, dx$$

$$\mu = \sqrt{M^2 - 1}\left[\left(p + \frac{i\omega M}{a_\infty(M^2 - 1)} \right)^2 + \frac{\omega^2}{a^2(M^2 - 1)^2} \right]^{\frac{1}{2}}$$

To reduce (4.2.33) to an ordinary differential equation in z, we take a Fourier transform with respect to y. Why would a Laplace transform be inappropriate? The result is:

$$\frac{d^2 \Phi^*}{dz^2} = (\mu^2 + \gamma^2)\Phi^* \tag{4.2.35}$$

where

$$\Phi^* \equiv F\Phi = \int_{-\infty}^\infty \Phi e^{-i\gamma y} \, dy$$

The solution to (4.2.34) is

$$\Phi^* = A \, \exp\left[+(\mu^2 + \gamma^2)^{\frac{1}{2}}z\right] + B \, \exp\left[-(\mu^2 + \gamma^2)^{\frac{1}{2}}z\right]$$

Selecting the appropriate solution for finiteness and/or radiation as $z \rightarrow +\infty$, we have

$$\Phi^* = B \, \exp\left[-(\mu^2 + \gamma^2)^{\frac{1}{2}}z\right] \tag{4.2.36}$$

Applying the body boundary condition (as transformed)

$$\left. \frac{d\Phi^*}{dz} \right|_{z=0} = W^* \tag{4.2.37}$$

we have from (4.2.36) and (4.2.37)

$$B = -\frac{W^*}{(\mu^2 + \gamma^2)^{\frac{1}{2}}}$$

and hence

$$\Phi^*_{z=0} = -\frac{W^*}{(\mu^2 + \gamma^2)^{\frac{1}{2}}}$$

Using the convolution theorem

$$\bar{\phi}(x, y, z = 0) = -\int_0^x \int_{-\infty}^{\infty} \bar{w}(\xi, \eta) \mathscr{L}^{-1} F^{-1} \frac{1}{(\mu^2 + \gamma^2)^{\frac{1}{2}}} \, d\xi \, d\eta \qquad (4.2.38)$$

Now let us consider the transform inversions. The Laplace inversion is essentially the same as for the two-dimensional case.

$$\mathscr{L}^{-1} \frac{1}{(\mu^2 + \gamma^2)^{\frac{1}{2}}} = \frac{\exp\left[-\dfrac{iM\omega x}{a_\infty(M^2-1)}\right]}{\sqrt{M^2-1}} J_0\left(\left[\frac{\omega^2}{a_\infty^2(M^2-1)^2} + \frac{\gamma^2}{(M^2-1)}\right]^{\frac{1}{2}} x\right)$$

To perform the Fourier inversion, we write

$$F^{-1}\left\{\mathscr{L}^{-1}\left\{\frac{1}{(\mu^2+\gamma^2)}\right\}\right\}$$

$$= \frac{\exp\left[-\dfrac{iM\omega x}{a_\infty(M^2-1)}\right]}{2\pi\sqrt{M^2-1}} \int_{-\infty}^{\infty} J_0\left(\left[\frac{\omega^2}{a_\infty^2(M^2-1)^2} + \frac{\gamma^2}{(M^2-1)}\right]^{\frac{1}{2}} x\right) e^{i\gamma y} \, d\gamma$$

$$= \frac{\exp\left[-\dfrac{iM\omega x}{a_\infty(M^2-1)}\right]}{\pi\sqrt{M^2-1}} \int_0^{\infty} J_0(\cdots) \cos \gamma y \, d\gamma$$

where the last line follows from the evenness of the integrand with respect to γ. The integral has been evaluated in Bateman, [7], p. 55.

$$\int_0^{\infty} J_0(\cdots) \cos \gamma y \, d\gamma = \left[\frac{x^2}{M^2-1} - y^2\right]^{-\frac{1}{2}} \cos\left[\frac{\omega}{a_\infty(M^2-1)^{\frac{1}{2}}} \left(\frac{x^2}{M^2-1} - y^2\right)^{\frac{1}{2}}\right]$$

$$\text{for} \quad |y| < \frac{x}{\sqrt{M^2-1}}$$

$$= 0 \quad \text{for} \quad |y| > \frac{x}{\sqrt{M^2-1}}$$

189

Thus finally

$$F^{-1}\mathscr{L}^{-1}\left\{\frac{1}{(\mu^2+\gamma^2)^{\frac{1}{2}}}\right\}$$

$$=\frac{1}{\pi}\frac{\exp\left[-\dfrac{iM\omega}{a_\infty(M^2-1)}\right]}{\sqrt{M^2-1}}\cos\frac{\left[\dfrac{\omega}{a_\infty(M^2-1)^{\frac{1}{2}}}\left(\dfrac{x^2}{M^2-1}-y^2\right)^{\frac{1}{2}}\right]}{\left[\dfrac{x^2}{M^2-1}-y^2\right]^{\frac{1}{2}}}$$

$$\text{for}\quad |y|<\frac{x}{\sqrt{M^2-1}}$$

$$=0\quad\text{for}\quad |y|>\frac{x}{\sqrt{M^2-1}}$$

Using the above in (4.2.37) and nondimensionalizing by $s\equiv$ wing semi-span and $b\equiv$ reference semi-chord,

$$\bar{\phi}(x^*,y^*,z=0)$$

$$=\frac{-s}{\pi}\int_0^{x^*}\int_{y^*-(2b/s)(x^*-\xi^*)/\beta}^{y^*+(2b/s)(x^*-\xi^*)/\beta}\bar{w}(\xi^*,\eta^*)\exp[-i\bar{\omega}(x^*-\xi^*)]\frac{\cos\dfrac{\bar{\omega}r^*}{M}}{r^*}\,d\xi^*\,d\eta^* \tag{4.2.39}$$

where

$$r^*\equiv[(x^*-\xi^*)^2-\beta^2\left(\frac{s}{2b}\right)^2(y^*-\eta^*)^2]^{\frac{1}{2}}$$

$$\beta\equiv\sqrt{M^2-1}$$

$$x^*,\xi^*\equiv x/2b,\xi/2b\qquad y^*,\eta^*\equiv y/s,\eta/s$$

$$k\equiv\frac{\omega 2b}{U_\infty},\qquad\bar{\omega}\equiv\frac{kM^2}{(M^2-1)}$$

If $\bar{w}$ is known everywhere in the region of integration then (4.2.39) is a solution to our problem. Unfortunately, in many case of interest, $\bar{w}$ is unknown over some portion of the region of interest. Recall that $\bar{w}$ is really $\dfrac{\partial\bar{\phi}}{\partial z}\bigg|_{z=0}$. In general this vertical fluid velocity is unknown off the wing. There are three principal exceptions to this:

(1) If we are dealing with a thickness problem then $\dfrac{\partial\phi}{\partial z}\bigg|_{z=0}=0$ everywhere off the wing and no further analysis is required.

(2) Certain wing geometries above a certain Mach number will have undisturbed flow off the wing even in the lifting case. For these so-called 'supersonic planforms', $\dfrac{\partial \phi}{\partial z}\Big|_{z=0} = 0$ off wing as well.

(3) Even in the most general case, there will be no disturbance to the flow ahead of the rearward facing Mach lines, $\eta = \pm \xi/\beta$, which originate at the leading most point of the lifting surface.

To make case (2) more explicit and in order to discuss what must be done for those cases where the flow off the wing is disturbed, let us consider the following figure; Figure 4.4. Referring first to case (2), we see that if the tangents of the forward facing Mach lines (integration limits of (4.2.39))

$$\eta = y \pm \frac{(x - \xi)}{\beta}$$

are sufficiently small, i.e., $\left|\dfrac{1}{\beta}\right| \rightarrow 0$ then the regions where $\bar{w}_a$ unknown, $\bar{\phi} = 0$

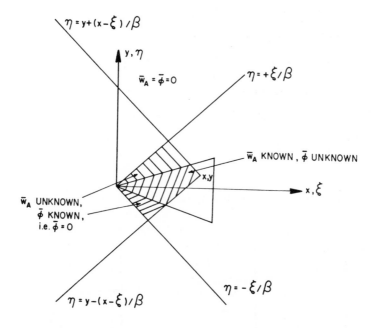

Figure 4.4 *Lifting surface geometry. A representative delta wing is shown.*

191

will vanish. This what we mean by a 'supersonic planform'. The mathematical problem for these planforms is essentially the same as for a 'thickness problem' whether or not lift is being produced.

Finally let us consider the most difficult case where we have a mixed boundary condition problem. In general analytical solutions are not possible and we resort to numerical methods. The technique which has received the largest amount of attention and development to date is the 'box' method. In this approach, the integral equation (4.2.39) is approximated by differences and sums, i.e.,

$$\frac{\bar{\phi}(x_i^*, y_j^*)}{U_\infty s} = \sum_{k=1}^{K} \sum_{l=1}^{L} A_{(ij)(kl)} \frac{\bar{w}_a(\xi_k^*, \eta_l^*)}{U_\infty} \tag{4.2.40}$$

where

$$A_{(ij)(kl)} \equiv -\frac{1}{\pi} \exp[-i\bar{\omega}(x_i^* - \xi_k^*)] \frac{\cos \frac{\bar{\omega}}{M} r_{(ij)(kl)}^*}{r_{(ij)(kl)}^*} \Delta\xi^* \Delta\eta^*$$

and

$$r_{(ij)(kl)}^* \equiv \left[(x_i^* - \xi_k^*)^2 - \beta^2 \left(\frac{s}{2b}\right)^2 (y_j^* - \eta_l^*)^2 \right]^{\frac{1}{2}}$$

$\Delta\xi^*, \Delta\eta^* \equiv$ dimensions of aerodynamic box

$A_{(ij)(kl)}$ aerodynamic influence coefficients; the velocity potential at point, ij, due to a unit 'downwash', $\bar{w}_a$, at point kl

Equation (4.2.40) can be written in matrix notation as:

$$\left\{ \begin{matrix} \bar{\phi} \\ \\ \end{matrix} \right\} = \left[A \right] \left\{ \begin{matrix} \bar{w}_a \\ \\ \end{matrix} \right\} \tag{4.2.41}$$

The system of linear equation may be separated as follows:

$$
\begin{matrix} N1 \\ \\ - \\ \\ N2 \end{matrix}
\left\{ \begin{matrix} \bar{\phi} \\ \text{unknown} \\ \underline{\quad\quad} \\ \bar{\phi} \\ \text{known} \end{matrix} \right\}
=
\left[\begin{matrix} A_1 & \bigg| & N2 \times N1 \\ N1 \times N1 & & A_2 \\ \hline N1 \times N2 & \bigg| & N2 \times N2 \\ A_3 & & A_4 \end{matrix} \right]
\left\{ \begin{matrix} \bar{w}_a \\ \text{known} \\ \underline{\quad\quad} \\ \bar{w}_a \\ \text{unknown} \end{matrix} \right\}
\begin{matrix} N1 \\ \\ \\ \\ N2 \end{matrix}
$$

where $N1$ number of boxes where $\bar{w}_a$ known, $\bar{\phi}$ unknown (on wing)

$N2$ number of boxes where $\bar{w}_a$ unknown, $\bar{\phi}$ known (off wing)

Using last $N2$ equations of (4.2.42)

$$\bar{\phi}_{known} = [A_3]\{\bar{w}_{a\,known}\} + [A_4]\{\bar{w}_{a\,unknown}\} \tag{4.2.43}$$

Solving for $\bar{w}_{a\,unknown}$,

$$\{\bar{w}_{a\,unknown}\} = [A_4]^{-1}\{\{\bar{\phi}_{known}\} - [A_3]\{\bar{w}_{a\,known}\}\}$$
$$= -[A_4]^{-1}[A_3]\{\bar{w}_{a\,known}\} \tag{4.2.44}$$

where we have noted that $\bar{\phi}_{known} = 0$. Using (4.2.44) in the first $N1$ equations of (4.2.42),

$$\{\bar{\phi}_{unknown}\} = [A_1]\{\bar{w}_{a\,known}\} + [A_2]\{\bar{w}_{a\,unknown}\}$$
$$= [[A_1] - [A_2][A_4]^{-1}[A_3]]\{\bar{w}_{a\,known}\} \tag{4.2.45}$$

Computer programs have been written to perform the various computations.* Also it should be pointed out that in the evaluation of the 'aerodynamic' influence coefficients it is essential to account for the singular nature of the integrand along the Mach lines. This requires an analytical integration of (4.2.39) over each box with $\bar{w}$ assumed constant and taken outside the integral.

Recent advances in this technique have been made to include more complicated geometries, e.g., nonplanar and multiple surfaces,* and also preliminary efforts have been made to include other physical effects.†

4.3 Subsonic flow‡

Subsonic flow is generally a more difficult problem area since all parts of the flow are disturbed due to the motion of the airfoil. To counter this difficulty an inverse method of solution has been evolved, the so-called 'Kernel Function' approach. To provide continuity with our previous development we shall formulate and solve the problem in a formal way through the use of Fourier Transforms. Historically, however, other methods were used. These will be discussed after we have obtained our formal solution. To avoid repetition, we shall treat the three-dimensional problem straight away.

---

* Many Authors, Oslo AGARD Symposium [8].

† Landahl and Ashley [9].

‡ BA, pp. 125–133; Landahl and Stark, [5], Williams [10].

Bernoulli's equation reads:

$$p = -\rho_\infty \left[\frac{\partial \phi}{\partial t} + U_\infty \frac{\partial \phi}{\partial x} \right] \tag{4.3.1}$$

It will prove convenient to use this relation to formulate our solution in terms of pressure directly rather than velocity potential.

Derivation of the integral equation by transform methods and solution by collocation

As before we will use the transform calculus. Since there is no limited range of influence in subsonic flow we employ Fourier transforms with respect to *x and y*. We shall also assume, as before, simple harmonic time dependent motion. Thus

$$\phi = \bar{\phi}(x, y, z) e^{i\omega t} \tag{4.3.2}$$

and transformed

$$\Phi^* = \int_{-\infty}^{\infty} \int \bar{\phi}(x, y, z) \exp(-i\alpha x - i\gamma y) \, \mathrm{d}x \, \mathrm{d}y \tag{4.3.3}$$

Hence (4.3.1) may be transformed

$$P^* = -\rho[i\omega + U_\infty i\alpha]\Phi^* \tag{4.3.4}$$

where

$$p = \bar{p}(x, y, z) e^{i\omega t}$$

$$P^* \equiv \int_{-\infty}^{\infty} \int \bar{p} \exp(-i\alpha x - i\gamma y) \, \mathrm{d}x \, \mathrm{d}y \tag{4.3.5}$$

As in supersonic flow we may relate the (transformed) velocity potential to the (transformed) 'upwash' (see (4.2.36) et. seq.)

$$\Phi^* \Big|_{z=0} = \frac{-W^*}{(\mu^2 + \gamma^2)^{\frac{1}{2}}} \tag{4.3.6}$$

Substituting (4.3.6) into (4.3.4),

$$P^* = \rho_\infty \frac{[i\omega + U_\infty i\alpha]}{(\mu^2 + \gamma^2)^{\frac{1}{2}}} W^*$$

or

$$\frac{\bar{W}^*}{U_\infty} = \frac{P^*}{\rho_\infty U_\infty^2} \frac{(\mu^2 + \gamma^2)^{\frac{1}{2}}}{\left[\dfrac{i\omega}{U_\infty} + i\alpha\right]} \tag{4.3.7}$$

Inverting

$$\frac{\bar{w}}{U_\infty}(x, y) = \int_{-\infty}^{\infty}\int K(x - \xi, y - \eta) \frac{\bar{p}}{\rho_\infty U_\infty^2}(\xi, \eta) \, d\xi \, d\eta \tag{4.3.8}$$

where

$$K(x, y) \equiv \frac{1}{(2\pi)^2} \int_{-\infty}^{\infty}\int \frac{(\mu^2 + \gamma^2)^{\frac{1}{2}}}{\left[\dfrac{i\omega}{U_\infty} + i\alpha\right]} \exp(i\alpha x + i\gamma y) \, d\alpha \, d\gamma$$

K is physically interpreted as the (non-dimensional) 'upwash', $\bar{w}/U_\infty$ at x, y due to a unit (delta-function) of pressure, $\bar{p}/\rho_\infty U_\infty^2$, at ξ, η. For lifting flow (subsonic or supersonic), $\bar{p} = 0$ off the wing; hence in (4.3.8) the (double) integral can be confined to the wing area. This is the advantage of the present formulation.

Now we are faced with the problem of extracting the pressure from beneath the integral in (4.3.9). By analogy to the supersonic 'box' approach we might consider approximating the integral equation by a double sum

$$\frac{\bar{w}_{ij}}{U_\infty} \simeq \Delta \xi \, \Delta \eta \sum_k \sum_l K_{(ij)(kl)} \frac{\bar{p}_{kl}}{\rho_\infty U_\infty^2} \tag{4.3.9}$$

In matrix notation

$$\left\{\frac{\bar{w}}{U_\infty}\right\} = [\tilde{K} \, \Delta \xi \, \Delta \eta] \left\{\frac{\bar{p}}{\rho_\infty U_\infty^2}\right\}$$

and formally inverting

$$\left\{\frac{\bar{p}}{\rho_\infty U_\infty^2}\right\} = \{K \, \Delta \xi \, \Delta \eta\}^{-1} \left\{\frac{\bar{w}}{U_\infty}\right\} \tag{4.3.10}$$

This solution is mathematically incorrect; worse, it is useless. The reason is that it is not unique unless an additional restriction is made, the so-called 'Kutta Condition'.* This restriction states that the pressure on the trailing edge of a thin airfoil must remain finite. For a lifting airfoil

---

* See Landahl and Stark or Williams, ibid.

this is tantamount to saying it must be zero. This constraint is empirical in nature being suggested by experiment. Other constraints such as zero pressure at the leading edge would also make the mathematical solution unique; however, this would not agree with available experimental data. Indeed these data suggest a pressure maxima at the leading edge; the theory with trailing edge Kutta condition gives a square root singularity at the leading edge.

Although, in principle, one could insure zero pressure at the trailing edge by using a constraint equation to supplement (4.3.9) and/or (4.3.10), another approach has gained favor in practice. In this approach the pressure is expanded in a series of (given) modes

$$\bar{p} = \sum_k \sum_l p_{kl} F_k(\xi) G_l(\eta) \tag{4.3.11}$$

where the functions $F_k(\xi)$ are chosen to satisfy the Kutta condition. (If the wing planform is other than rectangular, a coordinate transformation may need to be made in order to choose such functions readily.) The p_{kl} are, as yet, unknown.

Substituting (4.3.11) into (4.3.8) and integrating over the wing area

$$\frac{\bar{w}}{U_\infty}(x, y) = \sum_k \sum_l \frac{p_{kl}}{\rho_\infty U_\infty^2} \tilde{K}_{kl}(x, y) \tag{4.3.12}$$

where

$$\tilde{K}_{kl}(x, y) \equiv \iint K(x - \xi, y - \eta) F_k(\xi) G_l(\eta) \, d\xi \, d\eta$$

K is singular at $x = \xi$, $y = \eta$ (as we shall see later) and the above integral must be evaluated with some care.

The question remains how to evaluate the unknown coefficient, p_{kl}, in terms of $\bar{w}/U_\infty(x, y)$? The most common procedure is collocation. (4.3.12) is evaluated at a number of points, x_i, y_j, equal to the number of coefficients, p_{kl}. Thus (4.3.12) becomes

$$\frac{\bar{w}(x_i, y_j)}{U_\infty} = \sum_k \sum_l \frac{p_{kl}}{\rho_\infty U_\infty^2} \tilde{K}_{kl}(x_i, y_j) \tag{4.3.13}$$

Defining $\tilde{K}_{ijkl} \equiv \tilde{K}_{kl}(x_i, y_j)$, (4.3.13) becomes

$$\left\{ \frac{\bar{w}_{ij}}{U_\infty} \right\} = [\tilde{K}_{(ij)(kl)}] \left\{ \frac{p_{kl}}{\rho_\infty U_\infty^2} \right\}$$

Inverting

$$\left\{\frac{\bar{p}}{\rho_\infty U_\infty^2}\right\} = [\tilde{K}]^{-1}\left\{\frac{\bar{w}}{U_\infty}\right\} \qquad (4.3.14)$$

This completes our formal solution. Relative to the supersonic 'box' method, the above procedure, the so-called 'Kernal Function' method, has proven to be somewhat delicate. In particular, questions have arisen as to:

(1) the 'optimum' selection of pressure modes
(2) the 'best' method for computing $\tilde{K}$
(3) convergence of the method as the number of pressure modes becomes large

It appears, however, that as experience is acquired these questions are being satisfactorily answered at least on an 'ad hoc' basis.

More recently an alternative approach for solving (4.3.8) has gained popularity which is known as the 'doublet lattice' method. In this method the lifting surface is divided into boxes and collocation is used.*

An alternative determination of the Kernel Function using Green's Theorem

The transform methods are most efficient at least for formal derivations, however historically other approaches were first used. Many of these are now only of interest to history, however we should mention one other approach which is a powerful tool for nonsteady aerodynamic problems. This is the use of Green's Theorem.

First let us review the nature of Green's Theorem.† Our starting point is the Divergence Theorem or Gauss' Theorem.

$$\iiint \nabla \cdot \vec{b} \, dV = \iint \vec{b} \cdot \vec{n} \, dS \qquad (4.3.15)$$

S surface area enclosing volume V
$\vec{n}$ outward normal
$\vec{b}$ arbitrary vector

---

* Albano and Rodden [11]. The downwash is placed at the box three-quarters chord and pressure concentrated at the one-quarter chord. For two-dimensional steady flow this provides an exact solution which satisfies the Kutta condition. Lifanov, T. K. and Polanski, T. E., 'Proof of the Numerical Method of "Discrets Vortices" for Solving Singular Integral Equations', PMM (1975), pp. 742–746.
† References: Hildebrand [2] p. 312, Stratton [12], pp. 165–169.

Let $\vec{b} = \phi_1 \nabla \phi_2$ where ϕ_1, ϕ_2 are arbitrary scalars. Then (4.3.15) may be written as:

$$\iiint \nabla \cdot \phi_1 \nabla \phi_2 \, dV = \iint \vec{n} \cdot \phi_1 \nabla \phi_2 \, dS$$

Now use the vector calculus identity

$$\nabla \cdot c\vec{a} = c \nabla \cdot \vec{a} + \vec{a} \cdot \nabla c$$

c arbitrary scalar
$\vec{a}$ arbitrary vector

then $\nabla \cdot \phi_1 \nabla \phi_2 = \phi_1 \nabla^2 \phi_2 + \nabla \phi_2 \cdot \nabla \phi_1$ and (4.3.15) becomes

$$\iiint [\phi_1 \nabla^2 \phi_2 + \nabla \phi_2 \cdot \nabla \phi_1] \, dV = \iint \vec{n} \cdot \phi_1 \nabla \phi_2 \, dS \qquad (4.3.16)$$

This is the first form of Green's Theorem. Interchanging ϕ_1 and ϕ_2 in (4.3.16) and subtracting the result from (4.3.16) gives

$$\iiint [\phi_1 \nabla^2 \phi_2 - \phi_2 \nabla^2 \phi_1] \, dV = \iint \vec{n} \cdot (\phi_1 \nabla \phi_2 - \phi_2 \nabla \phi_1) \, dS$$

$$= \iint \left(\phi_1 \frac{\partial \phi_2}{\partial n} - \phi_2 \frac{\partial \phi_1}{\partial n} \right) dS \qquad (4.3.17)$$

This is the second (and generally more useful) form of Green's Theorem. $\partial/\partial n$ denotes a derivative in the direction of the normal. Let us consider several special but informative cases.

(a) $\phi_1 = \phi_2 = \phi$ in (4.3.16)

$$\iiint [\phi \nabla^2 \phi + \nabla \phi \cdot \nabla \phi] \, dV = \iint \phi \frac{\partial \phi}{\partial n} \, dS \qquad (4.3.18)$$

(b) $\phi_1 = \phi$, $\phi_2 = 1$ in (4.3.17)

$$\iiint \nabla^2 \phi \, dV = \iint \frac{\partial \phi}{\partial n} \, dS \qquad (4.3.19)$$

(c) $\nabla^2 \phi_1 = 0$, $\phi_2 = 1/r$, $r \equiv \sqrt{(x-y_1)^2(y-y_1)^2 + (z-z_1)^2}$ in (4.3.17)

$$\iiint \phi_1 \nabla^2 (1/r) \, dV = \iint \left[\phi_1 \frac{\partial}{\partial n} - \frac{\partial \phi_1}{\partial n} \right] \frac{1}{r} \, dS \qquad (4.3.20)$$

Now $\nabla^2(1/r) = 0$ everywhere except at $r = 0$. Thus

$$\iiint \phi_1 \nabla^2(1/r) \, dV = \phi_1(r = 0) \iiint \nabla^2(1/r) \, dV$$

$$= \phi_1(r = 0) \iiint \nabla \cdot \nabla \frac{1}{r} \, dV$$

from divergence theorem (4.3.15)

$$= \phi_1(r = 0) \iint \nabla(1/r) \cdot \frac{\nabla r}{|\nabla r|} \, dS$$

$$= \phi_1(r = 0) \int_0^{2\pi} \int_0^{\pi} -\frac{1}{r^2} r^2 \sin\theta \, d\theta \, d\phi$$

$$= -4\pi \phi_1(r = 0)$$

where we consider a small sphere of radius ε, say, in evaluating surface integral. Now

$$\phi_1(r = 0) = \phi_1(x_1 = x, y_1 = y, z_1 = z) = \phi_1(x, y, z)$$

Thus (4.3.20) becomes

$$\phi_1(x, y, z) = -\frac{1}{4\pi} \iint \left[\phi_1 \frac{\partial}{\partial n} - \frac{\partial \phi_1}{\partial n} \right] \frac{1}{r} \, dS \tag{4.3.21}$$

The choice of $\phi_2 = 1/r$ may seem rather arbitrary. This can be motivated by noting that

$$\frac{\nabla^2 \phi_2}{4\pi} = -\delta(x - x_1) \, \delta(y - y_1) \, \delta(z - z_1)$$

Hence we seek a ϕ_2 which is the response to a delta function. This is what leads to the simplification of the volume integral.

Incompressible, three dimensional flow

To simplify matters we will first confine ourselves to $M = 0$. However, similar, but more complex calculations subsequently will be carried out for $M \neq 0$.* For incompressible flow, the equation of motion is

$$\nabla^2 \phi = 0$$

or

$$\nabla^2 p = 0$$

---

* Watkins, Woolston and Cunningham [13], Williams [14].

where ϕ and p are (perturbation) velocity potential and pressure respectively. Hence we may identify ϕ_1 in (4.3.21) with either variable as may be convenient. To conform to convention in the aerodynamic theory literature, we will take the normal positive *into* the fluid and introduce a minus sign into (4.3.21) which now reads:

$$\phi_1(x, y, z) = \frac{1}{4\pi} \iint \left[\phi_1 \frac{\partial}{\partial n} - \frac{\partial \phi_1}{\partial n} \right] \frac{1}{r} dS \qquad (4.3.22)$$

For example for a planar airfoil surface

$$n \quad \text{on} \quad S \quad \text{at} \quad z_1 = 0^+ \quad \text{is} \quad +z_1$$
$$n \quad \text{on} \quad S \quad \text{at} \quad z_1 = 0^- \quad \text{is} \quad -z_1$$
$$dS = dx_1 \, dy_1$$

Note x, y, z is any given point, while x_1, y_1, z_1 are (dummy) integration variables. See following sketch.

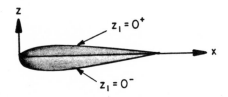

Let us identify the area S of the integral as enclosing the entire fluid, hence the area S is composed of two parts, the area of the airfoil plus wake, call it S_1, and the area of a sphere at infinity, call it S_2.

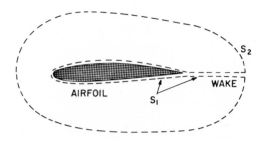

Figure 4.5 Airfoil and flow field geometry.

(i) *Thickness problem (nonlifting).*   Let $\phi_1 = \phi$, velocity potential. Because ϕ is bounded at $r \to \infty$, there is no contribution from S_2. Hence

$$\phi(x, y, z) = \frac{1}{4\pi} \iint \left[\phi \frac{\partial}{\partial z_1} - \frac{\partial \phi}{\partial z_1} \right] \frac{1}{r} \, dS$$

S_1 at $z_1 = 0^+$

$$+ \frac{1}{4\pi} \iint \left[\phi \left(-\frac{\partial}{\partial z_1} \right) - \left(-\frac{\partial \phi}{\partial z_1} \right) \right] \frac{dS}{r}$$

S_1 at $z_1 = 0^-$

Now $\phi_{z_1 = 0^+} = \phi_{z_1 = 0^-}$ for thickness problem and

$$\frac{\partial \phi}{\partial z_1} \bigg|_{z_1 = 0^+} = -\frac{\partial \phi}{\partial z_1} \bigg|_{z_1 = 0^-}$$

Thus

$$\phi(x, y, z) = -\frac{1}{2\pi} \iint \frac{\partial \phi}{\partial z_1} \bigg|_{z_1 = 0^+} \frac{dS}{r}$$

using body boundary condition

$$= -\frac{1}{2\pi} \iint w \frac{dS}{r} \tag{4.3.23}$$

where

$$w \equiv \frac{\partial z}{\partial t} + U_\infty \frac{\partial z}{\partial x}$$

Note this solution is valid for arbitrary time-dependent motion. Time only appears as a parameter in the solution $\phi(x, y, z) = \phi(x, y, z; t)$. This is a special consequence of $M \equiv 0$.

(ii) *Lifting problem.*   For the lifting problem it again will prove convenient to use pressure rather than velocity potential. (4.3.22) becomes

$$p(x, y, z) = \frac{1}{4\pi} \iint \left[(p_{z=0^+} - p_{z=0^-}) \frac{\partial}{\partial z_1} \left(\frac{1}{r} \right) \right.$$

$$\left. - \left(\frac{\partial p}{\partial z_1} \bigg|_{z_1 = 0^+} - \frac{\partial p}{\partial z_1} \bigg|_{z_1 = 0^-} \right) \frac{1}{r} \right] dS$$

Now

$$p_{z_1 = 0^+} = -p_{z_1 = 0^-}$$

for lifting problem and

$$\frac{\partial p}{\partial z_1}\bigg|_{z_1=0^+} - \frac{\partial p}{\partial z_1}\bigg|_{z_1=0^-} = 0$$

Thus

$$p(x, y, z) = \frac{1}{4\pi} \iint \Delta p \frac{\partial}{\partial z_1}\left(\frac{1}{r}\right) dS \qquad (4.3.24)$$

where

$$\Delta p \equiv p_{z=0^+} - p_{z=0^-}$$

(4.3.24) as it stands is not particularly helpful. We do not know either p or Δp. However we can relate p to something we do know, w. To simplify matters we shall specify simple harmonic motion,

$$p = \bar{p}e^{i\omega t}$$
$$\phi = \bar{\phi}e^{i\omega t}$$

hence from Bernoulli's equation

$$\bar{p} = -\rho_\infty\left[i\omega\bar{\phi} + U_\infty\frac{\partial\bar{\phi}}{\partial x}\right] \qquad (4.3.25)$$

Solving (4.3.25), by variation of parameters,

$$\bar{\phi}(x, y, z) = -\int_{-\infty}^{x} \frac{\bar{p}}{\rho_\infty U_\infty}(\lambda, y, z) \exp\left[i\frac{\omega}{U_\infty}(\lambda - x)\right] d\lambda \qquad (4.3.26)$$

and using (4.3.24), one has

$$\bar{\phi}(x, y, z) = -\int_{-\infty}^{x} \exp\left[i\frac{\omega}{U_\infty}(\lambda - x)\right]$$
$$\cdot \left\{\frac{1}{4\pi} \iint \frac{\Delta\bar{p}}{\rho_\infty U_\infty}(x_1, y_1, z_1 = 0)\frac{\partial}{\partial z_1}\left(\frac{1}{r(\lambda)}\right) dS\right\} d\lambda$$

where

$$r(\lambda) \equiv \sqrt{(\lambda - x_1)^2 + (y - y_1)^2 + (z - z_1)^2}$$
$$dS \equiv dx_1\, dy_1$$

Define

$$\xi \equiv \lambda - x_1,\, d\lambda = d\xi,\, \lambda = \xi + x_1$$

and interchange order of integration with respect to ξ and S, then

$$\bar{\phi}(x, y, z) = -\frac{1}{4\pi} \iint \frac{\Delta \bar{p}}{\rho_\infty U_\infty}(x_1, y_1, z_1 = 0)$$

$$\cdot \left\{ \int_{-\infty}^{x-x_1} \frac{\partial}{\partial z_1}\left(\frac{1}{r(\xi)}\right) \exp\left\{i\frac{\omega}{U_\infty}[\xi - (x - x_1)]\right\} d\xi \right\} dS$$

Compute $\partial\bar{\phi}/\partial z$ and set equal to $\bar{w}$ from body boundary condition, on $z = 0$.

$$\bar{w} = -\frac{1}{4\pi} \lim_{z \to 0} \iint \frac{\Delta \bar{p}}{\rho_\infty U_\infty}$$

$$\times \left\{ \frac{\partial}{\partial z} \int_{-\infty}^{x-x_1} \frac{\partial}{\partial z_1}\left(\frac{1}{r}\right) \exp\left\{i\frac{\omega}{U_\infty}[\xi - (x - x_1)]\right\} d\xi \right\} dS$$

Now

$$\frac{\partial}{\partial z}\left(\frac{1}{r}\right) = -\frac{\partial}{\partial z_1}\left(\frac{1}{r}\right)$$

therefore

$$\frac{\bar{w}}{U_\infty} = \iint \frac{\Delta \bar{p}}{\rho_\infty U_\infty^2}(x_1, y_1, z_1 = 0)K(x - x_1, y - y_1, 0)\, dx_1\, dy_1 \qquad (4.3.27)$$

where

$$K(x - x_1, y - y_1, 0) \equiv \frac{1}{4\pi} \lim_{z \to 0} \frac{\partial^2}{\partial z^2} \int_{-\infty}^{x-x_1} \frac{\exp\left\{\dfrac{i\omega}{U_\infty}[\xi - (x - x_1)]\right\}}{r} d\xi$$

and where

$$r \equiv \sqrt{\xi^2 + z^2 + (y - y_1)^2}$$

(4.3.27), of course, has the same form as we had previously derived by transform methods.

The expression for the Kernel function may be simplified.

$$K(x - x_1, y - y_1, 0) = \frac{\exp\left[-\dfrac{i\omega}{U_\infty}(x - x_1)\right]}{4\pi} \int_{-\infty}^{x-x_1} e^{+\frac{i\omega\xi}{U_\infty}} \lim_{z \to 0} \frac{\partial^2}{\partial z^2}\frac{1}{r} d\xi$$

Now

$$\frac{\partial^2}{\partial z^2}\frac{1}{r} = -\frac{1}{2}r^{-3}2 + (-1/2)(-3/2)r^{-5}(2z)^2$$

thus

$$\lim_{z \to 0} \frac{\partial^2}{\partial z^2} \frac{1}{r} = -[\xi^2 + (y - y_1)^2]^{-3/2}$$

and finally

$$K = -\frac{\exp\left[-\dfrac{i\omega}{U_\infty}(x - x_1)\right]}{4\pi} \int_{-\infty}^{x - x_1} \frac{\exp\left[+\dfrac{i\omega\xi}{U_\infty}\right]}{[\xi^2 + (y - y_1)^2]^{+3/2}} \, d\xi \qquad (4.3.28)$$

The integral in (4.3.28) must be evaluated numerically.

Compressible, three-dimensional flow

For the more general case of $M \neq 0$, we have an additional complication since

$$\nabla^2 \phi \neq 0$$

For simple harmonic motion, the equation of motion reads

$$\nabla^2 \bar{\phi} + \mathscr{L}\bar{\phi} = 0 \qquad (4.3.29)$$

where

$$\mathscr{L} \equiv -\frac{1}{a^2}\left[(i\omega) + U\frac{\partial}{\partial x}\right]^2$$

By making a transformation we may reduce the compressible equation to a simpler form. Defining

$$x^1 \equiv x, \ y^1 \equiv \beta y, \qquad z^1 \equiv \beta z$$

$$\beta \equiv \sqrt{1 - M^2}$$

$$\bar{\phi} \equiv \exp\left[i\frac{M^2}{(1 - M^2)}\frac{\omega}{U_\infty}x\right]\phi^* \ \dagger$$

The equation for ϕ^* is

$$\exp\left[i\frac{M^2}{(1 - M^2)}\frac{\omega}{U_\infty}x^1\right][\nabla^{1^2}\phi^* + k^2\phi^*] = 0 \qquad (4.3.30)$$

where

$$k \equiv \left[\frac{M}{(1 - M^2)}\right]\frac{\omega}{U_\infty}$$

† By assuming a transformation of the form $e^{\Omega x}\phi^* = \bar{\phi}$, one can always determine Ω such that (4.3.29) reduces to (4.3.31).

Note this equation is essentially the reduced wave equation in its form. We shall use Green's Theorem on ϕ^* and then transform back to $\bar{\phi}$. Let

$$\nabla^2\phi_1^* + k^2\phi_1^* = 0 \qquad (4.3.31)$$

$$\nabla^2\phi_2^* + k^2\phi_2^* = \delta(x^1 - x_1^1)\,\delta(y^1 - y_1^1)\,\delta(z^1 - z_1^1)$$

Solving for ϕ_2^*,

$$\phi_2^* = -\frac{e^{-ikr}}{4\pi r}$$

where

$$r = \sqrt{(x^1 - x_1^1)^2 + (y^1 - y_1^1)^2 + (z^1 - z_1^1)^2}$$

From (4.3.17)6,

$$\iiint [\phi_1^*(\delta - k^2\phi_2^*) - \phi_2^*(-k^2\phi_1^*)]\,dV = \iint \left[\phi_1^*\frac{\partial\phi_2}{\partial n} - \phi_2^*\frac{\partial\phi_1^*}{\partial n}\right]dS$$

$$(4.3.17a)$$

or

$$\phi_1^*(x, y, z) = -\frac{1}{4\pi}\iint\left[\phi_1^*\frac{\partial}{\partial n} - \frac{\partial\phi_1^*}{\partial n}\right]\frac{e^{-ikr}}{r}\,dS \qquad (4.3.21a)$$

or

$$\phi_1^*(x, y, z) = +\frac{1}{4\pi}\iint\left[\phi_1^*\frac{\partial}{\partial n} - \frac{\partial\phi_1^*}{\partial n}\right]\frac{e^{-ikr}}{r}\,dS \qquad (4.3.22a)$$

(if we redefine positive normal). Using symmetry and anti-symmetry properties of $\dfrac{\partial\phi_1^*}{\partial n}$ and ϕ_1^*

$$\phi_1^*(x, y, z) = \frac{1}{4\pi}\iint \Delta\phi_1^*\frac{\partial}{\partial z_1}\left\{\frac{e^{-ikr}}{r}\right\}dS \qquad (4.3.24a)$$

where

$$\Delta\phi_1^* = \phi_{1_{z_1=0^+}}^* - \phi_{1_{z_1=0^-}}^*$$

and

$$-\frac{\partial\phi_1^*}{\partial z_1}\bigg|_{z_1=0^+} + \frac{\partial\phi_1^*}{\partial z_1}\bigg|_{z_1=0^-} = 0$$

4 Nonsteady aerodynamics

Note $dS \equiv dx_1\, dy_1$ and

$$\left(\frac{\partial}{\partial z_1}\right) dx_1\, dy_1 = \left(\frac{\partial}{\partial z_1^1}\right) dx_1^1\, dy_1^1; \qquad x_1 = x_1^1$$

From (4.3.24a) and the definition of ϕ^*

$$
\begin{aligned}
\bar{\phi}_1 &= \exp\left[i\,\frac{M^2}{(1-M^2)}\,\frac{\omega}{U_\infty}\,x\right]\phi_1^*(x, y, z) \\
&= \frac{\exp\left[i\,\dfrac{M^2}{(1-M^2)}\,\dfrac{\omega}{U_\infty}\,x\right]}{4\pi} \\
&\qquad \times \iint \Delta\bar{\phi}_1\exp\left[-\frac{M^2}{(1-M^2)}\,\frac{\omega}{U_\infty}\,x_1\right]\frac{\partial}{\partial z_1}\left\{\frac{e^{-ikr}}{r}\right\}dS
\end{aligned}
\tag{4.3.32}
$$

Identifying $\bar{\phi}_1$ with $\bar{p}$ and using (4.3.32) in (4.3.26),

$$
\begin{aligned}
\bar{\phi}(x, y, z, \omega) &= -\frac{1}{4\pi}\int_{-\infty}^{x}\exp\left[i\,\frac{M^2}{(1-M^2)}\,\frac{\omega}{U_\infty}\,\lambda\right]\exp\left[i\,\frac{\omega}{U_\infty}\,(\lambda - x)\right] \\
&\qquad \cdot\left\{\iint \frac{\Delta\bar{p}}{\rho_\infty U_\infty}\exp\left[-i\,\frac{M^2}{(1-M^2)}\,\frac{\omega}{U_\infty}\,x_1\right]\frac{\partial}{\partial z_1}\left\{\frac{e^{-ikr}}{r}\right\}dS\right\}d\lambda
\end{aligned}
$$

Define $\xi \equiv \lambda - x_1$, $d\lambda = d\xi$, $\lambda = \xi + x_1$ and interchange order of integration with respect to ξ and S,

$$
\begin{aligned}
\bar{\phi}(x, y, z\,;\, \omega) &= -\frac{1}{4\pi}\iint_{z_1=0}\frac{\Delta\bar{p}}{\rho_\infty U_\infty}(x_1, y_1, z_1) \\
&\qquad \cdot\left\{\int_{-\infty}^{x-x_1}\frac{\partial}{\partial z_1}\left\{\frac{e^{-ikr}}{r}\right\}\exp\left[i\,\frac{M^2}{(1-M^2)}\,\frac{\omega}{U_\infty}\,\xi\right]\right. \\
&\qquad \left. \cdot\exp\left(i\,\frac{\omega}{U_\infty}\,\xi\right)\exp\left[-i\,\frac{\omega}{U_\infty}\,(x - x_1)\right]d\xi\right\}dS \\
&= -\frac{1}{4\pi}\iint_{z_1=0}\frac{\Delta\bar{p}}{\rho_\infty U_\infty}(x_1, y_1, z_1)\exp\left[-\frac{i\omega}{U_\infty}\,(x - x_1)\right] \\
&\qquad \cdot\left\{\int_{-\infty}^{x-x_1}\exp\left[i\,\frac{1}{(1-M^2)}\,\frac{\omega}{U_\infty}\,\xi\right]\frac{\partial}{\partial z_1}\left\{\frac{e^{-ikr}}{r}\right)d\xi\right\}dS
\end{aligned}
$$

Compute $\partial \bar{\phi}/\partial z$ and set equal to $\bar{w}$ from body boundary condition on $z = 0$, noting that

$$\frac{\partial}{\partial z}\left\{\frac{e^{-ikr}}{r}\right\} = -\frac{\partial}{\partial z_1}\left\{\frac{e^{-ikr}}{r}\right\}$$

The final result is

$$\frac{\bar{w}}{U_\infty} = \iint \frac{\Delta \bar{p}}{\rho_\infty U_\infty^2}(x_1, y_1, z_1 = 0)K(x - x_1, y - y_1, 0)\, dx_1\, dy_1 \qquad (4.3.33)$$

where

$$K(x, y) = \lim_{z \to 0} \frac{\exp\left(-i\dfrac{\omega}{U_\infty}x\right)}{4\pi} \int_{-\infty}^{x} \exp\left[\frac{i}{(1 - M^2)}\frac{\omega}{U_\infty}\xi\right]\frac{\partial^2}{\partial z^2}\left\{\frac{e^{-ikr}}{r}\right\} d\xi$$

$$r \equiv [\xi^2 + (1 - M^2)(y^2 + z^2)]^{\frac{1}{2}}$$

The expression for K may be simplified as follows: Define a new variable, τ, to replace ξ by

$$(1 - M^2)\tau \equiv \xi - Mr(\xi, y, z)$$

where one will recall

$$r(\xi, y, z) \equiv [\xi^2 + \beta^2(y^2 + z^2)]^{\frac{1}{2}}$$

and

$$\beta^2 \equiv 1 - M^2$$

After some manipulation one may show that

$$\left[\tau^2 + \frac{d\tau}{y^2} + z^2\right]^{\frac{1}{2}} = \frac{d\xi}{r}$$

and

$$\exp\left(+\frac{i\omega}{U_\infty}\frac{\xi}{(1 - M^2)}\right)e^{-ikr} = \exp\left[i\frac{\omega}{U_\infty}\tau\right]$$

Thus

$$K = \lim_{z \to 0} \frac{\exp\left(-i\dfrac{\omega x}{U_\infty}\right)}{4\pi}\frac{\partial^2}{\partial z^2}\int_{-\infty}^{[x - Mr(x,y,z)]/(1 - M^2)}\frac{\exp\left(\dfrac{i\omega\tau}{U_\infty}\right)}{[\tau^2 + y^2 + z^2]^{\frac{1}{2}}}\, d\tau \qquad (4.3.34)$$

207

Taking the second derivative and limit as indicated in (4.3.34) and using the identity

$$\left[\frac{Mx+r}{(x^2+y^2)}\right]^2 \equiv \frac{1}{\left[\dfrac{x-Mr}{(1-M^2)}\right]^2 + y^2}$$

one finally obtains

$$K = -\frac{1}{4\pi}\left\{\frac{M(Mx+r)}{r(x^2+y^2)} \exp\left[i\,\frac{\omega}{U_\infty}\frac{M}{(1-M^2)}(Mx-r)\right]\right.$$

$$\left. + \exp\left(-i\frac{\omega x}{U_\infty}\right) \int_{-\infty}^{(x-Mr)/(1-M^2)} \frac{\exp\left(i\dfrac{\omega\tau}{U_\infty}\right)}{[\tau^2+y^2]^{\frac{3}{2}}}\,d\tau\right\} \qquad (4.3.35)$$

This is one form often quoted in the literature. By expressing K in nondimensional form we see the strong singularity in K as $y \to 0$.

$$y^2 K(x,y) = -\frac{1}{4\pi}\left\{\frac{M(Mx/y+r/y)}{r/y[(x/y)^2+1]} \exp\left[i\,\frac{\omega y}{U_\infty}\frac{M}{(1-M^2)}\left(M\frac{x}{y}-\frac{r}{y}\right)\right]\right.$$

$$\left. + \exp\left(-i\frac{\omega x}{U}\right) \int_{-\infty}^{[x/y-M(r/y)]/(1-M^2)} \frac{\exp\left(\dfrac{i\omega y}{U_\infty}z\right)}{[z^2+1]^{\frac{3}{2}}}\,dz\right\}$$

$$z \equiv \tau/y$$

Note that the compressible Kernel, K, has the same strength singularity as for incompressible flow and is of no more fundamental complexity.

There is a vast literature on unsteady aerodynamics within the framework of linearized, potential flow models. Among recent references one may mention the work of A. Cunningham[*] on combined subsonic-supersonic Kernel Function methods including an empirical correction for transonic effects and also the work of Morino[†] using Green's Theorem in a more general form for both subsonic and supersonic flow. For a brief overview with authoritative suggestions for future work, the paper by Rodden[‡] is recommended. The reader who has mastered the material presented so far should be able to pursue this literature with confidence.

---

[*] Cunningham [15].

[†] Morino, Chen and Suciu [16].

[‡] Rodden [17], Ashley and Rodden [18].

Before turning to representative numerical results the historically important theory of incompressible, two-dimensional flow will be presented.

Incompressible, two dimensional flow

A classical solution is due to Theodorsen* and others. Traditionally, the coordinate system origin is selected at mid-chord with $b \equiv$ half-chord. The governing differential equation for the velocity potential, ϕ, is

$$\nabla^2 \phi = 0 \qquad (4.3.36)$$

with boundary conditions for a lifting, airfoil of

$$\left. \frac{\partial \phi}{\partial z} \right|_{z=0^{+,-}} = w_a \equiv \frac{\partial z_a}{\partial t} + U_\infty \frac{\partial z_a}{\partial x} \qquad (4.3.37)$$

on airfoil, $-b < x < b$, on $z = 0$ and

$$p = -\rho_\infty \left[\frac{\partial \phi}{\partial t} + U_\infty \frac{\partial \phi}{\partial x} \right] = 0 \qquad (4.3.38)$$

off airfoil, $x > b$ or $x < -b$, on $z = 0$ and

$$p, \phi \to 0 \quad \text{as} \quad z \to \infty \qquad (4.3.39)$$

From (4.3.36), (4.3.37) and (4.3.39) one may construct an integral equation,

$$w_a = \left. \frac{\partial \phi}{\partial z} \right|_{z=0} = -\frac{1}{2\pi} \int_{-b}^{\infty} \frac{\gamma(\xi, t)}{x - \xi} \, d\xi \qquad (4.3.40)$$

where

$$\gamma(x, t) \equiv \left. \frac{\partial \phi}{\partial x} \right|_U - \left. \frac{\partial \phi}{\partial x} \right|_L \qquad (4.3.41)$$

and

$$U \Rightarrow z = 0^+, \qquad L \Rightarrow z = 0^-$$

---

* Theodorsen [19]. Although this work is of great historical importance, the details are of less compelling interest today and some readers may wish to omit this section on a first reading. The particular approach followed here is a variation on Theodorsen's original theme by Marten Landahl.

4 Nonsteady aerodynamics

Further definitions include

$$\text{'Circulation'} \equiv \Gamma(x) \equiv \int_{-b}^{x} \gamma(\xi)\,d\xi \Rightarrow \frac{\partial \Gamma}{\partial x} = \gamma(x)$$

$$\Delta\phi \equiv \phi_L - \phi_U$$

$$C_p \equiv \frac{p}{\frac{1}{2}\rho_\infty U_\infty^2}$$

$$\Delta C_p \equiv C_{p_L} - C_{p_U}$$

From above, (4.3.41),

$$\Gamma(x, t) = \int_{-b}^{x} \gamma(\xi)\,d\xi = \int_{-b}^{x}\left[\frac{\partial \phi_U}{\partial \xi} - \frac{\partial \phi_L}{\partial \xi}\right]d\xi = -\Delta\phi(x), \qquad (4.3.42)$$

Note: $\Delta\phi(x = -b) = 0$. Also from (4.3.38) and (4.3.41),

$$\Delta C_p = -\frac{2}{U_\infty^2}\left[\frac{\partial \Delta\phi}{\partial t} + U_\infty \frac{\partial \Delta\phi}{\partial x}\right]$$

and using (4.3.42),

$$\Delta C_p = \frac{2}{U_\infty^2}\left[\frac{\partial \Gamma}{\partial t} + U_\infty \frac{\partial \Gamma}{\partial x}\right] \qquad (4.3.43)$$

Thus once γ (and hence Γ) is known, ΔC_p is readily computed. We therefore seek to solve (4.3.40) for γ. The advantage of (4.3.40) over (4.3.36)–(4.3.39) is that we have reduced the problem by one variable, having eliminated z. A brief derivation of (4.3.40) is given below.

Derivation of integral equation (4.3.40).   A Fourier transform of (4.3.36) gives

$$\frac{d^2\phi}{dz^2} - \alpha^2 \phi^* = 0 \qquad (4.3.36a)$$

where

$$\phi^*(\alpha, z, t) \equiv \int_{-\infty}^{\infty} \phi(x, z, t)e^{-i\alpha x}\,dx$$

(4.3.37) becomes

$$\left.\frac{d\phi^*}{dz}\right|_{z=0} = w_a^* \qquad (4.3.37a)$$

The general solution to (4.3.36a) is

$$\phi^* = A e^{+|\alpha|z} + B e^{-|\alpha|z} \tag{4.3.38a}$$

From the finiteness condition, (4.3.39), we see that one must require that $A = 0$ for $z > 0$ (and $B = 0$ for $z < 0$). Considering $z > 0$ for definiteness, we compute from (4.3.38a)

$$\left. \frac{d\phi^*}{dz} \right|_{z=0} = -|\alpha|\, B \tag{4.3.39a}$$

From (4.3.39a) and (4.3.37a),

$$B = -\frac{w_a^*}{|\alpha|} \tag{4.3.40a}$$

and from (4.3.38a) and (4.3.40a)

$$\left. \phi^* \right|_{z=0^+} = \frac{-w_a^*}{|\alpha|} \tag{4.3.41a}$$

From (4.3.41)

$$\gamma^* = \left(\frac{\partial\phi}{\partial x}\right)^* \Bigg|_{z=0^+} - \left(\frac{\partial\phi}{\partial x}\right)^* \Bigg|_{z=0^-}$$

and using (4.3.41a)

$$\gamma^* = -2i\alpha \frac{w_a^*}{|\alpha|} \tag{4.3.42a}$$

Re-arranging (4.3.42a),

$$w_a^* = -\frac{|\alpha|}{2i\alpha}\, \gamma^*$$

and inverting back to physical domain (using the convolution theorem) we obtain the desired result.

$$w_a = -\frac{1}{2\pi} \int_{-b}^{\infty} \frac{\gamma(\xi, t)}{x - \xi}\, d\xi \tag{4.3.40}$$

where

$$\frac{1}{2\pi} \int_{-\infty}^{\infty} -\frac{|\alpha|}{2i\alpha}\, e^{+i\alpha x}\, d\alpha = -\frac{1}{2\pi x}$$

The lower limit $x = -b$ in (4.3.40) follows from the fact that $p = 0$ for $x < -b$ (on $z = 0$) implies that $\phi = \phi_x = 0$ for $x < -b$. This will be made

211

more explicit when we consider $x > b$ where $p = 0$ does *not* imply $\phi = \phi_x = 0$! See discussion below.

Also one can calculate γ for $x > b$ in terms of γ for $b < x < b$ by using the condition that $\Delta C_p = 0$ (continuous pressure) for $x > b$. This is helpful in solving (4.3.40) for γ in terms of w_a. From (4.3.43)

$$\Delta C_p = 0 \Rightarrow \frac{\partial \Gamma}{\partial t} + U_\infty \frac{\partial \Gamma}{\partial x} = 0$$

$$\therefore \quad \Gamma = \Gamma\left(t - \frac{x}{U_\infty}\right) \tag{4.3.44}$$

Simple harmonic motion of an airfoil. For the special case of simple harmonic motion, one has

$$
\begin{aligned}
w_a(x, t) &= \bar{w}_a(x)e^{i\omega t} \\
\gamma(x, t) &= \bar{\gamma}(x)e^{i\omega t} \\
\Gamma &= \bar{\Gamma}e^{i\omega t} \quad \text{etc.}
\end{aligned}
\tag{4.3.45}
$$

(4.3.44) and (4.3.45) imply

$$\Gamma(x, t) = A \exp\left(i\omega[t - x/U_\infty]\right)$$

The (integration) constant A may be evaluated by considering the solution at $x = b$.

$$\Gamma(x = b, t) = A \exp\left(i\omega[t - b/U_\infty]\right)$$

$$\therefore \quad \Gamma(x, t) = \bar{\Gamma}(b) \exp\left\{i\omega[t - (x - b)/U_\infty]\right\}$$

and

$$\bar{\gamma} = \frac{\partial \bar{\Gamma}}{\partial x} = \frac{-i\omega}{U_\infty} \bar{\Gamma}(b) \exp\left[-i\omega(x - b)/U_\infty\right] \tag{4.3.46}$$

Introducing traditional nondimensionalization

$$x^* \equiv \frac{x}{b}, \qquad \xi^* \equiv \xi/b, \qquad k \equiv \frac{\omega b}{U_\infty}$$

a summary of the key relations is given below

$$\bar{w}_a(x^*) = -\frac{1}{2\pi} \int_{-1}^{\infty} \frac{\bar{\gamma}(\xi^*)}{x^* - \xi^*} \, d\xi^* \quad \text{from (4.3.40)}$$

where

$$\frac{\bar{\gamma}(x^*)}{U_\infty} = -ik \frac{\bar{\Gamma}(b)}{U_\infty b} \exp\left[-ik(x^*-1)\right]$$

for $x^* > 1$ from (4.3.46)

$$\frac{\bar{\Gamma}(x^*)}{U_\infty b} = \int_{-1}^{x^*} \frac{\bar{\gamma}(\xi^*)}{U_\infty} d\xi^* \quad \text{definition}$$

$$\Delta \bar{C}_p = 2\left[\frac{\bar{\gamma}(x^*)}{U_\infty} + ik \frac{\bar{\Gamma}(x^*)}{U_\infty b}\right] \quad \text{from (4.3.43)} \tag{4.3.47}$$

Special Case: steady motion. For simplicity let us first consider steady flow, $\omega \equiv 0$. From (4.3.46) or (4.3.47)

$$\gamma = 0 \quad \text{for} \quad x^* > 1$$

and hence we have

$$w_a(x^*) = -\frac{1}{2\pi} \int_{-1}^{1} \frac{\gamma(\xi^*)}{x^* - \xi^*} d\xi^* \tag{4.3.48}$$

To solve (4.3.48) for γ, we replace x^* by u, multiply both sides of (4.3.48) by the 'solving kernel'

$$\sqrt{\frac{1+u}{1-u}} \frac{1}{u-x^*}$$

and integrate $\int_{-1}^{1} \cdots du$. The result is

$$\int_{-1}^{1} \sqrt{\frac{1+u}{1-u}} \frac{w_a(u)}{u-x^*} du = -\frac{1}{2\pi} \int_{-1}^{1} \sqrt{\frac{1+u}{1-u}} \frac{1}{u-x^*} \int_{-1}^{1} \frac{\gamma(\xi^*)}{u-\xi^*} d\xi^* du$$

Now write $\gamma(\xi^*) = \gamma(x^*) + [\gamma(\xi^*) - \gamma(x^*)]$, then above may be written as

$$\int_{-1}^{1} \sqrt{\frac{1+u}{1-u}} \frac{w_a(u)}{u-x^*} du = -\frac{\gamma(x^*)}{2\pi}\left\{\int_{-1}^{1} \sqrt{\frac{1+u}{1-u}} \frac{1}{u-x^*} \int_{-1}^{1} \frac{d\xi^*}{u-\xi^*}\right\} du$$

$$-\frac{1}{2\pi}\left\{\int_{-1}^{1} \sqrt{\frac{1+u}{1-u}} \frac{1}{u-x^*} \int_{-1}^{1} \frac{(\xi^*-x^*)}{u-\xi^*} F(\xi^*, x^*) d\xi^* du\right\} \tag{4.3.49}$$

where

$$F(\xi^*, x^*) \equiv \frac{\gamma(\xi^*) - \gamma(x^*)}{\xi^* - x^*}$$

213

4 Nonsteady aerodynamics

To simplify (4.3.49) we will need to know several integrals. To avoid a diversion, these are simply listed here and are evaluated in detail at the end of this discussion of incompressible, two-dimensional flow.

$$I_0 \equiv \int_{-1}^{1} \frac{d\xi^*}{x^* - \xi^*} = \ln\left(\frac{1+x^*}{1-x^*}\right) \quad \text{for} \quad x^* < 1$$

$$= \ln\left(\frac{x^*+1}{x^*-1}\right) \quad \text{for} \quad x^* > 1$$

$$I_1 \equiv \int_{-1}^{1} \sqrt{\frac{1+u}{1-u}} \frac{du}{u-x^*} = \pi \quad \text{for} \quad x^* < 1$$

$$= \pi\left[1 - \sqrt{\frac{x^*+1}{x^*-1}}\right] \quad \text{for} \quad x^* > 1$$

$$I_2 \equiv \int_{-1}^{1} \sqrt{\frac{1+u}{1-u}} \ln\left|\frac{1-u}{1+u}\right| \frac{du}{u-x^*} = -\pi^2 \sqrt{\frac{1+x^*}{1-x^*}} \quad \text{for} \quad -1 < x^* < 1$$

$$(4.3.50)$$

Now we can proceed to consider the several terms on the RHS of (4.3.49)

1st term. Now

$$\int_{-1}^{1} \frac{d\xi^*}{u-\xi^*} = \ln\left|\frac{1+u}{1-u}\right| \quad \text{from} \quad I_0$$

$$\therefore \quad I_3 \equiv \oint_{-1}^{1} \sqrt{\frac{1+u}{1-u}} \frac{1}{u-x^*} \int_{-1}^{1} \frac{d\xi^*}{u-\xi^*} du$$

$$= \int_{-1}^{1} \sqrt{\frac{1+u}{1-u}} \frac{1}{u-x^*} \ln\left|\frac{1+u}{1-u}\right| du = -\pi^2 \sqrt{\frac{1+x^*}{1-x^*}} \quad \text{from} \quad I_2$$

$$\therefore \quad \text{1st term} = -\frac{\gamma I_3}{2\pi} = \frac{+\gamma(x^*)}{2} \pi \sqrt{\frac{1+x^*}{1-x^*}}$$

2nd term. Interchange order of integration;

$$I_4 \equiv \oint_{-1}^{1} [\xi^* - x^*] F(\xi^*, x^*) \int_{-1}^{1} \sqrt{\frac{1+u}{1-u}} \frac{du}{(u-x^*)(u-\xi^*)} d\xi^*$$

Now

$$\frac{1}{(u-x^*)(u-\xi^*)} = \frac{1}{x^*-\xi^*}\left[\frac{1}{u-x^*} - \frac{1}{u-\xi^*}\right]$$

partial fractions expansion

$$\therefore \quad I_4 = -\int_{-1}^{1} F(\xi^*, x^*)\left\{\int_{-1}^{1}\sqrt{\frac{1+u}{1-u}}\left[\frac{1}{u-x^*} - \frac{1}{u-\xi^*}\right]du\right\}d\xi^*$$

$$= -\oint_{-1}^{1} F(\xi^*, x^*)[\pi - \pi]^0 d\xi^* \quad \text{from} \quad I_1$$

Finally then, from above and (4.3.49),

$$\int_{1}^{1}\sqrt{\frac{1+u}{1-u}}\frac{w_a(u)}{u-x^*}\,du = -\frac{\pi}{2}\gamma(x^*)\sqrt{\frac{1+x^*}{1-x^*}}$$

$$\therefore \quad \gamma(x^*) = -\frac{2}{\pi}\sqrt{\frac{1-x^*}{1+x^*}}\int_{-1}^{1}\sqrt{\frac{1+u}{1-u}}\frac{w_a(u)}{u-x^*}\,du \qquad (4.3.51)$$

Note: Other 'solving kernels' exist, but they do not satisfy the Kutta condition, $\gamma(x^*)$ finite at $x^* = 1$, i.e., finite pressure at trailing edge.

One might reasonably inquire, how do we know what the solving kernel should be? Perhaps the most straightforward way to motivate the choice is to recognize that the solution for steady flow can be obtained by other methods. Probably the simplest of these alternative solution methods is to use the transformations $x^* = \cos\theta$, $\xi^* = \cos\phi$ and expand γ and w_a in Fourier series in ϕ and θ. See BAH, p. 216. Once the answer is known, i.e. (4.3.51), the choice of the solving kernel is fairly obvious. The advantage of the solving kernel approach over other methods is that it is capable of extension to unsteady airfoil motion where an analytical solution may be obtained as will be described below. On the other hand a method which is based essentially on the Fourier series approach is often employed to obtain numerical solutions for three-dimensional, compressible flow. This is the so-called Kernel Function approach discussed earlier.

In the above we have obtained the following integral relation: Given

$$f(x^*) = -\frac{1}{2\pi}\int_{-1}^{1}\frac{g(\xi^*)}{x^*-\xi^*}\,d\xi^*$$

with $g(1)$ finite or zero, then

$$g(x^*) = -\frac{2}{\pi} \sqrt{\frac{1-x^*}{1+x^*}} \int_{-1}^{1} \sqrt{\frac{1+\xi^*}{1-\xi^*}} \frac{f(\xi^*)}{\xi^* - x^*} \, d\xi^* \tag{4.3.52}$$

General case: Oscillating motion. We may employ the solving kernel approach to attack the unsteady problem also. Recall from (4.3.40), (4.3.43), (4.3.46) one has

$$\bar{w}_a(x^*) = -\frac{1}{2\pi} \int_{-1}^{1} \frac{\gamma(\xi^*)}{x^* - \xi^*} \, d\xi^* - \frac{1}{2\pi} \int_{1}^{\infty} \frac{\gamma(\xi^*)}{x^* - \xi^*} \, d\xi^* \tag{4.3.53}$$

$$\overline{\Delta C_p} = \frac{2\bar{\gamma}(x^*)}{U_\infty} + 2ik \frac{\bar{\Gamma}(x^*)}{U_\infty b} = 2\frac{\bar{\gamma}(x^*)}{U_\infty} + 2ik \int_{-1}^{x^*} \frac{\bar{\gamma}(\xi^*)}{U_\infty} \, d\xi^* \tag{4.3.54}$$

$$\frac{\bar{\gamma}(x^*)}{U_\infty} = -ik \frac{\bar{\Gamma}(1)}{U_\infty b} \exp[-ik(x^* - 1)] \quad \text{for} \quad x^* > 1 \tag{4.3.55}$$

Substitute (4.3.55) into (4.3.53).

$$\bar{w}_a(x^*) = -\frac{1}{2\pi} \int_{-1}^{1} \frac{\bar{\gamma}(\xi^*)}{x^* - \xi^*} \, d\xi^* + \bar{G}(x^*) \tag{4.3.56}$$

where

$$\bar{G}(x^*) \equiv \frac{ik\bar{\Gamma}(1)}{2\pi b} \int_{+1}^{\infty} \frac{\exp[-ik(\xi^* - 1)]}{x^* - \xi^*} \, d\xi^*$$

Invert (4.3.56) to determine $\gamma(x^*)$; recall the steady flow solution, (4.3.52).

$$\bar{\gamma}(x^*) = -\frac{2}{\pi} \sqrt{\frac{1-x^*}{1+x^*}} \int_{-1}^{1} \sqrt{\frac{1+\xi^*}{1-\xi^*}} \left\{ \frac{w_a(\xi^*) - \bar{G}(\xi^*)}{\xi^* - x^*} \right\} d\xi^*$$

$$= -\frac{2}{\pi} \sqrt{\frac{1-x^*}{1+x^*}} \int_{-1}^{1} \sqrt{\frac{1+\xi^*}{1-\xi^*}} \left\{ \frac{\bar{w}(\xi^*) - \dfrac{ik\bar{\Gamma}(1)}{2\pi b} \displaystyle\int_{1}^{\infty} \exp\left[\dfrac{-ik(u-1)}{\xi^* - u}\right] du}{\xi^* - x^*} \right\} d\xi^* \tag{4.3.57}$$

Interchanging the order of integration of the term involving $\bar{\Gamma}(1)$ on the RHS side of (4.3.57) we may evaluate the integral over ξ^* and obtain

$$\bar{\gamma}(x^*) = -\frac{2}{\pi} \sqrt{\frac{1-x^*}{1+x^*}} \left\{ \int_{-1}^{1} \sqrt{\frac{1+\xi^*}{1-\xi^*}} \frac{\bar{w}_a(\xi^*)}{(x^* - \xi^*)} \, d\xi^* \right.$$

$$\left. + ik \frac{\bar{\Gamma}(1)}{b} e^{ik} \int_{1}^{\infty} \frac{e^{-iku}}{x^* - u} \, du \right\} \tag{4.3.58}$$

(4.3.58) is not a complete solution until we determine $\bar{\Gamma}(1)$ which we do as follows. Integrating (4.3.58) with respect to x^* we obtain

$$\frac{\bar{\Gamma}(1)}{b} \equiv \int_{-1}^{1} \bar{\gamma}(x^*)\,dx^* = -2\int_{-1}^{1} \sqrt{\frac{1+\xi^*}{1-\xi^*}}\,\bar{w}_a(\xi^*)\,d\xi^*$$

$$- ik\frac{\bar{\Gamma}(1)}{b}\,e^{ik}\int_{1}^{\infty}\left[\sqrt{\frac{u+1}{u-1}} - 1\right]e^{-iku}\,du \quad (4.3.59)$$

where the integrals on the right hand side with respect to x^* have been evaluated explicitly. We may now solve (4.3.59) for $\bar{\Gamma}(1)$. Recognizing that

$$\int_{1}^{\infty}\left[\sqrt{\frac{u+1}{u-1}} - 1\right]e^{-iku}\,du = \frac{-\pi}{2}[H_1^{(2)}(k) + iH_0^{(2)}(k)] - \frac{e^{-ik}}{ik} \quad (4.3.60)$$

we determine from (4.3.59) and (4.3.60) that

$$\frac{\bar{\Gamma}(1)}{b} = 4\frac{e^{-ik}\displaystyle\int_{-1}^{1}\sqrt{\frac{1+\xi^*}{1-\xi^*}}\,\bar{w}_a(\xi^*)\,d\xi^*}{\pi ik[H_1^{(2)}(k) + iH_0^2(k)]} \quad (4.3.61)$$

$H_1^{(2)}$, $H_0^{(2)}$ are standard Hankel functions.[†] (4.3.58) and (4.3.61) constitute the solution for $\bar{\gamma}$ in terms of $\bar{w}_a$. From $\bar{\gamma}$, we may determine $\overline{\Delta C_p}$ by using

$$\overline{\Delta C_p} = 2\frac{\bar{\gamma}(x^*)}{U_\infty} + 2ik\int_{-1}^{x^*}\frac{\bar{\gamma}(\xi^*)}{U_\infty}\,ds^*$$

After considerable, but elementary, algebra

$$\overline{\Delta C_p} = \frac{4}{\pi}\sqrt{\frac{1-x^*}{1+x^*}}\int_{-1}^{1}\sqrt{\frac{1+\xi^*}{1-\xi^*}}\left\{\frac{\bar{w}_a(\xi^*)/U_\infty}{x^*-\xi^*}\right\}d\xi$$

$$+ \frac{4}{\pi}\,ik\sqrt{1-x^{*2}}\oint_{-1}^{1}\frac{W(\xi^*)\,d\xi^*}{U_\infty\sqrt{1-\xi^{*2}}(x^*-\xi^*)}$$

$$+ \frac{4}{\pi}[1-C(k)]\sqrt{\frac{1-x^*}{1+x^*}}\int_{-1}^{1}\sqrt{\frac{1+\xi^*}{1+\xi^*}}\,\frac{\bar{w}_a(\xi^*)}{U_\infty}\,d\xi^* \quad (4.3.62)$$

[†] Abramowitz and Stegun [20].

where

$$W(\xi^*) \equiv \int_{-1}^{\xi^*} \bar{w}_a(u)\, du$$

and

$$C(k) \equiv \frac{H_1^{(2)}}{[H_1^{(2)} + iH_0^{(2)}]}$$

is Theodorsen's well known Function.

The lift may be computed as the integral of the pressure.

$$
\begin{aligned}
\bar{L} &\equiv \frac{\rho U_\infty^2}{2}\, b \int_{-1}^{1} \overline{\Delta C_p}\, dx^* \\
&= \frac{\rho U_\infty^2}{2} b \left\{ -C(k) \int_{-1}^{1} \sqrt{\frac{1+\xi^*}{1-\xi^*}}\, \frac{w_a(\xi^*)}{U_\infty}\, d\xi^* \right. \\
&\quad \left. - ik \int_{-1}^{1} \sqrt{1-\xi^{*2}}\, \frac{\bar{w}_a(\xi^*)}{U_\infty}\, d\xi^* \right\}
\end{aligned}
\tag{4.3.63}
$$

Similarly for the moment about the point $x = ba$,

$$\bar{M}_y \equiv \frac{\rho U_\infty^2}{2}\, b^2 \int_{-1}^{1} \overline{\Delta C_p}[x^* - a]\, dx^* \tag{4.3.64}$$

In particular, for

$$z_a = -h - \alpha(x - ba)$$

$$\bar{z}_a = -\bar{h} - \bar{\alpha}(x - ba)$$

one has

$$\bar{w}_a = -i\omega\bar{h} - i\omega\bar{\alpha}(x - ba) - U_\infty\bar{\alpha} \tag{4.3.65}$$

Thus (4.3.65) in (4.3.63) and (4.3.64) give

$$
\begin{aligned}
\bar{L} &= \pi\rho b^2 [-\omega^2\bar{h} + i\omega U_\infty\bar{\alpha} + ba\omega^2\bar{\alpha}] \\
&\quad + 2\pi\rho U_\infty b C(k)[i\omega\bar{h} + U_\infty\bar{\alpha} + b(\tfrac{1}{2} - a)i\omega\alpha] \\
\bar{M}_y &= \pi\rho b^2 [-ba\omega^2\bar{h} - U_\infty b(\tfrac{1}{2} - a)i\omega\bar{\alpha} + b^2(\tfrac{1}{8} + a^2)\omega^2\bar{\alpha}] \\
&\quad + 2\pi\rho U_\infty b^2(\tfrac{1}{2} + a)C(k)[i\omega\bar{h} + U_\infty\bar{\alpha} + b(\tfrac{1}{2} - a)i\omega\bar{\alpha}]
\end{aligned}
\tag{4.3.66}
$$

Theodorsen's Function, $C(k) = F + iG$, is given below in Fig 4.6.

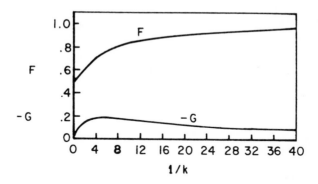

Figure 4.6 *The functions F and G against $\frac{1}{k}$. After Theodorsen [19].*

Transient motion. Using Fourier synthesis one may now obtain results for arbitrary time dependent motion from the simple harmonic motion results; using Fourier summation (integration) and (4.3.66),

$$L(t) = \frac{1}{2\pi} \int_{-\infty}^{\infty} \bar{L}(\omega)_{\text{due to } h} \bar{h}(\omega) e^{i\omega t} \, d\omega$$

$$+ \frac{1}{2\pi} \int_{-\infty}^{\infty} \bar{L}(\omega)_{\text{due to } \alpha} \bar{\alpha}(\omega) e^{i\omega t} \, d\omega$$

$$= \frac{1}{2\pi} \int_{-\infty}^{\infty} \{\pi\rho b^2(-\omega^2) + 2\pi\rho U_\infty b C(k)(i\omega)\} \cdot \bar{h}(\omega) e^{i\omega t} \, d\omega$$

$$+ \frac{1}{2\pi} \int_{-\infty}^{\infty} \{\pi\rho b^2(i\omega U_\infty + ba\omega^2) + 2\pi\rho U_\infty b C(k)(U_\infty + b(\tfrac{1}{2} - a)i\omega)\}$$

$$\bar{\alpha}(\omega) e^{i\omega t} \, d\omega \tag{4.3.67}$$

where

$$\bar{h}(\omega) = \int_{-\infty}^{\infty} h(t) e^{-i\omega t} \, d\omega$$

and

$$\bar{\alpha}(\omega) = \int_{-\infty}^{\infty} \alpha(t) e^{-i\omega t} \, d\omega \tag{4.3.68}$$

219

Now

$$\int_{-\infty}^{\infty} (i\omega)^n \bar{\alpha} e^{i\omega t}\, d\omega = \frac{d^n\alpha}{dt^n} \qquad n = 1, 2, \ldots \tag{4.3.69}$$

Thus

$$L = \pi\rho b^2 \left[\frac{d^2 h}{dt^2} + U_\infty \frac{d\alpha}{dt} - ba\frac{d^2\alpha}{dt^2}\right]$$
$$+ \rho U_\infty b \int_{-\infty}^{\infty} C(k) f(\omega) e^{i\omega t}\, d\omega$$

where

$$f(\omega) \equiv i\omega \bar{h}(\omega) + U_\infty \bar{\alpha}(\omega) + b(\tfrac{1}{2}-a) i\omega \bar{\alpha}(\omega)$$
$$= \int_{-\infty}^{\infty} \left[\frac{dh}{dt} + U_\infty \alpha + b(\tfrac{1}{2}-a)\frac{d\alpha}{dt}\right] e^{-i\omega t}\, dt \tag{4.3.70}$$

Physically,

$$\frac{dh}{dt} + U_\infty \alpha + b(\tfrac{1}{2}-a)\frac{d\alpha}{dt} = -w_a \quad \text{at} \quad x = b/2;$$

$x = b/2$ is $\frac{3}{4}$ chord of airfoil.

Similarly,

$$M_y = \pi\rho b^2 ba \left[\frac{d^2 h}{dt^2} - U_\infty b(\tfrac{1}{2}-a)\frac{d\alpha}{dt} - b^2(\tfrac{1}{8}+a^2)\frac{d^2\alpha}{dt^2}\right]$$
$$+ \rho U_\infty b^2(\tfrac{1}{2}+a) \int_{-\infty}^{\infty} C(k) f(\omega) e^{i\omega t}\, d\omega \tag{4.3.71}$$

Example I.   Step change in angle of attack.

$$h \equiv 0$$
$$\alpha = 0 \quad \text{for} \quad t < 0$$
$$\quad = \alpha_0 \equiv \text{constant} \quad \text{for} \quad t > 0$$
$$\therefore\ \frac{d\alpha}{dt} = \frac{d^2\alpha}{dt^2} = \frac{dh}{dt} = \frac{d^2 h}{dt^2} = 0 \quad \text{for} \quad t > 0$$
$$\therefore\ f(\omega) = U_\infty \alpha_0 \int_0^{\infty} e^{-i\omega t}\, dt$$
$$\quad = \frac{U_\infty \alpha_0}{-i\omega} e^{-i\omega t}\Big|_0^{\infty} = \frac{U_\infty \alpha_0}{i\omega}$$

$$\therefore \; L = \rho U_\infty^2 b \alpha_0 \int_{-\infty}^{\infty} \frac{C(k)}{i\omega} e^{i\omega t} \, d\omega$$

$$= \rho U_\infty^2 b \alpha_0 \int_{-\infty}^{\infty} \frac{C(k)}{ik} e^{iks} \, dk$$

where $s \equiv \dfrac{Ut}{b}$.   Finally,

$$L = 2\pi\rho U^2 b \alpha_0 \left\{ \frac{1}{2\pi i} \int_{-\infty}^{\infty} \frac{C(k)}{k} e^{iks} \, dk \right\} \qquad (4.3.72)$$

$\{\cdots\} \equiv \phi(s)$ is called the **Wagner Function**, see Figure 4.7. Note that if α is precisely a step function, then L has a singularity at $t = 0$ from (4.3.70). Also shown is the Kussner function, $\psi(s)$, to be discussed subsequently. Note also that ϕ is the lift of the airfoil due to step change in angle of attack or more generally due to step change in $-w_a/U_\infty$ at $\frac{3}{4}$ chord.

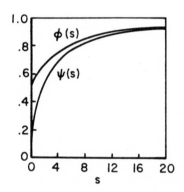

Figure 4.7   *Wagner's function $\phi(s)$ for indicial lift and Küssner's function $\psi(s)$ for lift due to a sharp-edged gust, plotted as functions of distance traveled in semichordlengths. After BAH, Fig. 5.21.*

Thus using Duhamel superposition formula

$$L(t) = \pi\rho b^2 \left[\frac{d^2 h}{dt^2} + U_\infty \frac{d\alpha}{dt} - ba \frac{d^2\alpha}{dt^2} \right]$$

$$- 2\pi\rho U_\infty b \left[w_{a_{\frac{3}{4}}}(0)\phi(s) + \int_0^s \frac{dw_{a_{\frac{3}{4}}}}{d\sigma}(\sigma)\phi(s-\sigma) \, d\sigma \right] \qquad (4.3.73)$$

221

Example II. Entrance into a sharp edged gust. In the primed coordinate system, i.e., fixed with respect to the atmosphere, one has

$$w_G = 0 \quad \text{for} \quad x' > 0$$
$$= w_0 \quad \text{for} \quad x' < 0$$

Note: The general transformation between fluid fixed and body fixed coordinate systems is

$$x' = x + b - U_\infty t, \qquad x + b = x' + U_\infty t'$$
$$t' = t \qquad\qquad\qquad t = t'$$

The leading edge enters the gust at $t = t' = 0$ at

$$t = 0, \qquad x' = x + b$$
$$t' = 0.$$

Thus in the coordinate system fixed with respect to the airfoil, one has

$$w_G = 0 \quad \text{for} \quad x + b > U_\infty t \quad \text{or} \quad \frac{x+b}{U_\infty} > t \qquad (4.3.74)$$

$$= w_0 \quad \text{for} \quad x + b < U_\infty t \quad \text{or} \quad \frac{x+b}{U_\infty} < t$$

$$\therefore \; w_G(\omega) \equiv \int_{-\infty}^{\infty} w_G e^{-i\omega t} \, dt$$

$$= w_0 \int_{(x+b)/U_\infty}^{\infty} e^{-i\omega t} \, dt$$

$$= \frac{w_0}{-i\omega} e^{-i\omega t} \Big|_{(x+b)/U_\infty}^{\infty}$$

$$= \frac{w_0}{i\omega} \exp\left[-i\omega \frac{(x+b)}{U_\infty} \right] = \frac{w_0}{i\omega} e^{-ik} e^{-ikx^*}$$

where

$$x^* \equiv x/b \qquad (4.3.75)$$

For

$$\bar{w}_a = -w_G \left(= -\frac{w_0}{i\omega} e^{-ik} e^{-ikx^*} \right)$$

one finds from the oscillating airfoil motion theory that

$$\bar{L} = 2\pi\rho U_\infty b\{C(k)[J_0(k) - iJ_1(k)] + iJ_1(k)\}\frac{w_0}{i\omega}e^{-ik}$$

and

$$\bar{M}_y = b(\tfrac{1}{2} + a)\bar{L}$$

$$\therefore \quad L(t) = \frac{1}{2\pi}\int_\infty^\infty \bar{L}(\omega)e^{i\omega t}\,d\omega$$

$$= \rho U_\infty bw_0 \int_{-\infty}^\infty \frac{\{\cdots\}}{ik}e^{-ik}e^{iks}\,dk \tag{4.3.76}$$

$$= 2\pi\rho U_\infty bw_0\psi(s)$$

where

$$\psi(s) \equiv \frac{1}{2\pi i}\int_{-\infty}^\infty \frac{\{\cdots\}}{k}\exp[ik(s-1)]\,dk \tag{4.3.77}$$

is called the Kussner Function and was shown previously in Figure 4.7. Finally then, using Duhamel's integral,

$$L = \pi\rho Ub\left\{w_G(0)\psi(s) + \int_0^s \frac{dw_G}{d\sigma}(\sigma)(s-\sigma)\,d\sigma\right\} \tag{4.3.78}$$

A famous controversy concerning the interpretation of Theodorsen's function for other than real frequencies (neutrally stable motion) took place in the 1950's. The issue has arisen again because of possible applications to feedback control of aeroelastic systems. For a modern view and discussion, the reader should consult Edwards, Ashley, and Breakwell [21]. Also see Sears, [10] in chapter 3.

Evaluation of integrals. I_0:
    For $x^* < 1$.

$$I_0 \equiv \oint_{-1}^1 \frac{d\xi^*}{x^* - \xi^*} = \lim_{\varepsilon\to 0}\left[\int_{-1}^{x^*-\varepsilon}\frac{d\xi^*}{x^* - \xi^*} + \int_{x^*+\varepsilon}^1 \frac{d\xi^*}{x^* - \xi^*}\right]$$

$$= \lim_{\varepsilon\to 0}\left[-\int_{-1}^{x^*-\varepsilon}\frac{d(x^*-\xi^*)}{(x^*-\xi^*)} - \int_{x^*+\varepsilon}^1 \frac{d(\xi^*-x^*)}{(\xi^*-x^*)}\right]$$

$$= -\ln(x^*-\xi^*)\Big|_{\xi^*=-1}^{x^*-\varepsilon} - \ln(\xi^*-x^*)\Big|_{x^*+\varepsilon}^1$$

$$= -[\ln\varepsilon - \ln(x^*+1)] - [\ln(1-x^*) - \ln\varepsilon] = \ln\left(\frac{1+x^*}{1-x^*}\right)$$

For $x^* > 1$, there is no need for a Cauchy Principal Value and

$$I_0 = \ln \left(\frac{x^* + 1}{x^* - 1} \right)$$

I_1: $I_1 \equiv \oint_{-1}^{1} \sqrt{\frac{1+u}{1-u}} \frac{du}{u - x^*}$

Use contour integration. Define $w \equiv u + iv$ (a complex variable whose real part is u) and

$$F(w) \equiv \left(\frac{w + 1}{w - 1} \right)^{\frac{1}{2}} \frac{1}{w - x^*}$$

Choose a contour as follows

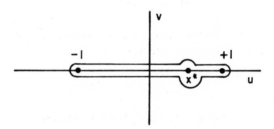

Figure 4.8   *Integral contour.*

Now

$$\frac{w + 1}{w - 1} = [R^2 + I^2]^{\frac{1}{2}} e^{-i\theta}$$

where

$$R \equiv \frac{(u + 1)(u - 1) + v^2}{(u - 1)^2 + v^2}, \qquad I \equiv \frac{-2v}{(u - 1)^2 + v^2}$$

$\theta = \tan^{-1} I / R$

(i) on top, $v = 0^+$, $u - 1 < 0$

$\therefore$ $R < 0$, $I = 0^- \Rightarrow \theta = -\pi$

(ii) on bottom, $v = 0^-$, $u - 1 < 0$

$\therefore$ $R < 0$, $I = 0^+ \Rightarrow \theta = +\pi$

$$\therefore \quad \left(\frac{w+1}{w-1}\right)^{\frac{1}{2}} = \sqrt{\frac{1+u}{1-u}}\, e^{-i\pi/2} \quad \text{on top}$$

$$= \sqrt{\frac{1+u}{1-u}}\, e^{+i\pi/2} \quad \text{on bottom}$$

Now $dw = du$ on top or bottom and $w - x^* = u - x^*$ except on arcs near $u = x^*$. On arcs $w - x^* = \varepsilon e^{i\theta}$, $dw = \varepsilon e^{i\theta} i\, d\theta$ where ε is radius of arc. Also

$$\left(\frac{w+1}{w-1}\right)^{\frac{1}{2}} = \sqrt{\frac{1+u}{1-u}}\,(-i) \quad \text{on top}$$

and $= \cdots (+i)$ on bottom. Thus

$$\zeta_1 \equiv \int_C F(w)\, dw = \overbrace{\int_{-1}^{x^*-\varepsilon} + \int_{x^*+\varepsilon}^{1}}^{\text{bottom}} i\sqrt{\frac{1+u}{1-u}}\,\frac{du}{u-x^*}$$

$$+ \overbrace{\int_{1}^{x^*+\varepsilon} + \int_{x^*-\varepsilon}^{-1}}^{\text{top}} -i\sqrt{\frac{1+u}{1-u}}\,\frac{du}{u-x^*} +$$

$+$ contributions from arcs which cancel each other

$$\therefore \quad \lim_{\varepsilon \to 0} \zeta_1 = 2i \int_{-1}^{1} \sqrt{\frac{1+u}{1-u}}\,\frac{du}{u-x^*} = 2iI_1$$

ζ_1 can be simply evaluated by Cauchy's Theorem. As $w \to \infty$, $F(w) \to 1/w$.

$$\therefore \quad \zeta_1 = \int_{\substack{\text{around} \\ \text{arc at } \infty}} \frac{dw}{w} = 2\pi i$$

$$\therefore \quad I_1 = \frac{\zeta_1}{2i} = \frac{2\pi i}{2i} = \pi$$

For $x^* > 1$, I_1 is still equal to $\zeta_1/2\pi i$; however, now $\zeta_1 = \int_{\text{arc at infinity}} F(w)\, dw - \text{Residue of } F \text{ at } x^*$

$$= 2\pi i - 2\pi i \sqrt{\frac{x^*+1}{x^*-1}}$$

$$\therefore \quad I_1 = \frac{\zeta_1}{2\pi i} = \pi\left[1 - \sqrt{\frac{x^*+1}{x^*-1}}\right]$$

A similar calculation gives I_2.

Evaluation of I_2.

$$I_2 \equiv \int_{-1}^{1} \sqrt{\frac{1+u}{1-u}} \ln \left| \frac{1+u}{1-u} \right| \frac{du}{u-x^*}$$

Define

$$w \equiv u + iv$$

and

$$F(w) \equiv \ln \left| \frac{w+1}{w-1} \right| \sqrt{\frac{w+1}{w-1}} \frac{1}{w-x^*}$$

The contour is the same as for I_1.

As before,

$$\left(\frac{w+1}{w-1} \right)^{\frac{1}{2}} = \sqrt{\frac{1+u}{1-u}} e^{-i\pi/2} \quad \text{on top}$$

$$= \sqrt{\frac{1+u}{1-u}} e^{+i\pi/2} \quad \text{on bottom}$$

Also

$$\ln \left(\frac{w+1}{w-1} \right) = \ln \sqrt{R^2 + I^2} + i\theta$$

$$= \ln \left| \frac{u+1}{u-1} \right| - \pi \quad \text{on top}$$

$$= \ln \left| \frac{u+1}{u-1} \right| + \pi \quad \text{on bottom}$$

Now $dw = du$ on top or bottom and $w - x^* = u - x^*$ *except* on arcs near $u = x^*$. On arcs $w - x^* = \varepsilon e^{i\theta}$, $dw = \varepsilon e^{i\theta} i \, d\theta$ where ε is radius of arc. Thus

$$\zeta_2 \equiv \int_C F_2(w) \, dw = \int_{-1}^{x^*-\varepsilon} + \int_{x^*+\varepsilon}^{1} i \left\{ \sqrt{\frac{1+u}{1-u}} \left[\ln \left| \frac{1+u}{1-u} \right| + \pi \right] \right\}$$

$$\times \frac{du}{u-x^*} \quad \text{bottom}$$

$$+ \int_{1}^{x^*+\varepsilon} + \int_{x^*-\varepsilon}^{-1} -i \left\{ \sqrt{\frac{1+u}{1-u}} \left[\ln \left| \frac{1+u}{1-u} \right| - \pi \right] \right\} \frac{du}{u-x^*} \quad \text{top}$$

$$+ \int_{0}^{\pi} i\pi \sqrt{\frac{1+x^*}{1-x^*}} \, d\theta + \int_{-\pi}^{0} i\pi \sqrt{\frac{1+x^*}{1-x^*}} \, d\theta \quad \text{arcs}$$

Note: ln terms cancel and thus are omitted in arc contributions. Cancelling π terms from bottom and top and adding arc terms, gives

$$\zeta_2 = \int_{-1}^{x^*-\varepsilon} + \int_{x^*+\varepsilon}^{1} i\sqrt{\frac{1+u}{1-u}}\ln\left|\frac{1+u}{1-u}\right|\frac{du}{u-x^*}$$

$$+ \int_{1}^{x^*+\varepsilon} + \int_{x^*-\varepsilon}^{-1} -i\sqrt{\frac{1+u}{1-u}}\ln\left|\frac{1+u}{1-u}\right|\frac{du}{u-x^*}$$

$$+ 2i\pi^2\sqrt{\frac{1+x^*}{1-x^*}}$$

Adding bottom and top terms,

$$\lim_{\varepsilon\to 0}\zeta_2 = 2i\int_{-1}^{1}\sqrt{\frac{1+u}{1-u}}\ln\left|\frac{1+u}{1-u}\right|\frac{du}{u-x^*} + 2i\pi^2\sqrt{\frac{1+x^*}{1-x^*}}$$

$$= 2iI_2 + 2i\pi^2\sqrt{\frac{1+x^*}{1-x^*}}$$

ζ_2 can be simply evaluated by Cauchy's Theorem. As $w\to\infty$, $F_2(w)\to 0$.

$$\therefore\quad \zeta_2 = 0 \Rightarrow I_2 = -\pi^2\sqrt{\frac{1+x^*}{1-x^*}}$$

4.4   Representative numerical results

Consider a flat plate airfoil, initially at zero angle of attack, which is given a step change in α, i.e.,

$$w = -U_\infty\alpha \quad \text{for} \quad t>0$$
$$= 0 \qquad \text{for} \quad t<0$$

Although most calculations in practice are carried out for sinusoidal time dependent motion, for our purposes examining aerodynamic pressures due to this step change leads to more insight into the nature of the physical system. Of course, *in principle*, the results for sinusoidal motion (or a step change) may be superposed to obtain results for arbitrary time dependent motion.

227

It is traditional to express the pressure in nondimensional form

$$\frac{p}{\dfrac{\rho_\infty U_\infty^2 \alpha}{2}} \equiv \frac{p}{q\alpha}$$

as a function of nondimensional time,

$$s \equiv \frac{tU_\infty}{c/2}$$

and M_∞. The results shown below are from an article by Lomax;[*] both subsonic and supersonic, two- and three dimensional results are displayed.

In Figure 4.9 the chord-wise pressure distribution for two-dimensional flow is shown at several times, s, for a representative subsonic Mach number. For $s = 0$, the result is given by piston theory (as in supersonic flow)[†]

$$p = \rho_\infty a_\infty w$$

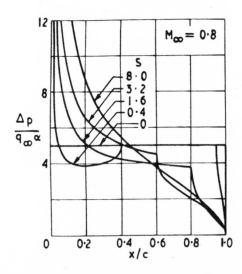

Figure 4.9   Chordwise lifting pressure distributions.

---

[*] Lomax [22].

[†] This can be shown by considering the transient analysis of Section 4.2 and noting it still applies for $t = 0^+$.

For a step change in α, piston theory gives

$$\frac{\Delta p}{\dfrac{\rho_\infty U_\infty^2 \alpha}{2}} = \frac{p_L - p_U}{\dfrac{\rho_\infty U_\infty^2 \alpha}{2}} = \frac{4}{M}$$

For $s \to \infty$, the result is also well known, with a square root singularity at the leading edge. Of course, the Kutta condition, $\Delta p = 0$, is enforced at the trailing edge for all s. As $s \to \infty$

$$\frac{\Delta p}{\dfrac{\rho_\infty U_\infty^2 \alpha}{2}} = \frac{4}{(1 - M^2)^{\frac{1}{2}}} \sqrt{\frac{c - x}{x}}$$

This result is implicit in the analysis of Section 4.3.

In Figure 4.10 the chord-wise pressure distribution is shown at several times, s, for a representative supersonic Mach number. For $s = 0$ the result is again that given by piston theory

$$\frac{\Delta p}{\dfrac{\rho_\infty U_\infty^2 \alpha}{2}} = \frac{4}{M}$$

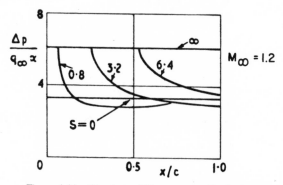

Figure 4.10   *Chordwise lifting pressure distribution.*

For $s \to \infty$, the result is (as previously cited in our earlier discussion, Section 4.2)

$$\frac{\Delta p}{\dfrac{\rho_\infty U_\infty^2 \alpha}{2}} = \frac{4}{(M^2 - 1)^{\frac{1}{2}}}$$

Indeed the pressure reaches this final steady state value at a finite s which can be determined as follows. All disturbances propagate *in the fluid* with the speed of sound, a_∞, but the airfoil moves faster with velocity $U_\infty > a_\infty$. Hence, the elapsed time for all disturbances (created by the step change of α for the airfoil) to move off the airfoil is the time required for a (forward propagating *in the fluid*) disturbance at the leading edge to move to the trailing edge, namely

$$t = c/(U_\infty - a_\infty)$$

or, in nondimensional form,

$$s \equiv \frac{tU_\infty}{c/2} = \frac{2M_\infty}{M_\infty - 1}$$

For

$$s > \frac{2M_\infty}{M_\infty - 1}$$

steady state conditions are obtained all along the airfoil. As can be seen from Figure 4.10 for $s = 0^+$ the leading edge pressure instantly reaches its final steady state value. As s increases the steady state is reached by increasing portions of the airfoil along the chord. Note that the initial results, $s = 0$, and steady state results,

$$s \geq \frac{2M_\infty}{M_\infty - 1}$$

have a constant pressure distribution; however, for intermediate s, the pressure varies along the chord.

The pressure distributions may be integrated along the chord to obtain the total force (lift) on the airfoil

$$L \equiv \int_0^c \Delta p \, dx,$$

$$C_{L_\alpha} \equiv \frac{L}{\dfrac{\rho_\infty U_\infty^2 c \alpha}{2}}, \text{ lift curve slope}$$

Again the $s = 0$ result is that given by piston theory

$$C_{L_\alpha} = \frac{4}{M}$$

and the steady-state result is

$$C_{L_\alpha} = \frac{4}{(M^2 - 1)^{\frac{1}{2}}} \quad \text{for} \quad M_\infty > 1$$

and it is also known that

$$C_{L_\alpha} = \frac{2\pi}{(1 - M_\infty^2)^{\frac{1}{2}}} \quad \text{for} \quad M_\infty < 1$$

see Section 4.3. Results for C_{L_α} are shown in Figure 4.11 for various Mach number.

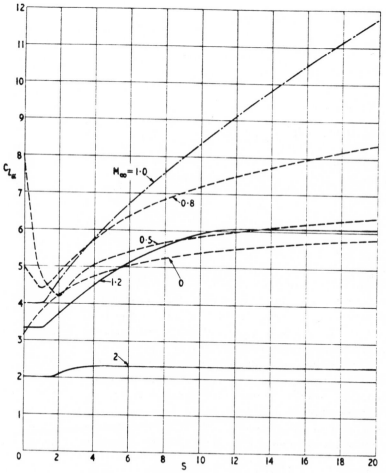

Figure 4.11   *Time history of lift curve slope.*

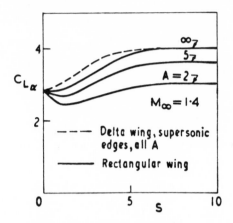

Figure 4.12   *Time history of lift curve slope.*

Finally some representative results for *three-dimensional*, supersonic flow are shown in Figure 4.12. The effect of three-dimensionality is to reduce the lift. For small aspect ratio, A, where

$A \equiv$ maximum span squared/wing area

it is known from slender body theory* (an asymptotic theory for $A \to 0$) that

$$C_{L_\alpha} = \frac{\pi}{2} A$$

for $s \to \infty$. Note however, that the $s = 0^+$ result is independent of A and is that given by piston theory.

Hence, piston theory gives the correct result for $s = 0^+$ for two- and three-dimensional flows, subsonic as well as supersonic. However, only for relatively high supersonic and nearly two-dimensional flows does it give a reasonable approximation for *all s*.

For subsonic flows, the numerical methods are in an advanced state of development and results have been obtained for rather complex geometries including multiple aerodynamic surfaces. In Figures 4.13 to 4.17 representative data are shown. These are drawn from a paper by Rodden,† *et al.*, which contains an extensive discussion of such data and the numerical techniques used to obtain them. Simple harmonic motion is

---

* See Lomax, for example [22].
† Rodden, Giesing and Kálmán [23].

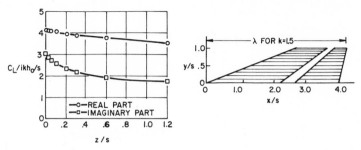

Figure 4.13 Lift coefficient of plunging wing-tail combination for various vertical separation distances; simple harmonic motion.

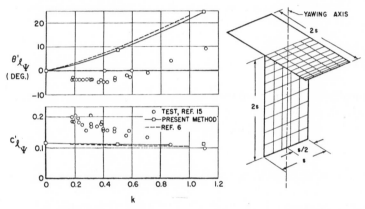

Figure 4.14 Rolling moment coefficient of horizontal stabilizer for simplified T-tail oscillating in yaw about fin mid-chord; simple harmonic motion.

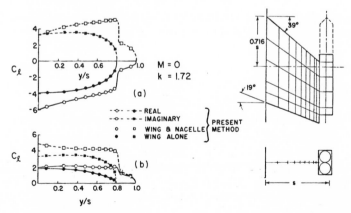

Figure 4.15 Distribution of span load for wing with and without engine nacelle. (a) *plunging* (b) *pitching; simple harmonic motion.*

233

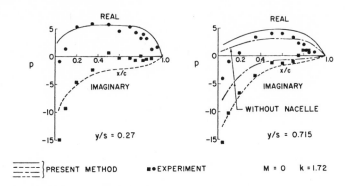

Figure 4.16   Comparison of experimental and calculated lifting pressure coefficient on a wing-nacelle combination in plunge; simple harmonic motion.

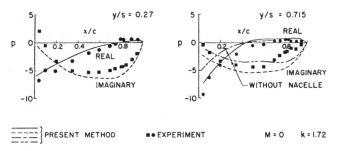

Figure 4.17   Comparison of experimental and calculated lifting pressure coefficient on a wing-nacelle combination in pitch; simple harmonic motion.

considered where k is a non-dimensional frequency of oscillation. Comparisons with experimental data are also shown.

4.5   Transonic flow

Major progress has been made in recent years on this important topic. Here we concentrate on the fundamental ideas and explore one simple approach to obtaining solutions using the same mathematical methods previously employed for subsonic and supersonic flow.

The failure of the classical linear, perturbation theory in transonic flow is well known and several attempts have been made to develop a theoretical model which will give consistent, accurate results. Among the more successful is the 'local linearization' concept of Spreiter which

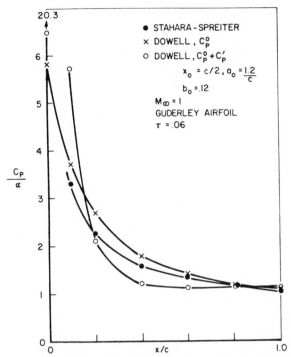

Figure 4.18 Pressure distribution for Guderley airfoil at constant angle of attack.

recently has been generalized to treat oscillating airfoils in transonic flow [24]. Another valuable method is that of parametric differentiation as developed by Rubbert and Landahl [25]. 'Local linearization' is an ad hoc approximation while parametric differentiation is a perturbation procedure from which the results of local linearization may be derived by making further approximations. Several authors [26–29] have attacked the problem in a numerical fashion using finite differences and results have been obtained for two-dimensional, high subsonic flow. Cunningham [30] has suggested a relatively simple, empirical modification of the classical theory.

In the present section a rational approximate method* is discussed which is broadly related to the local linearization concept. It has the advantages of (1) being simpler than the latter and (2) capable of being systematically improved to obtain an essentially exact solution to the governing transonic equation. Although the method has been developed

---

* This section is a revised version of Dowell [31]. A list of nomenclature is given at the end of this section.

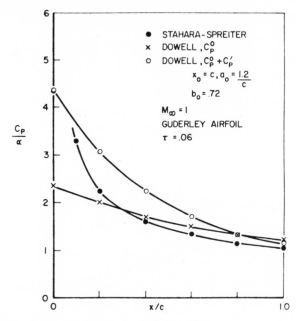

Figure 4.19   Pressure distribution for Guderley airfoil at constant angle of attack.

for treating infinitesimal dynamic motions of airfoils of finite thickness, it may also be employed (using the concept of parametric differentiation) to obtain solutions for nonlinear, steady nonlifting flows. This is the problem for which 'local linearization' was originally developed.

First, the basic idea will be explained for an infinitesimal steady motion of an airfoil of finite thickness in two-dimensional flow. Results will also be given for dynamic motion. The aerodynamic Green's functions for three-dimensional flow have also been derived. These are needed in the popular Mach Box and Kernel Function methods [32]. Using the Green's functions derived by the present method, three-dimensional calculations are effectively no more difficult than for the classical theory.

Analysis

From (4.1.21), Section 4.1, the full nonlinear equation for ϕ is

$$a^2 \nabla^2 \phi - \left[\frac{\partial}{\partial t} (\nabla \phi \cdot \nabla \phi) + \frac{\partial^2 \phi}{\partial t^2} + \nabla \phi \cdot \nabla \left(\frac{\nabla \phi \cdot \nabla \phi}{2} \right) \right] = 0$$

In cartesian, scalar notation and re-arranging terms

$$\phi_{xx}(a^2 - \phi_x^2) + \phi_{yy}(a^2 - \phi_y^2) + \phi_{zz}(a^2 - \phi_z^2)$$

$$-2\phi_{yz}\phi_y\phi_z - 2\phi_{xz}\phi_x\phi_z - 2\phi_{xy}\phi_x\phi_y$$

$$-\frac{\partial}{\partial t}(\phi_x^2 + \phi_y^2 + \phi_z^2) - \frac{\partial^2\phi}{\partial t^2} = 0 \tag{4.5.1}$$

Also we previously determined that ((4.1.22), Section 4.1)

$$\frac{a^2 - a_\infty^2}{\gamma - 1} = \frac{U_\infty^2}{2} - \left(\frac{\partial\phi}{\partial t} + \frac{\nabla\phi \cdot \nabla\phi}{2}\right) \tag{4.5.2}$$

Now let $\phi = U_\infty x + \hat{\phi}$, then (4.5.2) becomes

$$\frac{a^2 - a_\infty^2}{\gamma - 1} = -\left[\frac{\frac{\partial\hat{\phi}}{\partial t} + 2U_\infty\frac{\partial\hat{\phi}}{\partial x} + \frac{\partial\hat{\phi}}{\partial x} + \left(\frac{\partial\hat{\phi}}{\partial x}\right) + \left(\frac{\partial\hat{\phi}}{\partial y}\right) + \left(\frac{\partial\hat{\phi}}{\partial z}\right)^2}{2}\right]$$

$$\cong -\left[\frac{\partial\hat{\phi}}{\partial t} + U_\infty\frac{\partial\hat{\phi}}{\partial x}\right]$$

or

$$a^2 \cong a_\infty^2 - (\gamma - 1)\left[\frac{\partial\hat{\phi}}{\partial t} + U_\infty\frac{\partial\hat{\phi}}{\partial x}\right] \tag{4.5.3}$$

(4.5.1) becomes

$$\hat{\phi}_{xx}\left(a_\infty^2 - (\gamma - 1)\left[\frac{\partial\hat{\phi}}{\partial t} + U_\infty\frac{\partial\hat{\phi}}{\partial x}\right] - U_\infty^2 - 2U_\infty\frac{\partial\hat{\phi}}{\partial x}\right)$$

$$+ \hat{\phi}_{yy}a_\infty^2 + \hat{\phi}_{zz}a_\infty^2 - \frac{\partial}{\partial t}\left(2U_\infty\frac{\partial\hat{\phi}}{\partial x}\right) - \frac{\partial^2\hat{\phi}}{\partial t^2} \cong 0 \tag{4.5.4}$$

where obvious higher order terms have been neglected on the basis of $\hat{\phi}_x$, $\hat{\phi}_y$, $\hat{\phi}_z \ll U_\infty$ and a_∞.

The crucial distinction in transonic perturbation theory is in the coefficient of $\hat{\phi}_{xx}$. In the usual subsonic or supersonic small perturbation theory one approximates it as simply

$$a_\infty^2 - U_\infty^2$$

However if $U_\infty = a_\infty$ or nearly so then the terms retained above become important. The time derivative term in the coefficient of $\hat{\phi}_{xx}$ may still be neglected compared to the next to last term in (4.5.4), but no further simplification is possible, in general. Hence, (4.5.4) becomes (dividing by a_∞^2)

$$\hat{\phi}_{xx}[1 - M_L^2] + \hat{\phi}_{yy} + \hat{\phi}_{zz} - \frac{1}{a_\infty^2}\left[2U_\infty \frac{\partial^2 \hat{\phi}}{\partial x\,\partial t} + \frac{\partial^2 \hat{\phi}}{\partial t^2}\right] = 0 \tag{4.5.5}$$

where

$$M_L^2 \equiv M_\infty^2\left[1 + \frac{(\gamma+1)\hat{\phi}_x}{U_\infty}\right], \; M_\infty \equiv U_\infty/a_\infty$$

It may be shown that M_L is the consistent transonic, small perturbation approximation to the local (rather than free stream) Mach number. Hence, the essence of *transonic* small perturbation theory is the allowance for variable, local Mach number rather than simply approximating the local Mach number by M_∞ as in the usual subsonic and supersonic theories.

We digress briefly to show that in (4.5.4) the term

$$\hat{\phi}_{xx}\left[-(\gamma-1)\frac{\partial\hat{\phi}}{\partial t}\right] \tag{4.5.6}$$

may be neglected compared to

$$-2U_\infty \frac{\partial^2 \hat{\phi}}{\partial t\,\partial x} \tag{4.5.7}$$

This is done both for its interest in the present context as well as a prototype for estimation of terms in analyses of this general type.

We assume that a length scale, L, and a time scale, T, may be chosen so that

$$x^* \equiv x/L \quad \text{'is of order one'}$$
$$t^* \equiv t/T \quad \text{'is of order one'}$$

Hence, derivatives with respect to x^* or t^* do *not*, by assumption, change the order or size of a term. Thus (4.5.6) and (4.5.7) may be written (ignoring constants of order one like $\gamma - 1$ and 2) as

$$\frac{\hat{\phi}_{x^*x^*}}{L^2}\frac{\hat{\phi}_{t^*}}{T} \tag{4.5.6}$$

and

$$U_\infty \frac{\hat{\phi}_{t^*x^*}}{TL} \tag{4.5.7}$$

Hence

$$\frac{(A)}{(B)} \sim 0\left[\frac{\hat{\phi}}{U_\infty L}\right]$$

This ratio however, is much less than one by our original assumption of a small perturbation, viz.

$$\phi = U_\infty L x^* + \hat{\phi}$$

In the beginning we have assumed

$$\frac{\hat{\phi}}{U_\infty L x^*} \ll 1$$

Hence (4.5.6) may be neglected compared to (4.5.7).

(4.5.5) is a nonlinear equation even though we have invoked small perturbation ideas. One may develop a linear theory by considering a steady flow due to airfoil shape, $\hat{\phi}_s$, and an infinitesimal time dependent motion of the airfoil superimposed, $\hat{\phi}_d$. For definiteness, one may consider ϕ_s as due to an airfoil of symmetric thickness at zero angle of attack. Thus let

$$\hat{\phi}(x, y, z, t) = \hat{\phi}_s(x, y, z) + \hat{\phi}_d(x, y, z, t) \tag{4.5.6}$$

and substitute into (4.5.5). The equation for ϕ_s is (by definition)

$$\hat{\phi}_{s_{xx}}[1 - M_{L_s}^2] + \hat{\phi}_{s_{yy}} + \hat{\phi}_{s_{zz}} = 0 \tag{4.5.7}$$

where

$$M_{L_s}^2 \equiv M_\infty^2 \frac{[1 + (\gamma + 1)\hat{\phi}_{s_x}]}{U_\infty}$$

The equation for $\hat{\phi}_d$ (neglecting products of $\hat{\phi}_d$ and its derivatives which is acceptable for sufficiently small time dependent motions) is

$$\hat{\phi}_{d_{zz}} + \hat{\phi}_{d_{yy}} - \frac{1}{a_\infty^2}\hat{\phi}_{d_{tt}} - 2\frac{U_\infty}{a_\infty^2}\hat{\phi}_{d_{xt}} - b\hat{\phi}_{d_{xx}} - a\hat{\phi}_{d_x} = 0 \tag{4.5.8}$$

where

$$b \equiv \left[M_\infty^2 - 1 + (\gamma + 1)\frac{\hat{\phi}_{s_x}}{U_\infty}\right]$$

$$a \equiv (\gamma + 1)M_\infty^2\frac{\hat{\phi}_{s_{xx}}}{U_\infty}$$

239

4 Nonsteady aerodynamics

From Bernoulli's equation

$$C_{p_{ms}} \equiv \frac{\hat{p}_s}{\frac{\rho_\infty U_\infty^2}{2}} = -\frac{2\hat{\phi}_{s_x}}{U_\infty}$$

Hence, a and b may be written as

$$b \equiv \left[M_\infty^2 - 1 - \frac{(\gamma+1)M_\infty^2 C_{p_{ms}}(x)}{2} \right]$$

$$a \equiv -(\gamma+1)\frac{M_\infty^2}{2}\frac{dC_{p_{ms}}(x)}{dx}$$

ϕ is velocity potential due to the infinitesimal motion (henceforth $\hat{}$ and d are dropped for simplicity). $C_{p_{ms}}$ is the mean steady pressure coefficient due to airfoil finite thickness and is taken as known. In general, it is a function of x, y, z and the method to be described will, in principle, allow for such dependence. However, all results have been obtained ignoring the dependence on y and z. See Refs. [24], [25] and [33] for discussion of this point.

The (perturbation) pressure, p, is related to ϕ by the Bernoulli relation

$$p = -\rho_\infty \left[\frac{\partial \phi}{\partial t} + U_\infty \frac{\partial \phi}{\partial x} \right]$$

and the boundary conditions are

$$\frac{\partial \phi}{\partial z}\bigg|_{z=0} = w \equiv \frac{\partial f}{\partial t} + U_\infty \frac{\partial f}{\partial x}$$

on airfoil where

$f(x, y, t) \equiv$ vertical displacement of point x, y on airfoil
$w \qquad \equiv$ upwash velocity

and

$$p|_{z=0} = 0 \text{ off airfoil}$$

plus appropriate finiteness or radiation conditions as $z \to \infty$.

Note that equation (4.5.7) is nonlinear in $\hat{\phi}_s$. If one linearizes, as for example in the classical *supersonic* theory, one would set $M_L = M_\infty$ and obtain as a solution to (4.5.7)

$$\hat{p}_s = \frac{\rho_\infty U_\infty^2}{(M_\infty^2 - 1)^{\frac{1}{2}}} \frac{\partial f}{\partial x}$$

240

where $\partial f/\partial x$ is the slope of airfoil shape. As $M_\infty \to 1$, $\hat{p} \to \infty$ which is a unrealistic physical result of the linear theory. On the other hand if one uses

$$M_L = M_\infty \left[1 + (\gamma - 1) \frac{\hat{\phi}_{s_x}}{U_\infty} \right]^{\frac{1}{2}}$$

a finite result is obtained for $\hat{p}_s$ as $M_\infty \to 1$ which is in reasonable agreement with the experimental data.*

Equation (4.5.7) with the full expression for M_L is a nonlinear partial differential equation which is much more difficult to solve than its linear counterpart. However two types of methods have proven valuable, the numerical finite difference methods† and various techniques associated with the name 'local linearization' as pioneered by Oswatitsch and Spreiter [34].

Once $\hat{\phi}_s$ is known (either from theory or experiment) (4.5.8) may be used to determine $\hat{\phi}_d$. (4.5.8) is a linear differential equation with variable coefficients which depend upon $\hat{\phi}_s$. Hence, the solution for the lifting problem, $\hat{\phi}_d$, depends upon the thickness solution, $\hat{\phi}_s$, unlike the classical linear theory where the two may be calculated separately and the results superimposed. Again either finite difference methods or 'local linearization' may be employed to solve (4.5.8). Here we pursue an improved analytical technique to determine $\hat{\phi}_d$, which has been recently developed in the spirit of 'local linearization' ideas [31].

To explain the method most concisely, let $\phi_y = \phi_t = 0$ in equation (4.5.1), i.e., consider two-dimensional, steady flow.

Assume

$$a = \sum_{m=0}^{\infty} a_m (x - x_0)^m$$

$$b = \sum_{n=0}^{\infty} b_n (x - x_0)^n \ddagger$$

and $\phi = \phi^0 + \phi'$ where, by *definition*,

$$\phi_{zz}^0 - b_0 \phi_{xx}^0 - a_0 \phi_x^0 = 0 \qquad (4.5.8a)$$

---

* Spreiter [34].
† Ballhaus, Magnus and Yoshihara [35].
‡ We expand in a *power* series about $x = x_0$; however, other series might be equally or more useful for some applications. Results suggest the details of a and b are unimportant.

and ϕ^0 satisfies any nonhomogeneous boundary conditions on ϕ. The equation for ϕ' is thus from (4.5.8) and using the above

$$\phi'_{zz} - b_0\phi'_{xx} - a_0\phi'_x = \sum_{n=1}^{\infty} b_n(x-x_0)^n[\phi^0_{xx} + \phi'_{xx}]$$

$$+ \sum_{m=1}^{\infty} a_m(x-x_0)^m[\phi^0_x + \phi'_x] \quad (4.5.8b)$$

with homogeneous boundary conditions on ϕ'.

If $\phi' \ll \phi^0$, i.e., ϕ^0 is a good approximation to the solution, then ϕ' may be computed from (4.5.8b) by neglecting ϕ' in the right hand side. The retention of a_0 (but not b_1!) in (4.5.8a) is the key to the method, even though this may seem inconsistent at first.

We begin our discussion with steady airfoil motion in a two-dimensional flow. This is the simplest case from the point of view of computation, of course; however, it is also the most critical in the sense that, as Landahl [33] and others have pointed out, unsteadiness and/or three-dimensionality alleviate the nonlinear transonic effects. Indeed, if the flow is sufficiently unsteady and/or three-dimensional, the classical linear theory gives accurate results transonically.

Steady airfoil motion in two-dimensional, 'supersonic' $(b_0 > 0)$ flow

Solution for ϕ^0. For $b_0 > 0$, x is a time-like variable and the flow is undisturbed ahead of the airfoil (as far as ϕ^0 is concerned). Hence, solutions may be obtained using a Laplace transform with respect to x. Defining

$$\phi^{0*} \equiv \int_0^\infty \phi^0(x, z)e^{-px} \, dx$$

(4.5.8a) becomes

$$\phi^{0*}_{zz} - \mu^2\phi^{0*} = 0 \tag{4.5.9}$$

with

$$\mu^2 \equiv [b_0 p^2 + a_0 p]$$

Solving (4.5.9)

$$\phi^{0*} = A^0_1 e^{-\mu z} + A^0_2 e^{+\mu z} \tag{4.5.10}$$

In order to satisfy finiteness/radiation condition at infinity, one selects $A^0_2 \equiv 0$. A^0_1 is determined from the (transformed) boundary condition,

$$\phi^{0*}_z\big|_{z=0} = w^* \tag{4.5.11}$$

From (4.5.10) and (4.5.11),

$$\phi^0 *|_{z=0} = \frac{-w^*}{\mu}$$ (4.5.12)

Inverting (4.5.12),

$$\phi^0|_{z=0} = -\int_0^x b_0^{-\frac{1}{2}} \exp\left(\frac{-a_0\xi}{2b_0}\right) I_0\left[\frac{a_0\xi}{2b_0}\right] w(x-\xi)\, d\xi$$ (4.5.13)

It is of interest to note two limiting cases. As $a_0\xi/2b_0 \to 0$,

$$\phi^0|_{z=0} = -\int_0^x b_0^{-\frac{1}{2}} w(x-\xi)\, d\xi$$ (4.5.14)

the classical result. But, more importantly, as $a_0\xi/2b_0 \to \infty$,

$$\phi^0|_{z=0} = -\int_0^x (\pi a_0\xi)^{-\frac{1}{2}} w(x-\xi)\, d\xi$$ (4.5.15)

Hence, even when the effective Mach number at $x = x_0$ is transonic, i.e., $b_0 \equiv 0$, the present model gives a finite result. Before computing the correction, ϕ', to the velocity potential we shall exploit ϕ^0 to obtain several interesting results. For this purpose we further restrict ourselves to an airfoil at angle of attack, $w = -U_\infty\alpha$. From (4.5.15),

$$\frac{\phi^0_{z=0}}{U_\infty\alpha} = \frac{2b_0^{\frac{1}{2}}}{a_0} \tilde{x} e^{-x}[I_0(\tilde{x}) + I_1(\tilde{x})]; \qquad \tilde{x} \equiv \frac{a_0 x}{2b_0}$$ (4.5.16)

and the pressure on the *lower* aerodynamic surface is

$$\frac{C_p}{\alpha} \equiv \frac{p^0}{\frac{\rho_\infty U_\infty^2 \alpha}{2}} = \frac{2\phi^0_x}{U_\infty\alpha}\bigg|_{z=0} = 2b_0^{-\frac{1}{2}} e^{-x} I_0(\tilde{x})$$ (4.5.17)

The lift, moment and center of pressure may be computed.

$$L^0 \equiv \int_0^C 2p^0\, dx = \rho_\infty U_\infty^2 \alpha c 4(\pi a_0 c)^{-\frac{1}{2}} \tilde{L}^0$$

$$\tilde{L}^0 \equiv (\pi/2)^{\frac{1}{2}} \tilde{c}^{\frac{1}{2}} e^{-\tilde{c}}[I_0(\tilde{c}) + I_1(\tilde{c})]; \qquad \tilde{c} \equiv \frac{a_0 c}{2b_0}$$ (4.5.18)

$$M^0 = \int_0^c 2p^0 x\, dx = L^0 c - \rho_\infty U_\infty^2 c^2 \tfrac{8}{3}(\pi a_0 c)^{-\frac{1}{2}} \tilde{M}^0$$

$$\tilde{M}_0 \equiv \tfrac{3}{4}(2\pi)^{\frac{1}{2}}\{e^{-\tilde{c}} I_1(\tilde{c})[\tilde{c}^{-\frac{1}{2}} + \tfrac{2}{3}\tilde{c}^{\frac{1}{2}}] + \tfrac{2}{3} e^{-\tilde{c}} I_2(\tilde{c}) \tilde{c}^{\frac{1}{2}}\}$$ (4.5.19)

The center of pressure may be obtained from L^0 and M^0 in the usual way. We shall use and discuss these results for a particular airfoil later. But first let us consider the computation of ϕ'.

Solution for ϕ'. For *simplicity*, we shall consider only a *linear* variation in mean pressure, $C_{p_{ms}}$, along the airfoil chord. Hence, a_0, b_0 and b_1 are not zero and $b_1 = a_0$. All other a_m and b_n are zero. Assuming $\phi' \ll \phi^0$, the equation for ϕ' is

$$\phi'_{zz} - a_0 \phi'_x - b_0 \phi'_{xx} = b_1 (x - x_0) \phi^0_{xx} \tag{4.5.20}$$

Taking a Laplace transform of (4.5.20),

$$\phi'^*_{zz} - \mu^2 \phi'^* = -b_1 \left[2p\phi^{0*} + p^2 \frac{d\phi^{0*}}{dp} + x_0 p^2 \phi^{0*} \right] \tag{4.5.21}$$

A particular solution of (4.5.21) is

$$\phi'^*_p = (C_0 z + C_1 z^2) e^{-\mu z} \tag{4.5.22}$$

where

$$C_0 \equiv b_1 \left[\frac{A}{2\mu} + \frac{B}{4\mu^2} \right]; \qquad C_1 \equiv \frac{b_1}{4\mu} B$$

$$A \equiv \frac{-2pw^*}{\mu} + \frac{p^2 w^*}{\mu^3} \frac{[2b_0 p + a_0]}{2} - x_0 \frac{p^2 w^*}{\mu} - \frac{p^2}{\mu} \frac{dw^*}{dp}$$

$$B \equiv \frac{p^2 w^*}{\mu^2} \frac{[2b_0 p + a_0]}{2}$$

The homogeneous solution for ϕ' is of the same form as for ϕ^0. After some calculation, applying homogeneous boundary conditions to ϕ', we determine

$$\phi'^* \big|_{z=0} = \frac{C_0}{\mu} \tag{4.5.23}$$

Inverting (4.5.23) using the definitions of C_0, A, B above, and assuming $w = -U_\infty \alpha$ for simplicity, we have

$$\frac{\phi'}{U_\infty \alpha} \bigg|_{z=0} = \frac{b_0^{\frac{1}{2}}}{a_0} \left\{ 2e^{-\tilde{x}} \tilde{x} I_1(\tilde{x}) - \left[\frac{d^2}{d\tilde{x}^2} + \frac{d}{d\tilde{x}} \right] [e^{-\tilde{x}} \tilde{x}^2 I_2(\tilde{x})] \right.$$

$$\left. + \frac{\tilde{c} 2 x_0}{c} e^{-\tilde{x}} \tilde{x} [I_0(\tilde{x}) + I_1(\tilde{x})] \right\}; \qquad \tilde{c} \equiv \frac{a_0 c}{2b_0} \tag{4.5.24}$$

The pressure coefficient corresponding to ϕ' is given by

$$C_p' = C_{p_1}' + C_{p_2}'$$

where

$$\frac{b_0^{\frac{1}{2}} C_{p_1}'}{\alpha} \equiv e^{-\tilde{x}}\{(2I_1 - I_0)(\tilde{x} - \tilde{x}^2) + I_2 \tilde{x}^2\} \qquad (4.5.25)$$

$$\frac{b_0^{\frac{1}{2}} C_{p_2}'}{\alpha} \equiv \bar{c} \frac{2x_0}{c} e^{-\tilde{x}}\{2\tilde{x}(I_1 - I_0) + I_0\}$$

As may be seen C_{p_1}' is always a small correction to C_p^0; however, C_{p_2}' may be large or small (particularly near the leading edge as $\tilde{x} \to 0$) depending on the size of

$$\frac{x_0}{c} \frac{a_0 c}{2b_0}$$

Since we are free to choose x_0 in any application, it is in our interest to choose it so that

$$C_{p_2}' \ll C_p^0$$

More will be said of this in the following section.

We note that *higher* terms in the power series for a and b may be included and a solution for ϕ' obtained in a similar manner. The algebra becomes more tedious, of course.

Results and comparisons with other theoretical and experimental data

We have calculated two examples, a Guderley airfoil and a parabolic arc airfoil, both of 6% thickness ratio, τ, and for Mach numbers near one. These were chosen because they have smooth mean steady pressure distributions (at least for some Mach number range) and because other investigators have obtained results for these airfoils. These two airfoils and their mean, steady pressure distributions are shown in [24]. The Guderley airfoil has a linear mean pressure variation while the parabolic arc has a somewhat more complicated variation including a (theoretical) logarithmic singularity at the leading edge. For $M_\infty = 1$, when $C_{p_{ms}} = 0$ the local Mach number along the chord equals one and if one expanded about that point then $b_0 \equiv 0$, and our procedure would fail in that $\phi' \gg \phi^0$. Hence, one is led to believe that one should choose x_0 as far away from the sonic point, $C_{p_{ms}} \equiv 0$ at $M_\infty \equiv 1$, as possible. To fix this idea more concretely, we first consider the Guderley airfoil.

245

4 Nonsteady aerodynamics

Guderley airfoil. We have calculated C_p^0 and $C_p^0 + C_p'$ for $M_\infty = 1$. Two different choices of x_0 were used, $x_0 = c/2$ (Figure 4.18) and c (Figure 4.19). Results from Stahara and Spreiter [24] are also shown for reference. As can be seen for $x_0 = c/2$, the 'correction' term, C_p', dominates the basic solution, C_p^0, as $x/c \to 0$. For $x_0 = c$, on the other hand, the correction term is much better behaved, in agreement with our earlier speculation about choosing x_0 as far as possible from the sonic point. Note that if, for example, we choose $x_0 = 0$ this would also work in principle, but now $b_0 < 0$, and a 'subsonic' solution would have to be obtained for ϕ^0.

Parabolic arc airfoil. Similar results have been obtained and are displayed in Figure 4.20 ($x_0 = c/2$) and Figure 4.21 ($x_0 = c_0$). Both of these solutions are well behaved in the sense that $C_p' < C_p^0$, though again the results for $x_0 = c$ appear to be better than those for $x_0 = c/2$. The relatively better behavior of the $x_0 = c/2$ results for the parabolic arc as

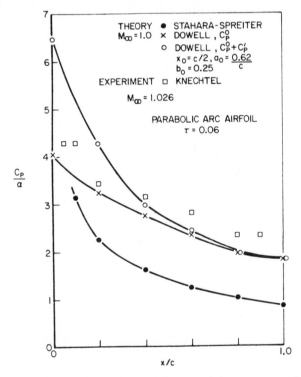

Figure 4.20   Pressure distribution for parabolic arc airfoil at constant angle of attack.

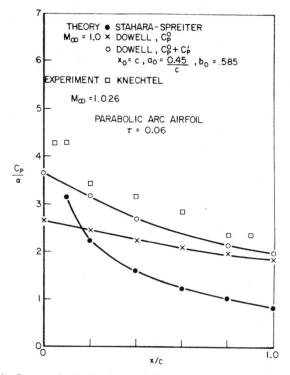

Figure 4.21   Pressure distribution for parabolic arc airfoil at constant angle of attack.

compared with the Guderley airfoil is probably related to the sonic point being further ahead of $x_0 = c/2$ for the former than the latter. See [24]. Also shown in Figures 4.20 and 4.21 are the theoretical results of Stahara–Spreiter [24] and the experimental data of Knechtel [36]. Knechtel indicates the effective Mach number of his experiments should be reduced by approximately 0.03 due to wall interference effects. Also he shows that the measured mean steady pressure distributions at zero angle of attack, $C_{p_{ms}}$, agree well with the theoretical results of Spreiter [24, 37] for $M_\infty \geq 1$. However, for $M_\infty \leq 1$, $C_{p_{ms}}$ deviates from that theoretically predicted; see Figure 4.22 taken from [36]. The change in slope for $C_{p_{ms}}$ near the trailing edge may be expected to be important for computing the lifting case. In Figure 4.23 results are shown for $M_\infty = 0.9$ which dramatically make this point. Shock induced separation of the boundary layer is the probable cause of the difficulty.

Finally, we present a graphical summary of lift curve slope and center

247

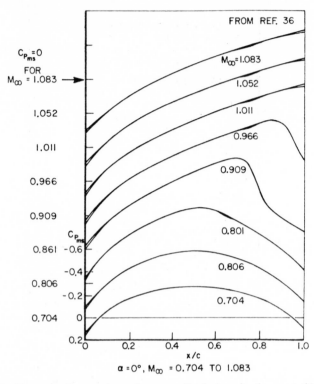

Figure 4.22 Representative experimental pressure distributions for 6-percent-thick circular-arc airfoil with roughness elements near the leading edge.

of pressure for the parabolic arc airfoil comparing results of Knechtel's experimental data and the present analysis. See Figure 4.24.

All things considered the agreement between theory and experiment is rather good; however, it is clear that if $C_{p_{ms}}$ varies in a complicated way one must go beyond the straight line approximation used in obtaining the present results. In principle this can be done; how much effort will be required remains to be determined.

Nonsteady airfoil motion in two-dimensional, 'supersonic' $b_0 > 0$ flow

Solution for ϕ^0. Again taking a Laplace transform with respect to x of (4.5.8) (for $\phi_{yy} \equiv 0$ and $a = a_0$, $b = b_0$) we obtain

$$\phi_{zz}^{0*} - \mu^2 \phi^{0*} = 0 \tag{4.5.26}$$

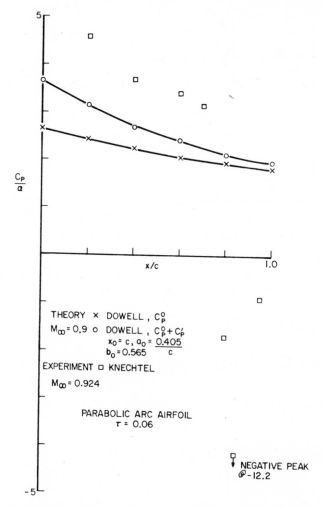

Figure 4.23 Pressure distribution for parabolic arc airfoil at constant angle of attack.

where $\mu \equiv [b_0 p^2 + \tilde{a}_0 p - d]^{\frac{1}{2}}$; b_0—as before

$$\tilde{a}_0 \equiv a_0 + \frac{2U_\infty}{a_\infty^2} i\omega; \qquad d \equiv \left(\frac{\omega}{a_\infty}\right)^2$$

and we have assumed simple harmonic motion in time. Solving (4.5.26) subject to the boundary condition, (4.5.11), and appropriate finiteness

249

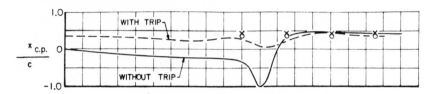

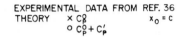

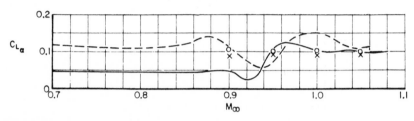

Figure 4.24   Effect of boundary-layer trip on the variation with Mach number of lift-curve slope and center of pressure of the circular-arc airfoil at $\alpha_0 \cong 0°$.

and/or radiation condition at infinity we have (after inversion)

$$\phi^0\big|_{z=0} = -\int_0^x b_0^{-\frac{1}{2}} \exp\left(\frac{-\bar{a}_0 \xi}{2b_0}\right) I_0\left\{\left[\left(\frac{\bar{a}_0}{2b_0}\right)^2 + \frac{d}{b_0}\right]^{\frac{1}{2}} \xi\right\} w(x-\xi)\,d\xi \quad (4.5.27)$$

The perturbation pressure on the lower surface is given by

$$p^0 = \rho_\infty[\phi_t^0 + U_\infty \phi_x^0] \quad (4.5.28)$$

which may be evaluated from (4.5.27) directly using Leibnitz' rule

$$C_p^0 = \frac{p^0}{\dfrac{\rho_\infty U_\infty^2}{2}} = -2b_0^{-\frac{1}{2}}\left\{\exp\left(\frac{-\bar{a}_0 x}{2b_0}\right) I_0\left[\left(\frac{\bar{a}_0}{2b_0}\right)^2 + \frac{d}{b}\right]^{\frac{1}{2}} x\right\} \frac{w(0)}{U_\infty}$$

$$+ \int_0^x \exp\left(\frac{-\bar{a}_0 \xi}{2b_0}\right) I_0\left\{\left[\left(\frac{\bar{a}_0}{2b_0}\right)^2 + \frac{d}{b_0}\right]^{\frac{1}{2}} \xi\right\}$$

$$\cdot \left[i\omega \frac{w(x-\xi)}{U_\infty^2} + \frac{w'(x-\xi)}{U_\infty}\right] d\xi$$

where

$$w'(x) \equiv \frac{dw}{dx} \quad (4.5.29)$$

250

An alternative form for C_p^0 may be obtained by first interchanging the arguments x and $x - \xi$ in (4.5.27). For $a_0 \equiv 0$, $b_0 \equiv M_\infty^2 - 1$ the above reduces to the classical result. For any a_0 and b_0 and $k \equiv \omega c/U_\infty$ large the results approach those of the classical theory and for $k \to \infty$ approach the 'piston' theory [32]. For the specific case of an airfoil undergoing vertical translation, $w = -h_t$, where h is vertical displacement and h_t is the corresponding velocity, we have the following results,

$$\phi^0\big|_{z=0} = h_t b_0^{-\frac{1}{2}} \left[\left(\frac{\tilde{a}_0}{2b_0}\right)^2 + \frac{\mathrm{d}}{b_0} \right]^{-\frac{1}{2}} e^{-e\tilde{x}} \left\{ I_0(\tilde{x}) + \frac{I_1(\tilde{x})}{e} \right\}$$

where

$$\tilde{x} \equiv \left[\left(\frac{\tilde{a}_0}{2b_0}\right)^2 + \frac{\mathrm{d}}{b_0} \right]^{\frac{1}{2}} x$$

$$e \equiv \frac{\tilde{a}_0}{2b_0} \left[\left(\frac{\tilde{a}_0}{2b_0}\right)^2 + \frac{\mathrm{d}}{b_0} \right]^{-\frac{1}{2}} \qquad (4.5.30)$$

In the limit as $b_0 \to 0$, (corresponding to $M_\infty \to 1$ in the classical theory)

$$\left[\left(\frac{\tilde{a}_0}{2b_0}\right)^2 + \frac{\mathrm{d}}{b_0} \right]^{\frac{1}{2}} \to \frac{\tilde{a}_0}{2b_0}; \qquad e \to 1$$

and

$$\phi^0\big|_{z=0} \to h_t 2 \left(\frac{x}{\tilde{a}_0 \pi}\right)^{\frac{1}{2}} \qquad (4.5.31)$$

Using (4.5.30) or (4.5.31) in (4.5.28) gives the perturbation pressure. The latter form is particularly simple

$$\frac{C_p}{ik\dfrac{\bar{h}}{c} e^{i\omega t}} \equiv \frac{\dfrac{p}{\rho_\infty U_\infty^2}}{ik\dfrac{\bar{h}}{c} e^{i\omega t}} = (\pi \tilde{a}_0 c)^{-\frac{1}{2}} \left[2(x/c)^{-\frac{1}{2}} + i4k\left(\frac{x}{c}\right)^{\frac{1}{2}} \right] \qquad (4.5.32)$$

where

$$h \equiv \bar{h} e^{i\omega t}; \qquad k \equiv \frac{\omega c}{U_\infty}$$

Solution for ϕ'.   Park [38] has computed ϕ' and made comparisons with available experimental and theoretical data. It is well-known, of course,

251

that for sufficiently large k the classical theory itself is accurate transonically [33]. Hence, we also expect the present theory to be more accurate for increasing k.

Results and comparison with other theoretical data

We have calculated a numerical example for the Guderley airfoil for $M = 1$ and $k = 0.5$ in order to compare with the results of Stahara–Spreiter [24]. We have chosen $x_0 = c/2$ for which

$$b_0 = 0.12; \qquad a_0 = 1.2/c$$

For such small b_0, we may use the asymptotic form for $b_0 \to 0$, (4.5.32), and the results are plotted in Figures 4.25 and 4.26 along with the results of

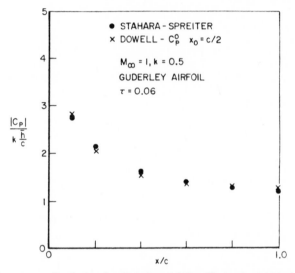

Figure 4.25   Pressure distribution for Guderley airfoil oscillating in rigid body translation.

[24]. As $k \to 0$, the phase angle, Φ, is a constant at 90° and the pressure coefficient amplitude is the same as that of Figure 4.18. Presumably somewhat more accurate results could be obtained by choosing $x_0 = c$ and computing the correction, C'_p. However, the agreement is already good between the present results and those of [24].

As Stahara–Spreiter [24] point out even for k as large as unity there are still substantial quantitative differences between their results (and hence the present results) and those of the classical theory. However, for

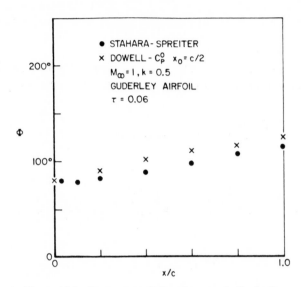

Figure 4.26   Pressure-translation phase angle distribution.

$k \gg 1$, one may expect the present theory and that of [24] to give results which approach those of the classical theory.

Nonsteady airfoil motion in three-dimensional 'supersonic' $(b_0 > 0)$ flow

Solution for ϕ^0. We begin with (4.5.1) and take a Fourier transform with respect to y,

$$\phi^\dagger \equiv \int_{-\infty}^{\infty} \phi e^{-i\gamma y} \, dy \tag{4.5.33}$$

and a Laplace transform with respect to x,

$$(\phi^\dagger)^* \equiv \int_{0}^{\infty} (\phi^\dagger) e^{-px} \, dx \tag{4.5.34}$$

(4.5.1) becomes

$$\phi_{zz}^{0*\dagger} - \mu^2 \phi^{0*\dagger} = 0 \tag{4.5.35}$$

where

$$\mu \equiv [b_0 p^2 + \tilde{a}_0 p - \tilde{d}]^{\frac{1}{2}}; \qquad b_0, \tilde{a}_0 \quad \text{as before}$$

253

and

$$\tilde{d} \equiv (\omega/a_\infty)^2 - \gamma^2$$

Solving (4.5.35) subject to the boundary condition, (4.5.11), and appropriate boundary finiteness/radiation conditions at infinity we have (after inversion)

$$\phi^0\big|_{z=0} = \int_0^x \int_{-\infty}^\infty A(x-\xi, y-\eta) w(\xi, \eta) \, d\xi \, d\eta \tag{4.5.36}$$

where

$$A(x, y) = \frac{-\exp\left(\dfrac{-\tilde{a}_0}{2b_0} x\right)}{\pi} r^{-1} \cosh\left\{ \left[\left(\frac{\omega}{a_\infty}\right)^2 + \left(\frac{\tilde{a}_0}{2b_0}\right)^2 \right]^{\frac{1}{2}} r \right\}$$

$$\text{for} \quad r^2 > 0, \quad \text{i.e.}$$
$$0 < |y| < x b^{-\frac{1}{2}}$$
$$= 0 \qquad \text{for} \quad r^2 < 0, \quad \text{i.e.}$$
$$x b_0^{-\frac{1}{2}} < |y|$$

and

$$r^2 \equiv x^2 - b_0^2 y^2 \tag{4.5.37}$$

A is the aerodynamic Green's function required in the Mach Box numerical lifting surface method [32].

For $b_0 \to M_\infty^2 - 1$; $a_0 \to 0$; $\tilde{a}_0 \to 2(i\omega U_\infty/a_\infty^2)$ and A reduces to the classical result. For $b_0 \to 0$, $R_e \tilde{a}_0 > 0$,

$$A \to -\frac{1}{2\pi x} e^{-\tilde{a}_0 y^2/4x} \quad \text{for} \quad x > 0; \quad |y| < \infty \tag{4.5.38}$$

For $b_0 \to 0$, $R_e \tilde{a}_0 < 0$,

$$A \to -\frac{1}{2\pi x} \exp\left(\frac{-\tilde{a}_0 x}{b_0}\right) \quad \text{for} \quad x > 0; \quad |y| < \infty \tag{4.5.39}$$

Non-steady airfoil motion in three-dimensional 'subsonic' ($b_0 < 0$) flow

Solution for ϕ^0. We begin with (4.5.1), assuming simple harmonic motion,

$$-b_0\phi_{xx}^0 - \tilde{a}_0\phi_x^0 + d\phi^0 + \phi_{yy}^0 + \phi_{zz}^0 = 0 \tag{4.5.40}$$

where $\tilde{a}_0$, b_0, d as before.

To put (4.5.40) in canonical form by eliminating the term ϕ_x, we introduce the new dependent variable, Φ

$$\phi^0 \equiv e^{\Omega x} \Phi \qquad (4.5.41)$$

where Ω is determined to be

$$\Omega = -\bar{a}_0/2b_0 \qquad (4.5.42)$$

and the equation for Φ is

$$B\Phi_{xx} + \Phi\left[\frac{-a^2}{4B} + d\right] + \Phi_{yy} + \Phi_{zz} = 0 \qquad (4.5.43)$$

and

$$B \equiv -b_0 > 0$$

We further define new independent variables,

$$x' \equiv x, \qquad y' \equiv B^{\frac{1}{2}}y, \qquad z' \equiv B^{\frac{1}{2}}z \qquad (4.5.44)$$

then (4.5.43) becomes

$$\Phi_{x'x'} + \Phi_{y'y'} + \Phi_{z'z'} + \tilde{k}^2\Phi = 0 \qquad (4.5.45)$$

where

$$\tilde{k}^2 \equiv \left(d - \frac{a^2}{4B}\right)\bigg/B$$

We are now in a position to use Green's theorem

$$\iiint [\Phi\nabla^2\psi - \psi\nabla^2\Phi]\,dV = \iint_S \left[\Phi\frac{\partial\psi}{\partial n} - \psi\frac{\partial\Phi}{\partial n}\right]dS \qquad (4.5.46)$$

V volume enclosing fluid
S surface area of volume indented to pass over airfoil surface and wake
$\overset{\cdot}{n}$ outward normal.

We take Φ to be the solution we seek and choose ψ as

$$\psi \equiv \left(\frac{e^{-i\tilde{k}r}}{r}\right) \qquad (4.5.47)$$

where

$$r \equiv [(x' - x_1')^2 + (y' - y_1')^2 + (z' - z_1')^2]$$

255

Note that

$$[\nabla^2 + \tilde{k}^2]\left(\frac{e^{-ikr}}{r}\right) = -4\pi\delta(x'-x_1')\delta(y'-y_1')\delta(z'-z_1') \qquad (4.5.48)$$

Thus the LHS of (4.5.7) becomes $-4\pi\Phi(x', y', z')$. On the RHS, there is no contribution from the surface area of sphere at infinity. Thus (4.5.46) becomes

$$4\pi\Phi(x, y, z) = \iint_{\substack{S \\ \text{airfoil} \\ \text{plus wake}}} \left[(\Phi_U - \Phi_L)\frac{\partial}{\partial z_1}\left(\frac{e^{-ikr}}{r}\right)\right.$$

$$\left. - \left(\frac{e^{-ikr}}{r}\right)\frac{\partial}{\partial z_1}(\Phi_U - \Phi_L)\right] dx_1\, dy_1 \qquad (4.5.49)$$

where

Φ_U, Φ_L upper, lower surface

$$\frac{\partial}{\partial n} = \frac{-\partial}{+\partial z_1} \quad \text{on} \quad \begin{array}{l}\text{upper} \\ \text{lower}\end{array} \text{ surface}$$

and we have returned to the original independent variables, x, y, z and x_1, y_1, z_1. Since Φ is an odd function of z, z_1,

$$\frac{\partial}{\partial z_1}(\Phi_U - \Phi_L) = 0 \qquad (4.5.50)$$

Also

$$\frac{\partial}{\partial z_1}\left(\frac{e^{-ikr}}{r}\right) = \frac{\partial}{\partial z}\frac{e^{-ikr}}{r}(-1) \qquad (4.5.51)$$

Thus (4.5.49) becomes, re-introducing the original dependent variable, ϕ^0,

$$\phi^0(x, y, z) = \frac{-e^{-\Omega x}}{4\pi}\iint \Delta\phi\, e^{-\Omega x_1}\frac{\partial}{\partial z}\left\{\frac{e^{-ikr}}{r}\right\} dx_1\, dy_1 \qquad (4.5.52)$$

where

$$\Delta\phi \equiv \phi_U^0 - \phi_L^0$$

Up to this point we have implicitly identified ϕ^0 with the velocity potential. However, *within the approximation*, $a = a_0$, $b = b_0$, $\phi = \phi^0$, $p = p^0$, ϕ and p satisfy the same equation, (4.5.40); hence, we may use

(4.5.54) with ϕ^0 replaced by p^0. Further using Bernoulli's equation, (4.5.5), we may relate ϕ^0 to p^0

$$\phi^0(x, y, z) = -\int_{-\infty}^{x} \frac{p^0(\lambda, y, z)}{\rho_\infty U_\infty} \exp\left[\frac{i\omega(\lambda - x)}{U_\infty}\right] d\lambda \qquad (4.5.53)$$

Substituting (4.5.52) into (4.5.53) (where (4.5.52) is now expressed in terms of p^0); introducing a new variable $\xi \equiv \lambda - x_1$ and interchanging the order of integration with respect to ξ and x_1, y_1; gives

$$\phi(x, y, z) = \frac{1}{4\pi} \iint \frac{\Delta p}{\rho_\infty U_\infty}(x_1, y_1) \exp\left[\frac{-i\omega(x - x_1)}{U_\infty}\right]$$

$$\cdot \left\{\int_{-\infty}^{x-x_1} \exp\left(\frac{[\Omega + i\omega]\xi}{U_\infty}\right) \frac{\partial}{\partial z}\left\{\frac{e^{-i\bar{k}r}}{r}\right\} d\xi\right\} dx_1 \, dy_1 \qquad (4.5.54)$$

Finally, computing from (4.5.56)

$$w = \frac{\partial \phi}{\partial z}\bigg|_{z=0}$$

we obtain

$$\frac{w(x, y)}{U_\infty} = \iint \frac{\Delta p}{\rho_\infty U_\infty^2}(x_1, y_1) K(x - x_1, y - y_1) \, dx_1 \, dy_1 \qquad (4.5.55)$$

where

$$K \equiv \lim_{z \to 0} \frac{\exp\left[\dfrac{-i\omega(x - x_1)}{U_\infty}\right]}{4\pi} \int_{-\infty}^{x-x_1} \exp\left(\left[\Omega + \frac{i\omega}{U_\infty}\right]\xi\right) \frac{\partial^2}{\partial z^2}\left\{\frac{e^{i\bar{k}r}}{r}\right\} d\xi$$

$$(4.5.56)$$

and

$$r^2 \equiv [\xi^2 + B(y - y_1)^2]$$

The above derivation, though lengthy, is entirely analogous to the classical one. For $a_0 \to 0$, $B \to 1 - M_\infty^2$ we retrieve the known result [32].

It should be noted that in the above derivation we have assumed $\mathrm{Re}\, \tilde{a}_0 > 0$ and thus $\mathrm{Re}\, \Omega < 0$. This permits both the radiation and finiteness conditions to be satisfied as $z \to \pm\infty$. For $\mathrm{Re}\, \tilde{a}_0 < 0$ one may not satisfy both conditions and one must choose between them.

Asymmetric mean flow. In the above derivations we have assumed a mean flow about symmetrical airfoils at zero angle of attack and considered small motions of that configuration. It is of interest to generalize

257

this to a mean flow about asymmetrical airfoils at nonzero angles of attack. First consider the Mach box form of the integral relation between velocity potential and downwash, cf. equation (4.5.36),

$$\phi_U = \int\int A_U(x-\xi, y-\eta) w_U(\xi, \eta) \, d\xi \, d\eta \qquad (4.5.57)$$

Here we have written the relation as though we knew w_U everywhere on $z = 0^+$. We do not, of course, and thus the need for the Mach Box procedure [32]. Here A_U is that calculated using upper surface parameters, ignoring the lower surface. A similar relation applies for the lower surface with A_U replaced by $-A_L$. Hence, we may compute from (4.5.57) (for lifting motion where $w_U = w_L \equiv w$ on and off the airfoil)

$$\phi_U - \phi_L = \int\int A(x-\xi, y-\eta) w(\xi, \eta) \, d\xi \, d\eta \qquad (4.5.58)$$

where

$$A \equiv A_U + A_L$$

is the desired aerodynamic influence function. Note that A_U and A_L are the same basic function, but in one the upper surface parameters are used and in the other the lower surface parameters.

Using the Kernel Function approach the situation is somewhat more complicated. Here we have, cf. equation (4.5.55),

$$w_U = \int\int K_U(x-\xi, y-\eta) p_U(\xi, \eta) \, d\xi \, d\eta \qquad (4.5.59)$$

Note $K_U = 2K_{\Delta p}$ where $K_{\Delta p}$ is the Kernel Function for Δp when the lower surface mean flow parameters are the same as those of the upper surface.

A similar equation may be written for w_L and p_L with K_U replaced by $-K_L$. Again we note $w_L = w_U \equiv w$. These two integral equations must be solved simultaneously for p_U and p_L with given w. Hence, the number of unknowns one must deal with is doubled for different upper and lower surface parameters. This poses a substantial additional burden on the numerics.

There is a possible simplification, however. Define

$$K \equiv \frac{K_U + K_L}{2}; \qquad \Delta K \equiv \frac{K_U - K_L}{2} \qquad (4.5.60)$$

If $(\Delta K/K)^2 \ll 1$, then one may simply use K, i.e., the average of the upper and lower surface kernel functions. Formally, one may demonstrate this using perturbation ideas as follows.

Using (4.5.59) (and its counterpart for the lower surface) and (4.5.60) one may compute

$$w_U + w_L \equiv 2w = \iint [K(p_U - p_L) + \Delta K(p_U + p_L)]\, d\xi\, d\eta$$

and

$$w_U - w_L \equiv 0 = \iint [K(p_U + p_L) + \Delta K(p_U - p_L)]\, d\xi\, d\eta \qquad (4.5.61)$$

From the second of these equations, the size of the terms may be estimated.

$$\frac{p_U + p_L}{p_U - p_L} \sim 0\!\left(\frac{\Delta K}{K}\right)$$

Thus in the first of (4.5.61) the two terms on the right hand side are of order

$$K(p_U - p_L) \quad \text{and} \quad \frac{(\Delta K)^2}{K}(p_U - p_L)$$

The second of these terms may be neglected if

$$(\Delta K/K)^2 \ll 1 \qquad (4.5.62)$$

and (4.5.61) may be approximated as

$$w(x, y) \approx \iint \frac{K}{2}(x - \xi, y - \eta)\Delta p(\xi, \eta)\, d\xi\, d\eta \qquad (4.5.63)$$

where

$$\Delta p \equiv p_U - p_L$$

(4.5.62) would not appear to be an unduly restrictive condition for some applications.

The development in this section is *not* dependent upon the particular method used to compute K_U and/or K_L elsewhere in the text. The crucial assumptions are that (1) the oscillating motion is a small perturbation to the mean flow and (2) the difference between the upper and lower surface Kernel functions is small compared to either.

Concluding remarks

A relatively simple, reasonably accurate and systematic procedure has been developed for transonic flow. A measure of the simplicity of the

259

method is that all numerical results presented herein were computed by hand and analytical forms have been obtained for general 'supersonic' Mach number and airfoil motion for two-dimensional flow. For three-dimensional flow the relevant Green's functions have been determined which may be used in the Kernel Function and Mach Box numerical lifting surface methods.

This approach has recently been extended to include a more accurate form of Bernoulli's equation and airfoil boundary condition. Also numerical examples are now available for two dimensional airfoils in transient motion and three dimensional steady flow over a delta wing. Finally a simple correction for shock induced flow separation has been suggested.*

For a recent, highly readable survey of transonic flow, the reader should consult the paper, by Spreiter and Stahara [40].

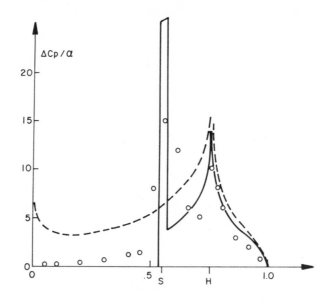

IN PHASE PRESSURE
o o o TIJDEMAN, M_∞ = .875, k = .12
– – – CLASSICAL LINEAR THEORY
——— WILLIAMS
NACA 64A006
α = 1°

Figure 4.27a

---

* Dowell [39].

Also important recent advances in finite differences and finite element solutions are discussed in the following papers (all presented at the AIAA Dynamic Specialists Conference, San Diego, March 1977): Chan and Chen [41], Ballhaus and Goorjian [42] and Isogai [43].

In an important, but somewhat, neglected paper Eckhaus [44] gave a transonic flow model including shock waves which considered a constant supersonic Mach number ahead of the shock and a constant subsonic Mach number behind it. An obvious next step is to combine the Eckhaus and Dowell models. M. H. Williams [45], in work recently published, has extended Eckhaus' results by utilizing a somewhat broader theoretical formulation and obtaining more accurate and extensive solutions. He has compared his results to those of Tijdeman and Schippers [46] (experiment) and Ballhaus and Goorjian [42] (finite difference solutions) and obtained good agreement. The comparison with experiment is shown here in Figure 4.27 for a NACA 64 A006 airfoil with a trailing edge quarter chord oscillating flap. The measured steady state shock strength and location for no flap oscillation is used as an input to the theoretical model. Since the flap is downstream of the shock, the theory predicts no

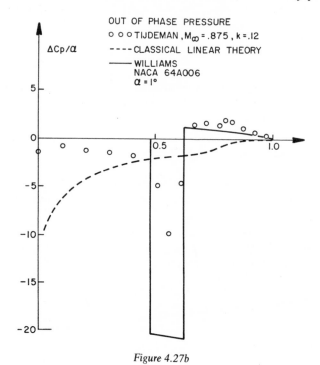

Figure 4.27b

4 Nonsteady aerodynamics

disturbance upstream of the shock. The experiment shows the upstream effect is indeed small. Moreover the agreement on the pressure peaks at the shock and at the flap hinge line is most encouraging. It would appear that the transonic airfoil problem is finally yielding to a combination of analytical and numerical methods. As Tijdeman and others have emphasized, however, the effects of the viscous boundary layer may prove significant for some applications. In particular the poorer agreement between theory and experiment for the imaginary pressure peak at the shock in Figure 4.27 is probably due to the effects of viscosity. The same theoretical model has also been studied by Goldstein, et al. for cascades with very interesting results [47]. Rowe, a major contributor to subsonic aerodynamic solution methods, has in the same spirit discussed how the classical boundary conditions and Bernoulli's equation can be modified to partially account for transonic effects as the airfoil critical mach number is approached [48].

For a recent, broad-ranging survey of unsteady fluid dynamics including a discussion of linear potential theory, transonic flow, unsteady boundary layers, unsteady stall, vortex shedding and the Kutta–Joukowski trailing edge condition the paper by McCroskey [49] is recommended.

Nomenclature

A	aerodynamic influence function; see equation (4.5.36)
a, b	see definitions following equation (4.5.1)
a_m, b_n	see equation (4.5.5)
$\tilde{a}_0$	see equation (4.5.26)
a_∞	free stream speed of sound
B	$\equiv -b_0$
C_p	$\equiv 2(p - p_\infty)/\rho_\infty U_\infty^2$; pressure coefficient due to airfoil motion
$C_{p_{ms}}$	mean steady pressure coefficient due to airfoil finite thickness at zero angle of attack
C_{L_α}	lift curve slope per degree
c	airfoil chord
d	see equation (4.5.26)
$\tilde{d}$	see equation (4.5.35)
e	see equation (4.5.30)
f	vertical airfoil displacement
h	rigid body translation of airfoil
Im	imaginary part
i	$\equiv (-1)^{\frac{1}{2}}$

K	aerodynamic kernel function; see equation (4.5.55)
k	$\equiv \dfrac{\omega c}{U_\infty}$
$\tilde{k}$	see equation (4.5.45)
L	lift
M	pitching moment about leading edge
M_∞	free stream Mach number
p	perturbation pressure; also Laplace Transform variable
Re	real part
r	see equations (4.5.37), (4.5.47) and (4.5.56)
t	time
U_∞	free stream velocity
w	downwash; see equation (4.5.4)
x, y, z	spatial coordinates
x', x', z'	see equation (4.5.44)
$\tilde{x}$	$\equiv \dfrac{a_0 x}{2b_0}$
$x_{c.p.}$	center of pressure; measured from leading edge
x_0	see equation (4.5.5)
α	angle of attack
γ	ratio of specific heats; also Fourier transform variable
Φ	see equation (4.5.41)
ϕ	velocity potential
ψ	see equation (4.5.47)
ρ_∞	free stream density
Ω	see equation (4.5.42)
ω	frequency of airfoil oscillation
ξ, λ, η	dummy integration variables for x, x, y

Superscripts

0	basic solution
'	correction to basic solution
*	Laplace Transform
†	Fourier Transform

Subscripts

U, L	upper, lower surfaces

References for Chapter 4

[1] Liepmann, H. W. and Roshko, A., *Elements of Gasdynamics*, John Wiley, 1957.

[2] Hildebrand, F. B., *Advanced Calculus for Engineers*, Prentice-Hall, Inc., 1961.

[3] van der Vooren, A. I., *Two Dimensional Linearized Theory*, Vol. II, Chapter 2, AGARD Manual on Aeroelasticity.

[4] Lomax, H., Heaslet, M. A., Fuller, F. B. and Sluder, L., 'Two- and Three-dimensional Unsteady Lift Problems in High Speed Flight', *NACA Report 1077*, 1952.

[5] Landahl, M. T. and Stark, V. J. E., 'Numerical Lifting Surface Theory—Problems and Progress', *AIAA Journal* (November 1968) pp. 2049–2060.

[6] Watkins, C. E., *Three Dimensional Supersonic Theory*, Vol. II, Chapter 5, AGARD Manual on Aeroelasticity.

[7] Bateman, H., *Table of Integral Transforms*, McGraw-Hill, 1954.

[8] Many Authors, *Oslo AGARD Symposium Unsteady Aerodynamics for Aeroelastic Analyses of Interfering Surfaces*, Tønsberg, Oslofjorden, Norway (Nov. 1970).

[9] Landahl, M. T. and Ashley, H., *Thickness and Boundary Layer Effects*, Vol. II, Chapter 9, AGARD Manual on Aeroelasticity.

[10] Williams, D. E., *Three-Dimensional Subsonic Theory*, Vol. II, Chapter 3, AGARD Manual on Aeroelasticity.

[11] Albano, E. and Rodden, W. P., 'A Doublet-Lattice Method for Calculating Lift Distributions on Oscillating Surfaces in Subsonic Flows', *AIAA J.* (February 1969) pp. 279–285.

[12] Stratton, J. A., *Electromagnetic Theory*, McGraw-Hill, 1941.

[13] Watkins, C. E., Woolston, D. S. and Cunningham, J. J., 'A Systematic Kernel Function Procedure for Determining Aerodynamic Forces on Oscillating or Steady Finite Wings at Subsonic Speeds', *NASA Technical Report TR*-48, 1959.

[14] Williams, D. E., *Some Mathematical Methods in Three-Dimensional Subsonic Flutter Derivative Theory*, Great Britain Aeronautical Research Council, R&M 3302, 1961.

[15] Cunningham, A. M., Jr., 'Further Developments in the Prediction of Oscillatory Aerodynamics in Mixed Transonic Flow', *AIAA Paper 75–99* (January 1975).

[16] Morino, L., Chen, L. T. and Suciu, E. O., 'Steady and Oscillatory Subsonic and Supersonic Aerodynamics Around Complex Configurations', *AIAA Journal* (March 1975), pp. 368–374.

[17] Rodden, W. P., 'State-of-the-Art in Unsteady Aerodynamics', *AGARD Report No. 650*, 1976.

[18] Ashley, H. and Rodden, W. P., 'Wing-Body Aerodynamic Interaction', *Annual Review of Fluid Mechanics*, Vol. 4, 1972, pp. 431–472.

[19] Theodorsen, T., 'General Theory of Aerodynamic Instability and the Mechanism of Flutter', *NACA Report 496*, 1935.

[20] Abramowitz, M. and Stegun, I. A., *Handbook of Mathematical Functions*, National Bureau of Standards, U.S. Printing Office, 1965.

[21] Edwards, J. V., Ashley, H. and Breakwell, J. B., 'Unsteady Aerodynamics Modeling for Arbitrary Motions', *AIAA Paper 77–451*, AIAA Dynamics Specialist Conference, San Diego, March 1977.

[22] H. Lomax, *Indicial Aerodynamics*, Vol. II, Chapter 6, AGARD Manual on Aeroelasticity, 1960.

[23] Rodden, W. P., Giesing, J. P. and Kálmán, T. P., 'New Developments and Applications of the Subsonic Doublet-Lattice Method for Non-Planar Configurations',

References for chapter 4

AGARD Symposium on Unsteady Aerodynamics for Aeroelastic Analyses of Interfering Surfaces, Tønsberg, Oslofjorden, Norway (November 3–4) 1970.

[24] Stahara, S. S. and Spreiter, J. R., 'Development of a Nonlinear Unsteady Transonic Flow Theory', *NASA CR-2258* (June 1973).

[25] Rubbert, P. and Landahl, M., 'Solution of the Transonic Airfoil Problem through Parametric Differentiation', *AIAA Journal* (March 1967), pp. 470–479.

[26] Beam, R. M. and Warming, R. F., 'Numerical Calculations of Two-Dimensional, Unsteady Transonic Flows with Circulation', *NASA TN D-7605*, (February 1974).

[27] Ehlers, F. E., 'A Finite Difference Method for the Solution of the Transonic Flow Around Harmonically Oscillating Wings', *NASA CR-2257* (July 1974).

[28] Traci, R. M., Albano, E. D., Farr, J. L., Jr. and Cheng, H. K., 'Small Disturbance Transonic Flow About Oscillating Airfoils', *AFFDL-TR-74-37* (June 1974).

[29] Magnus, R. J. and Yoshihara, H., 'Calculations of Transonic Flow Over an Oscillating Airfoil', *AIAA Paper 75-98* (January 1975).

[30] Cunningham, A. M., Jr., 'Further Developments in the Prediction of Oscillatory Aerodynamics in Mixed Transonic Flow', *AIAA Paper 75-99* (January 1975).

[31] Dowell, E. H. A Simplified Theory of Oscillating Airfoils in Transonic Flow', *Proceedings of Symposium on Unsteady Aerodynamics*, pp. 655–679, University of Arizona (July 1975).

[32] Bisplinghoff, R. L. and Ashley, H., *Principles of Aeroelasticity*, John Wiley and Sons, Inc., New York, 1962.

[33] Landahl, M., *Unsteady Transonic Flow*, Pergamon Press, London, 1961.

[34] Spreiter, J. R., 'Unsteady Transonic Aerodynamics—An Aeronautics Challenge', *Proceedings of Symposium on Unsteady Aerodynamics*, pp. 583–608, University of Arizona (July 1975).

[35] Ballhaus, W. F., Magnus, R. and Yoshihara, H., 'Some Examples of Unsteady Transonic Flows Over Airfoils', *Proceedings of a Symposium on Unsteady Aerodynamics*, pp. 769–792, University of Arizona (July 1975).

[36] Knechtel, E. D., 'Experimental Investigation at Transonic Speeds of Pressure Distributions Over Wedge and Circular-Arc Airfoil Sections and Evaluation of Perforated-Wall Interference', *NASA TN D-15* (August 1959).

[37] Spreiter, J. R. and Alksne, A. Y., 'Thin Airfoil Theory Based on Approximate Solution of the Transonic Flow Equation', *NACA TN 3970*, 1957.

[38] Park, P. H., 'Unsteady Two-Dimensional Flow Using Dowell's Method', *AIAA Journal* (October 1976) pp. 1345–1346. Also see Isogai, K., 'A Method for Predicting Unsteady Aerodynamic Forces on Oscillating Wings with Thickness in Transonic Flow Near Mach Number 1', National Aerospace Laboratory Technical Report NAL-TR-368T, Tokyo, Japan, June 1974. Isogai, using a modified local linearization procedure, obtains aerodynamic forces comparable to Park's and these provide significantly better agreement with transonic flutter experiments on parabolic arc airfoils.

[39] Dowell, E. H., 'A Simplified Theory of Oscillating Airfoils in Transonic Flow: Review and Extension', *AIAA Paper 77-445*, presented at AIAA Dynamic Specialists Conference, San Diego (March 1977).

[40] Spreiter, J. R. and Stahara, S. S., 'Developments in Transonic Steady and Unsteady Flow Theory', *Tenth Congress of the International Council of the Aeronautical Sciences*, Paper No. 76-06 (October 1976).

[41] Chan, S. T. K. and Chen, H. C., 'Finite Element Applications to Unsteady Transonic Flow', *AIAA Paper 77-446*.

[42] Ballhaus, W. F. and Goorjian, P. M., 'Computation of Unsteady Transonic Flows by the Indicial Method', *AIAA Paper 77-447.*

[43] Isogai, K., 'Oscillating Airfoils Using the Full Potential Equation', *AIAA Paper 77-448.* Also see NASA TP1120, April 1978.

[44] Eckhaus, W., 'A Theory of Transonic Aileron Buzz, Neglecting Viscous Effects', *Journal of the Aerospace Sciences* (June 1962) pp. 712–718.

[45] Williams, M. H. 'Unsteady Thin Airfoil Theory for Transonic Flow with Embedded Shocks', Princeton University MAE Report No. 1376, May 1978.

[46] Tijdeman, H. and Schippers, P., *Results of Pressure Measurements on an Airfoil with Oscillating Flap in Two-Dimensional High Subsonic and Transonic Flow,* National Aerospace Lab. Report TR 730780, The Netherlands (July 1973).

[47] Goldstein, M. E., Braun, W., and Adamczyk, J. J., 'Unsteady Flow in a Supersonic Cascade with Strong In-Passage Shocks', *J. Fluid Mechanics,* Vol. 83, 3 (1977) pp. 569–604.

[48] Rowe, W. S., Sebastian, J. D., and Redman, M. C., 'Recent Developments in Predicting Unsteady Airloads Caused by Control Surfaces', *J. Aircraft* (December 1976) pp. 955–963.

[49] McCroskey, W. J., 'Some Current Research in Unsteady Fluid Dynamics—The 1976 Freeman Scholar Lecture', *Journal of Fluids Engineering* (March 1977) pp. 8–39.

5

Stall flutter

As the name implies, stall flutter is a phenomenon which occurs with partial or complete breakaway of the flow from the airfoil during at least a part of every cycle of oscillation. As contrasted with the so-called classical flutter (i.e., flow attached at all times) the mechanism for energy transfer from the airstream to the oscillating airfoil does not rely on elastic and/or aerodynamic coupling between two modes, nor upon a phase lag between a displacement and its aerodynamic reaction. These latter effects are necessary in a linear system to account for an airstream doing positive work on a vibrating wing. The essential feature of stall flutter is the *nonlinear* aerodynamic reaction to the motion of the airfoil/structure. Thus, although coupling and phase lag may alter the results somewhat, the basic instability and its principal features are explicable in terms of nonlinear normal force and moment characteristics.

5.1 Background

Stall flutter of aircraft wings and empennages is associated with very high angles of attack. Large incidence is necessary to induce separation of the flow from the suction surface. This type of operating condition and vibratory response was observed as long ago as World War I at which time stall flutter occurred during sharp pull-up maneuvers in combat. The surfaces were usually monoplane without a great deal of effective external bracing. The cure was to stiffen the structure and avoid the dangerous maneuvers whenever possible.

Electric power transmission cables of circular cross-section, or as modified by bundling or by ice accretion, etc., and structural shapes of various description are classified as bluff bodies. As such they do not require large incidence for flow separation to occur. In fact incidence is

chiefly an orientation parameter for these 'airfoils' rather than an indication of steady aerodynamic loading. Again, largely attributable to the
nonlinearity in the force and moment as a function of incidence, such
structures are prone to stall flutter vibrations. These are sometimes called
'galloping' as in the case of transmission lines. The number and classes of
structures that potentially could experience stall flutter are very great, and
include such diverse examples as suspension bridges, helicopter rotors
and turbomachinery blades. More mundane examples are venetian blind
slats and air deflectors or spoilers on automobiles.

5.2   Analytical formulation

Although analysis of stall flutter based on purely theoretical considerations
is not quantitatively useful, it is nevertheless instructive to couch the
problem in analytical terms so as to discriminate clearly the actual
mechanism of instability.* We will consider two important cases: bending
and twisting.

In the case of bending, or plunging displacement of a two-
dimensional 'typical section', let us assume that the force coefficient,
including penetration well into the stall regime, is given by a polynomial
approximation in α,

$$-C_n = \sum_{n=0}^{\nu} a_n(\alpha_{ss})\alpha^n \qquad a_0 \cong -C_{nss}(\alpha_{ss})$$

where α is the departure from the steady state value of angle of attack,
α_{ss}, attributable to vibration of the airfoil. This method of expressing the
normal force characteristic gives a good local fit with a few terms.
However, the coefficients, a_n, are functions of the mean angle of attack,
α_{ss}. Force has been taken to be positive in the same direction as positive
displacement h. (In the usual (static) theory of thin unstalled and uncam-
bered profiles $-C_n = \pi \sin 2\alpha_{ss}$. The a_n could then be obtained by
deriving the Maclaurin series expansion of $\pi \sin 2(\alpha_{ss} + \alpha)$ considered as a
function of α). In general the $-C_n$ function is an empirically determined
function, or characteristic, when stall occurs on a cambered airfoil, but
the procedure is still the same. The a_n are in fact given by the slope and
higher order derivatives according to

$$a_n = -\frac{1}{n!}\frac{d^n C_n}{d\alpha}\bigg|_{\alpha=0}$$

* Sisto [1].

We next consider a small cosinusoidal bending oscillation

$$h = h_0 \cos \omega t$$

to exist and enquire as to the stability of that motion: Will it amplify or decay?

Under these circumstances, it is possible to interpret the instantaneous angle of attack perturbation to be given by

$$\alpha = \arctan\left(\tan \alpha_{ss} + \frac{\dot{h}}{V \cos \alpha_{ss}}\right) - \alpha_{ss}$$

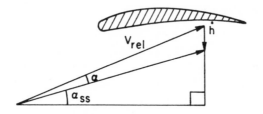

with Maclaurin series expansion in powers of $\dot{h}$ as follows

$$\alpha = \cos \alpha_{ss}\left(\frac{\dot{h}}{V}\right) - \tfrac{1}{2}\sin 2\alpha_{ss}\left(\frac{\dot{h}}{V}\right)^2 - \tfrac{1}{3}\cos 3\alpha_{ss}\left(\frac{\dot{h}}{V}\right)^3 + \tfrac{1}{4}\sin 4\alpha_{ss}\left(\frac{\dot{h}}{V}\right)^4 + \cdots$$

It should be noted that this incidence is relative to a coordinate system fixed to the airfoil. The dynamic pressure also changes periodically with time in this coordinate system according to

$$q_{rel} = \tfrac{1}{2}\rho V_{rel}^2 = \tfrac{1}{2}\rho V^2\left[1 + 2\sin \alpha_{ss}\left(\frac{\dot{h}}{V}\right) + \left(\frac{\dot{h}}{V}\right)^2\right]$$

It is assumed for simplicity that the single static characteristic of normal force coefficient versus angle of attack continues to be operative in the dynamic application described above. Thus, the expanded equation for the normal force $N = q(2b)C_n$ is given by

$$N = -\tfrac{1}{2}\rho V^2(2b)\left[1 + 2\sin \alpha_{ss}\left(\frac{\dot{h}}{V}\right) + \left(\frac{\dot{h}}{V}\right)^2\right]$$

$$\cdot \sum_{n=0}^{\nu} a_n(\alpha_{ss})\left[\cos \alpha_{ss}\left(\frac{\dot{h}}{V}\right) - \tfrac{1}{2}\sin 2\alpha_{ss}\left(\frac{\dot{h}}{V}\right)^2 - \tfrac{1}{3}\cos 3\alpha_{ss}\left(\frac{\dot{h}}{V}\right)^3\right.$$

$$\left. + \tfrac{1}{4}\sin 4\alpha_{ss}\left(\frac{\dot{h}}{V}\right)^4 + \cdots\right]^n$$

with

$$\frac{\dot{h}}{V} = -\frac{\omega h_0}{V}\sin \omega t = -k\frac{h_0}{b}\sin \omega t$$

A slight concession to the dynamics of stalling may be introduced by the inclusion of a time delay, ψ/ω, in the oscillatory velocity term appearing in the C_n expansion, i.e., within the summation of the previous equation, but not in the development of q_{rel}. The latter is assumed to respond instantaneously to α or $\dot{h}$.

5.3   Stability and work flow

As is common with single degree of freedom systems such as that postulated above, the question of amplification or subsidence of the amplitude of the initial motion can easily be decided on the basis of the work done by this force acting on the displacement. Thus

$$\text{Work/Cycle} = \int_0^T N\dot{h}\,dt = \frac{1}{\omega}\int_0^{2\pi} N\dot{h}\,d(\omega t)$$

and since the frequency is effectively the number of cycles per unit time, the power may be expressed as

$$\mathbb{P} = \text{Power} = (\text{Work/Cycle})(\text{Cycles/Second}) = \frac{1}{2\pi}\int_0^{2\pi} N\dot{h}\,d(\omega t)$$

Using the previous expression for N and $\dot{h}$, it is clear that only even powers of $\sin \omega t$ in the integrand of the power integral will yield nonzero contributions. Also, terms of the form $\sin^n \omega t \cos \omega t$ will integrate to zero for any integer value of n including zero. Restricting the series expansions for $-C_n$ and α to their leading terms such that the power integral displays terms of vibratory amplitude up to the sixth power (i.e., up to h_0^6) results in

$$\mathbb{P} = \tfrac{1}{2}\rho V^3 b[A(\omega h_0/V)^2 + B(\omega h_0/V)^4 + C(\omega h_0 V)^6 + \cdots]$$

where

$$A = -2a_0 \sin \alpha_{ss} - a_1 \cos \alpha_{ss} \cos \psi$$
$$B = -\tfrac{1}{4}a_1[-(\cos \alpha_{ss} - \cos 3\alpha_{ss})(1 + \tfrac{1}{2}\cos 2\psi)$$
$$+ (3\cos \alpha_{ss} - \cos 3\alpha_{ss})\cos \psi]$$
$$-\tfrac{1}{4}a_2[(\sin \alpha_{ss} + \sin 3\alpha_{ss})(1 - \tfrac{3}{2}\cos \psi + \tfrac{1}{2}\cos 2\psi)]$$
$$-\tfrac{3}{16}a_3[(3\cos \alpha_{ss} + \cos 3\alpha_{ss})\cos \psi]$$

$$C = -\tfrac{1}{16}a_1[(\cos 3\alpha_{ss} - \cos 5\alpha_{ss})(\tfrac{3}{2} + \cos 2\psi)$$
$$-\tfrac{1}{16}(3\cos 3\alpha_{ss} - 2\cos 5\alpha_{ss})\cos \psi - \tfrac{1}{3}\cos 3\alpha_{ss}\cos 3\psi] - \cdots$$

The cubic dependence on V is a consequence of the dimensions of power, or work per unit time.

5.4 Bending stall flutter

The analytical expression for the aerodynamic power in a sinusoidal bending vibration is too cumbersome for easy physical interpretation. However, for very small amplitudes of motion, as might be triggered by turbulence in the fluid, or other 'noise' in the system, it is clear that the sign of the work flow will be governed by the coefficient of $(\omega h_0/V)^2$. Assuming a small to moderate positive mean incidence, α_{ss}, the coefficient a_0 will be positive. With $\cos \psi$ near unity, a positive power can only occur if a_1 is sufficiently negative, i.e., if the $-C_n$ vs α characteristic has a negative slope at the static operating incidence. More precisely, if $|\psi| < 90°$ and

$$a_1 < -2a_0 \tan \alpha_{ss} \sec \psi$$

the small amplitude vibration is unstable and the work flow will be such as to feed energy into the vibration and increase its amplitude.

In the previous expression for the power,

$$\mathbb{P}/(\tfrac{1}{2}\rho V^3 b) = A(\omega h_0/V)^2 + B(\omega h_0/V)^4 + C(\omega h_0/V)^6$$

the coefficients A, B and C are complicated functions of ψ, α_{ss} and the a_n, the coefficients of the power series representation of the normal force characteristic. For example in the highly simplified case of $\alpha_{ss} = \psi = 0$, we obtain

$$A = a_1 = \left.\frac{dC_n}{d\alpha}\right|_{\alpha=0}, \quad B = \left.\frac{1}{2}\frac{dC_n}{d\alpha}\right|_{\alpha=0} + \left.\frac{1}{8}\frac{d^3C_n}{d\alpha^3}\right|_{\alpha=0}$$

and

$$C = \left.\frac{1}{12}\frac{dC_n}{d\alpha}\right|_{\alpha=0} + \left.\frac{1}{192}\frac{d^5C_n}{d\alpha^5}\right|_{\alpha=0}$$

In the general case A, B and C individually may be either positive, zero or negative. The several possible cases are of fundamental interest in describing possible bending stall flutter behavior.

    I. $A < 0$, $B < 0$, $C < 0$ No flutter is possible.

    II. $A > 0$, $B > 0$, $C > 0$ Flutter amplitude grows from zero to very large values.

III. $A > 0$, $B < 0$, $C < 0$ Flutter amplitude grows smoothly from zero to a finite amplitude given by

$$(\omega h_0 / V)_{\text{III}}^2 = (-|B| + \sqrt{B^2 + 4A\,|C|})/2\,|C|$$

At this amplitude the power once again becomes zero.

IV. $A < 0$, $B > 0$, $C > 0$ No flutter at small amplitudes; if an external 'triggering' disturbance carries the system beyond a certain critical vibratory amplitude

$$(\omega h_0 / V)_{\text{IV}}^2 = (-B + \sqrt{B^2 + 4\,|A|\,C})/2C$$

the flutter will continue to grow beyond that amplitude up to very large values. At the critical amplitude the power is zero.

V. $A > 0$, $B > 0$, $C < 0$ This is similar to case III except that the finite amplitude, or equilibrium, flutter amplitude

$$(\omega h_0 / V)_{\text{V}}^2 = (B + \sqrt{B^2 + 4A\,|C|})/2\,|C|$$

might be expected to be somewhat larger.

VI. $A > 0$, $B < 0$, $C > 0$ This is similar to case IV except that the critical vibratory amplitude beyond which flutter may be expected to grow

$$(\omega h_0 / V)_{\text{VI}}^2 = (|B| + \sqrt{B^2 + 4\,|A|\,C})/2C$$

is perhaps a larger value.

VII. $A > 0$, $B < 0$, $C > 0$ This case has behavior similar to case II if B is very small and similar to case III if C is very small and also very large amplitudes are excluded from consideration.

VIII. $A < 0$, $B > 0$, $C < 0$ This case behavior similar to case I if B is very small and similar to case IV if C is very small and also very large amplitudes are excluded from consideration.

5.5 Nonlinear mechanics description

A number of these variations of power dependency on amplitude have been sketched in Figure 5.1. Case II is an example of what may be termed 'soft flutter'; given an airstream velocity V, incidence α_{ss} and time delay ψ/ω such as produce values of A, B and C according to case II, the vibratory amplitude of flutter might be expected to grow smoothly from zero.

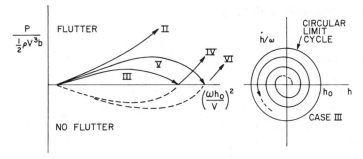

Figure 5.1   Power vs. amplitude.

Cases III and V similarly are examples of soft flutter; in these cases however, the amplitude of vibration reaches a steady value and does not increase further. An equilibrium flutter amplitude is attained after a period of time and maintained thereafter. If, in either of these cases, one were to plot h versus $\dot{h}/\omega$ with time as a parameter, it would be found that the 'trajectory' of the 'characteristic point' would be a spiral around the origin, beginning at the origin at $t = 0$ and asymptotically approaching a circle of radius h_0 for very large time. In the parlance of nonlinear mechanics the circular path is a 'limit cycle' and hence most instances of stall flutter may be termed limit cycle vibrations.

Case IV, or alternatively case VI, describes a type of behavior which may be termed 'hard flutter'. In this situation when flutter appears as a self-sustaining oscillation, the amplitude is immediately a large finite value. Here the motion spirals away from the circular limit cycle to either larger or smaller amplitudes in the phase plane (i.e., the h, $\dot{h}/\omega$ plane). This example is an instance in which the limit cycle is unstable. The slightest perturbation from an initially purely circular path, either to larger or smaller radii, will result in monotonic spiralling away from the limit cycle. The previous example of case III illustrated the case of a stable limit cycle.

The origin of the phase plane is also a degenerate limit cycle in the sense that the limit of a circle is a point in which case only path radii larger than zero have physical meaning. However, the origin may be an unstable limit cycle (soft flutter) or a stable limit cycle (hard flutter).

It is clear from a consideration of cases VII and VIII that more than two limit cycles may obtain; it is a theorem of mechanics that the concentric circles which are limit cycles of a given system are alternately stable and unstable.

273

5.6   Torsional stall flutter

With pure twisting motion of the profile, the analytical formulation is more complex stemming from the fact that the dynamic angle of incidence is compounded of two effects: the instantaneous angular displacement and the instantaneous linear velocity in a direction normal to the chord. The magnitude of the first effect is a constant independent of chordal position and frequency; the second magnitude is linearly dependent upon the distance along the chord from the elastic axis and upon the frequency of vibration. Both components, of course, vary harmonically with the frequency ω. Thus, assuming a displacement $\theta_0 \cos \omega t$ the 'local' angle of attack becomes

$$\alpha = \theta_0 \cos \omega t + \arctan \left[\tan \alpha_{ss} - \frac{(x - x_0)\omega\theta_0}{V \cos \alpha_{ss}} \sin \omega t \right] - \alpha_{ss}$$

and the relative dynamic pressure becomes

$$q_{\text{rel}} = \tfrac{1}{2}\rho V_{\text{rel}}^2 = \tfrac{1}{2}\rho V^2 \left[1 + 2 \sin \alpha_{ss} \frac{\dot{\theta}(x - x_0)}{V} + \left(\frac{\dot{\theta}(x - x_0)}{V} \right)^2 \right]$$

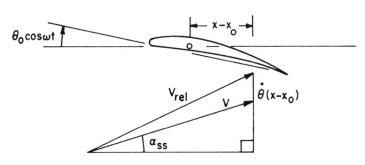

Figure 5.2   Geometry.

Since the local incidence varies along the chord in the torsional case, it is not possible to formulate the twisting problem in a simple and analogous manner to the bending case unless a single 'typical' incidence is chosen. From incompressible, potential flow, thin airfoil theory, it is known[*] that the three-quarter chord point is 'most representative' in relating changes in incidence to changes in aerodynamic reaction for an unstalled thin airfoil with parabolic camber. Replacing $x - x_0$ by a *constant*, say eb, for

---

[*] Durand [2]. Also see equation (4.3.70) in Section 4.3, Incompressible Two-dimensional Flow and discussion in Section 3.4.

simplicitly, one has by analogy with bending

$$\alpha = \theta_0 \cos \omega t + \cos \alpha_{ss}(-ek\theta_0) \sin \omega t - \tfrac{1}{2} \sin 2\alpha_{ss}(-ek\theta_0)^2$$
$$\cdot \sin^2 \omega t - \tfrac{1}{3} \cos 3\alpha_{ss}(-ek\theta_0)^3 \sin^3 \omega t + \tfrac{1}{4} \cdots$$

where α is, again, the departure in angle of attack from α_{ss}. The constant e will normally be of order unity for an elastic axis location forward of midchord.

From this point onward, the illustrative analysis involves the substitution of α into an analytical approximation for the aerodynamic moment coefficient

$$C_m = \sum_{n=0}^{\nu} b_n(\alpha_{ss})\alpha^n$$

In this equation, the b_n may be associated with the slope and higher order derivatives of the characteristic

$$b_n = \frac{1}{n!} \frac{d^n C_m}{d\alpha^n}\bigg|_{\alpha=0}$$

at the mean incidence point, in a manner analogous to the role of the a_n in the normal force coefficient.

The work done by the aerodynamic moment acting on the torsional displacement is given by

$$\text{Work/Cycle} = \int_0^T M\dot\theta \, dt = \frac{1}{\omega} \int_0^{2\pi} M\dot\theta \, d(\omega t)$$

and hence, the work flow, or power is

$$\mathbb{P} = \frac{1}{2\pi} \int_0^{2\pi} M\dot\theta \, d(\omega t)$$

Using the previously derived expressions contributing to the moment $M = q(2b)^2 C_m$ leads to

$$M = \tfrac{1}{2}\rho V^2 (2b)^2 \left[1 + 2\sin\alpha_{ss}\left(\frac{\dot\theta e b}{V}\right) + \left(\frac{\dot\theta e b}{V}\right)^2\right]$$

$$\cdot \sum_{m=0}^{\nu} b_n(\alpha_{ss})[\theta_0 \cos\omega t - \cos\alpha_{ss}(ek\theta_o)\sin\omega t$$

$$- \tfrac{1}{2}\sin 2\alpha_{ss}(ek\theta_0)^2 \sin^2\omega t + \tfrac{1}{3}\cos 3\alpha_{ss}(ek\theta_0)^3 \sin^3\omega t + \cdots]^n$$

and this expression, in turn inserted into the integrand of the power integral, will allow an analytical expression to be derived by quadrature.

275

At this stage in the development of torsional stall flutter, a key difference emerges more clearly when compared to bending stall flutter; a fundamental component of the moment coefficient appears ($b_1\theta_0 \cos \omega t$) which is out of phase with the torsional velocity ($\dot{\theta} = -\omega\theta_0 \sin \omega t$). Noting that $\dot{\theta}$ is the second factor in the integrand, it is seen that the final integrated expression for the power will have terms similar in nature to the expression derived for the bending case, and in addition may have terms proportional to

$$b_1\theta_0, \ b_2\theta_0^2, \ b_2\theta_0, \ b_3\theta_0^3, \ b_3\theta_0^2, \text{ etc.}$$

It is not particularly instructive to set out this result in full detail.

However, let us consider briefly the case of very slow oscillations, so that terms proportional to higher powers of the frequency can be ignored. Then

$$\mathbb{P}_1 = -\tfrac{1}{2}\rho V^2 (2b)^2 \frac{\omega\theta_0}{2\pi} \sum_{n=0}^{\nu} b_n\theta_0^n \int_0^{2\pi} \cos^n(\omega t - \psi) \sin \omega t \, d(\omega t)$$

$$= -\tfrac{1}{2}\rho V^2 (2b)^2 \frac{\omega}{2\pi} \sin \psi \sum_{n=\text{odd}}^{\nu} b_n\theta_0^{n+1} \int_0^{2\pi} \cos^{n+1}\Omega \, d\Omega$$

$$= -\tfrac{1}{2}\rho V^3 (4b)k \sin \psi \sum_{n=\text{odd}}^{\nu} b_n\theta_0^{n+1} \frac{1 \cdot 3 \cdot 5 \cdots n}{2 \cdot 4 \cdot 6 \cdots (n+1)}$$

We conclude from this equation that the work flow again will be proportional to a sum of terms in even powers of the vibratory amplitude, but in this instance, the low frequency torsional stall flutter is critically dependent on the time lag ψ/ω between the oscillatory motion and the response of the periodic aerodynamic moment.

Torsional stall flutter is thus seen to be a much more complex phenomenon, with a greater dependence on time lag and exhibiting very strong dependence on the location of the elastic axis. For example, if the elastic axis were artificially moved rearward on an airfoil such as to reduce the effective value of the parameter e to zero, the airfoil flutter behavior would be governed by exactly the same specialization of the analysis as was just termed 'low frequency'. Exactly the same terms would be eliminated from consideration. In qualitative terms one may also conclude that the actual behavior in torsional flutter in the general case (with $e \neq 0$) is some intermediate state between the low frequency behavior (critical dependence on $\sin \psi$) and a type of behavior characteristic of bending stall flutter (critical dependence on the slope of a dynamic characteristic at the mean incidence).

5.7   Concluding comments

An interesting by-product of the nonlinear nature of stall flutter is the ability, in principle, to predict the final equilibrium amplitude of the vibration. This is in contradistinction to classical flutter in which only the stability boundary is usually determined. The condition for constant finite flutter amplitude is that the work, or power flow, again be zero. As we have seen this can be discerned when the power equation is set equal to zero; the resulting quadratic equation is solved for the squared flutter amplitude, either $(h_0/b)^2$ or θ_0^2 as the case may be. Since all the a_n or b_n coefficients are functions of α_{ss}, the two types of flutter are displayed in Figure 5.3 as *presumed* functions of this parameter. Hard flutter displays a sudden jump to finite amplitude as a critical parameter is varied and a lower 'quench' value of that parameter where the vibration suddenly disappears. The two effects conspire to produce the characteristic hysteresis loop indicated by arrows.

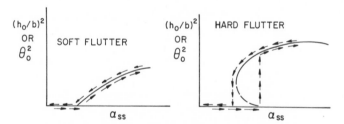

Figure 5.3   Flutter amplitude vs steady state angle of attack.

In summary then, stall flutter is associated with nonlinearity in the aerodynamic characteristic; the phenomenon may occur in a single degree of freedom and the amplitude of vibratory motion will often be limited by the aerodynamic nonlinearities. Although structural material damping have not been considered explicitly, it is clear that since damping is an absorber of energy its presence will serve to limit the flutter amplitudes to smaller values; damping limited amplitudes will obtain when the positive power flow from airstream to airfoil equals the power conversion to heat in the mechanical forms of damping.

It is also clear that motion in a third degree of freedom is also possible. Oscillatory surging of the airfoil in the chordwise direction can be related to a nonlinear behavior in the drag acting on the profile. However, airfoils are usually very stiff structurally in the chordwise direction and the drag/surging mechanism would normally be of importance only for bluff structural shapes such as bundles of electric power conductors suspended between towers, etc.

Under certain circumstances such as the example noted directly above, stall flutter in more than one degree of freedom may occur. In these cases, the dynamic characteristics of normal force, aerodynamic moment (and drag) become functions of an effective incidence compounded of many effects: plunging velocity, torsional displacement, torsional velocity and surging velocity. The resultant power equation will also contain cross-product terms in the various displacement amplitudes, and hence the equation cannot be used to predict stability or equilibrium flutter amplitudes without additional information concerning the vibration modes.

Perhaps the greatest deficiency in the theory, however, is the fact that even in pure bending motion or pure torsional motion, the dynamic force and moment are in fact frequency dependent: $a_n = a_n(\alpha_{ss}, k)$ and $b_n = b_n(\alpha_{ss}, k)$. And in general $a_0 \neq -C_{nss}$ and $b_0 \neq C_{mss}$. In analogy with classical flutter it may be shown that even this dependence is deficient in that the characteristics may be in practice double valued. That is, for the same value of effective incidence α, the characteristic may have different values depending upon whether α is decreasing or increasing with time. Such a hysteretic characteristic is usually more pronounced at high frequencies of oscillation; an airfoil may have two lift or moment coefficients at a particular angle of attack even in the static case, depending upon how the operating point was approached.

It is for these reasons that practical stall flutter prediction is at best a semi-empirical process, and often entirely empirical. A model is oscillated in torsion, or bending, in a wind tunnel under controlled conditions with parametric variation of reduced frequency, mean incidence and oscillatory amplitude. Various elastic axis locations also may be studied. Data which is taken may vary from instantaneous normal force and moment down to the actual time-dependent pressure distribution on the profile. Data reduction consists essentially of cross-plotting the various data so that flutter prediction for prototype application is largely a matter of interpolation in model data using dimensionless groups. Specific representative data will be taken up in subsequent chapters where stall flutter applications are studied.

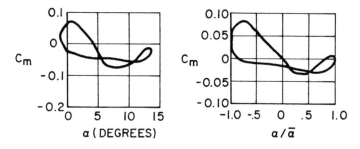

Figure 5.4   Dynamic moment loops.

One exception to the previous reliance on experimental data has recently been introduced. This new theory* postulates that the departure of the normal force and/or aerodynamic moment from the classical (attached flow) values can be modelled by considering a flat plate with separated flow on the suction side. As the plate oscillates harmonically in time, the position of the separation point is also considered to move periodically with the same frequency as the oscillation. The movement of the separation point along the suction surface is between two arbitrarily specified upstream and downstream limits and with an arbitrarily specified phase angle with respect to the oscillatory motion.

Under these circumstances, it is possible to solve the unsteady flow problem (analogous to the classical Theodorsen solution for attached flow) including the effects of separation. In effect the appropriate dynamic force and moment characteristics are generated for each function specifying the separation point movement and airfoil motion. The empirical part of the flutter prediction technique then resides in correlation of the separation point behavior as a function of the airfoil attitude and oscillatory motion. At the present writing experiments are underway attempting to close the loop on this promising new method. To illustrate the potential of the technique, two moment loops from the reference are shown below. The one on the left is from an experimental program† the one on the right is from the previously quoted reference. Although the variation of moment with torsional displacement is remarkably similar, it must be emphasized that the particular choice of elastic axis location is different in experiment and theory, and the assumed separation point behavior in the theory was reasonable, but quite arbitrary and unrelated to the unknown separation point behavior in the experiment.

---

* Sisto and Perumal [3].
† Gray and Liiva [4].

279

5 Stall flutter

References for Chapter 5

[1] Sisto, F., 'Stall Flutter in Cascades', *Journal of the Aeronautical Sciences*, v. 20, n. 9 (September 1953) pp. 598–604.

[2] Durand, W. F. (Ed.) *Aerodynamic Theory*, Sec. E., Vol. II, Dover Publications, New York, 1963, p. 50.

[3] Sisto, F. and Perumal, P. V. K., 'Lift and Moment Prediction For An Oscillating Airfoil With a Moving Separation Point', *Trans. ASME, J. Engrg. for Power*, v. 96, Series A, n. 4 (October 1974) pp. 372–278.

[4] Gray, Lewis and Liiva, J., *Windtunnel Tests of Thin Airfoils Oscillating Near Stall*, Vol. II, USAAVLABS Technical Report 68–89B, 1969.

6

Aeroelastic problems of civil engineering structures

In recent decades the designs and methods of fabrication of new engineering structures have been such as to render them more flexible and not as highly damped as their counterparts of the past. As a result the sensitivity of such structures to the natural wind has become an object of considerable attention, and many studies in wind engineering have been occasioned. In these not only the character of the wind alone, as loading function, has been taken into account, but also the interaction between wind and consequent structural motion has required consideration. It is the latter group of *aeroelastic* studies which will be examined particularly in the present chapter.

Although the problems that arise are associated directly with specific structures such as towers, tall buildings and bridges, it has been judged preferable to discuss them below according to phenomenological type rather than structure type. In following out this method of presentation, the phenomena that will be considered are *divergence, galloping, vortex-shedding, flutter* and *buffeting*. The discussion of each of these in turn will include some commentary on the particular structures that undergo the phenomena, and methods of prevention will be mentioned where appropriate.

6.1 Divergence

This phenomenon has already been discussed in Sections 2.1–2.3, relative to airfoils and lifting surfaces. It is, in general, a rather high-velocity phenomenon and so occurs relatively infrequently in civil structures at the normal speeds of the natural wind. These, except in tornadoes, rarely exceed 125 mph (1 mph = 0.447 m/sec). Hence the aerodynamics involved remains in the incompressible flow range. Only relatively weak structures risk divergence, but it should nonetheless be considered for all structures of importance.

281

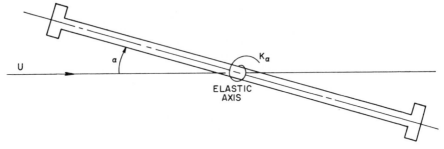

Figure 6.1   Cross-section of bridge deck, with torsional restraint.

The phenomenon can be illustrated in the case of a long, torsionally flexible suspension bridge. In terms of a single representative section of the bridge the action of torsional divergence is completely analogous to that described in Sections 2.1–2.3 for airfoils and lifting surfaces. Consider the bridge section shown in Figure 6.1.

There, horizontal wind at velocity U is blowing against the structural section, the angle of twist of which is α. The section is restrained by a structural spring of elastic characteristic K_α. If increased twist α implies increased moment M_α acting on the section, there will be, for some high wind velocity U, sufficient torsional moment on the section to overcome the structural restraint, and the section will rotate to destruction (diverge). The phenomenon is not unlike that of structural buckling, and its details are very similar to those already given in Sections 2.1–2.3.

Consider now, to be more explicit, a full bridge span for which the section conditions resemble these in Figure 6.1 but for which the deflection at the representative section is a function of loads both there and elsewhere. To this end let D_T be the matrix of torsional influence coefficients of the deck at various spanwise stations i. An element d_{ij} of D_T will represent the angle of attack at spanwise station i due to a unit torsional moment applied at spanwise station j.

Let C_M be the mean (i.e., time-averaged) aerodynamic moment coefficient about the rotation point of a representative section; then

$$M = \tfrac{1}{2}\rho U^2 B^2 C_M \tag{6.1.1}$$

is the aerodynamic moment per unit span applied by the wind at the section.* B is a reference length and is often taken to be the full width (chord) of the bridge section.

---

* Strip theory aerodynamics, previously discussed in Sections 2.2 and 3.4, will be employed here.

If the bridge span L is conceptually divided into equal subspans of length ΔL, the moment at section i is

$$M_i = \tfrac{1}{2}\rho U^2 B^2 C_{M_i} \Delta L \tag{6.1.2}$$

and the column of discrete torsional deformations of the bridge $\{\alpha_i\}$ is expressible as

$$\{\alpha_i\} = D_T\{M_i\} \tag{6.1.3}$$

which will have a definite solution for any assigned wind speed U.

This solution will now be considered. As is well known, the coefficient C_M is found experimentally to be a function of α. See Figure 6.2 for an example of such a coefficient for the experimental bridge deck section shown. [1].* Suppose, for example, that C_M is expressible in the form

$$C_M(\alpha) = c_0 + c_1\alpha + c_2\alpha^2 + \cdots + c_n\alpha^n \tag{6.1.4}$$

that is, the C_M curve analogous to that in Figure 6.2 is fitted by a polynomial in α. Then, if $\tfrac{1}{2}\rho U^2$ is written as q, M_i takes the form

$$M_i = qB^2 \Delta L[c_0 + c_1\alpha_i + \cdots + c_n\alpha_i^n] \tag{6.1.5}$$

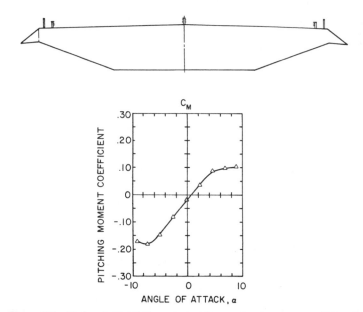

Figure 6.2 *Modern bridge deck section and its torsional moment coefficient [1].*

---

*Numbers in brackets refer to references at the end of this chapter.

and equation (6.1.3) becomes

$$\{\alpha_i\} = qB^2 \, \Delta L D_T \{\{c_0\} + c_1\{\alpha_i\} + \cdots + c_n\{\alpha_i^n\}\} \tag{6.1.6}$$

This problem may be solved iteratively for an assumed value of q. For example, the value of (α_i) is first estimated by taking $c_1 \cdots c_n$ equal to zero in (6.1.3); then the result (α_i) is employed on the right of (6.1.3) to recalculate (α_i). The process is repeated with the latest value of (α_i) until convergence occurs for a given q. With a new value of q assigned, the iteration scheme is then repeated.

The divergence problem concerns the determination of the highest value of q for which a solution $\{\alpha_i\}$ remains statically stable, that is, the limiting value of q for which convergence of the above-described process will occur. This depends, of course, upon the exact form of the function $C_M(\alpha)$; and equation (6.1.6) must, in general, be solved approximately.

A simple form of the problem occurs if the expression (6.1.4) for $C_M(\alpha)$ is linear:

$$C_M(\alpha) = c_0 + c_1\alpha \tag{6.1.7}$$

where c_0 is the value of C_M for $\alpha = 0$ and

$$c_1 = \frac{dC_M}{d\alpha} = C'_m \tag{6.1.8}$$

Letting

$$\frac{1}{p} \equiv qB^2 \, \Delta L \tag{6.1.9}$$

the problem reduces to the form

$$[pI - C'_M D_T]\{\alpha_i\} = D_T\{c_0\} \tag{6.1.10}$$

This has infinite (divergent) solutions for α_i when the determinant of the leading matrix vanishes:

$$|pI - C'_M D_T| = 0 \tag{6.1.11}$$

where I here is the identity matrix.

In the further special case that $c_0 = 0$ (moment coefficient is zero for $\alpha = 0$), the classic eigenvalue form for the problem is obtained:

$$[pI - C'_M D_T]\{\alpha_i\} = \{0\} \tag{6.1.12}$$

The largest eigenvalues p from either (6.1.11) or (6.1.12) permit the determination from (6.1.9) of the velocity U at which the divergent instability occurs.

Note that the necessary data, namely the evolution of C_M as a function of α (Figure 6.2) for this problem must be provided by wind tunnel test, for example on a geometrically similar model of a section of the bridge deck.

Many variants on the procedures described above can be devised, but that given serves to illustrate the character of the problem. The phenomenom can, in principle, occur for actual suspension bridge decks, though usually the calculated velocity for divergence of a normally designed bridge is considerably higher than expected wind speeds. Clearly, the most effective means of prevention of torsional divergence— or assuring that it will not occur until very high velocities are reached— are (1) a section moment coefficient with low slope and/or low overall value, and (2) high torsional stiffness of the structure.

6.2 Galloping

This is a relatively low-frequency oscillatory phenomenon of elongated, bluff bodies acted upon by a wind stream. Circular-section bodies do not generally gallop across-wind, but they may gallop in other modes, for example, if located in a shear flow. In the discussion which ensues, two illustrative types of galloping, across-wind and wake-induced, will be discussed.

When a bluff object is located in a cross flow, it trips off turbulence, some of which passes first over the object itself, causing time-varying pressures. In certain cases, which will be more completely discussed under the topic of vortex shedding, the turbulent wake organizes itself into a pattern of coherent, alternating vortices. Accompanying these are alternating pressures on the bluff object. In the usual discussion of the galloping phenomenon the natural structural frequency at which the bluff object in question responds is much lower than the frequencies of the vortex shedding or turbulence which take place around it. It is in this sense that galloping may be considered a low-frequency phenomenon.

In this circumstance, it becomes possible to formulate a theory of across-wind galloping that utilizes only the average lift and drag that develop over the body, and to disregard the rapid changes in these quantities that necessarily accompany the detailed local flow over it. The usual theories developed to describe galloping may be called 'quasi-steady' theories in light of these observations.

Across-wind galloping

One of the early manifestations of large-amplitude, across-wind oscillation was the galloping of ice-coated power lines under wind. Often a full
catenary span would exhibit such a vibration in its fundamental mode,
with center amplitudes approximating the cable sag. This phenomenon
was examined by Den Hartog [2]. A body having an undistorted circular
cross-section cannot gallop but various cross-sectional shapes other than
purely circular are gallop-prone. For example, the net configuration in
which a fairly heavy cable is suspended from a smaller diameter,
strapped-on supporting cable has been observed to gallop under strong,
steady winds in the field. Some large-diameter power cables made up of
wire strands had a configuration of wire lay that was conducive to
galloping [3]. Square and D-section rods (the latter with flat face to the
wind) have been demonstrated by tests to be gallop-prone.

The phenomenon will be examined here in terms of the time-
averaged or steady aerodynamic properties of the cross section of the
galloping object. Consider the cross section shown in Figure 6.3. Let
$C_D(\alpha)$ and $C_L(\alpha)$ be the experimentally determined steady drag and lift
coefficients, respectively, of the section given as functions of the geometric angle of attack α of the wind.

The effective, or relative, angle of attack of the wind will depend
upon the attitude of the fixed body or upon the across-wind velocity of
the moving body. If its across-wind displacement is denoted as y (positive
downward) the velocity in question is $\dot{y}$. The angle of attack of the

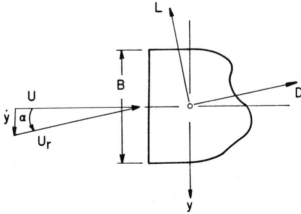

Figure 6.3 *Wind and motion components, with resultant lift and drag, on a bluff cross
section.*

relative wind to the body moving across-wind will then be

$$\alpha = \tan^{-1} \frac{\dot{y}}{U} \tag{6.2.1}$$

and the relative wind velocity U_r will have the value

$$U_r = [U^2 + \dot{y}^2]^{\frac{1}{2}} \tag{6.2.2}$$

that is

$$U_r = U \sec \alpha \tag{6.2.3}$$

The average drag force on the section will be

$$D = \tfrac{1}{2}\rho U_r^2 B C_D(\alpha) \tag{6.2.4}$$

and the corresponding lift will be

$$L = \tfrac{1}{2}\rho U_r^2 B C_L(\alpha) \tag{6.2.5}$$

The projection of these components in the across-wind or y direction gives the across-wind force

$$F_y(\alpha) = - D(\alpha) \sin \alpha - L(\alpha) \cos \alpha \tag{6.2.6}$$

$F_y(\alpha)$ may alternately be expressed in the form

$$F_y(\alpha) = \tfrac{1}{2}\rho U^2 B C_{F_y}(\alpha) \tag{6.2.7}$$

where

$$C_{F_y}(\alpha) = -[C_L(\alpha) + C_D(\alpha) \tan \alpha] \sec \alpha \tag{6.2.8}$$

If the section of the galloping body is assumed to be elastically supported so that it can vibrate across wind, the equation of motion of this section can be written

$$m[\ddot{y} + 2\zeta\omega_1 \dot{y} + \omega_1^2 y] = F_y \tag{6.2.9}$$

where m is the mass per unit span, ζ is the damping ratio-to-critical of the mechanical system ($\zeta = \delta/2\pi$, where δ is the logarithmic decrement of the system) and ω_1 is the undamped natural circular frequency of the system.

A basic criterion for galloping arises if the situation of incipient motion or initial, small departure from rest is considered; that is, the condition of very small angle of attack α. Then

$$F_y \cong \frac{\partial F_y}{\partial \alpha} \bigg|_{\alpha=0} \alpha \tag{6.2.10}$$

This leads to examination of $dC_{Fy}/d\alpha$, which from (6.2.8) is found to have the following value at $\alpha = 0$:

$$\frac{dC_{Fy}}{d\alpha}\bigg|_{\alpha=0} = -\left(\frac{dC_L}{d\alpha} + C_D\right) \tag{6.2.11}$$

Thus, for small motion, the galloping is governed by:

$$m[\ddot{y} + 2\zeta\omega_1\dot{y} + \omega_1^2 y] = -\tfrac{1}{2}\rho U^2 B\left[\frac{dC_L}{d\alpha} + C_D\right]\frac{\dot{y}}{U} \tag{6.2.12}$$

The system is then simply a linear oscillator with damping term ($\dot{y}$ term) having the coefficient

$$d = 2m\zeta\omega_1 + \tfrac{1}{2}\rho UB\left[\frac{dC_L}{d\alpha} + C_D\right] \tag{6.2.13}$$

According to linear theory the system is then stable if $d > 0$ and marginally or totally unstable if $d \leq 0$.

Since the mechanical damping ζ is always positive, the condition

$$\frac{dC_L}{d\alpha} + C_D < 0 \tag{6.2.14}$$

is seen to be necessary for galloping. This is known as the Den Hartog criterion [4]. In many cases examination of this criterion alone suffices to assess the galloping proclivities of a structure. For example, Figure 6.4 [5] illustrates the properties of a section of an octagonal lamp standard that exhibited susceptibility to galloping at certain azimuth angles of the wind because $dC_L/d\alpha$ was strongly negative at those angles.

A more complete investigation of the galloping problem reveals, however, that it is in fact strongly nonlinear. Parkinson [6–11], Novak [12, 13] and others have investigated these aspects of galloping.

The force F_y, being a function of α as given by (6.2.7), (6.2.8), can, in virtue of (6.2.1), be expressed as a polynomial in $\dot{y}$. This is done, for example, by Novak [12] who expressed F_y in the form:

$$F_y = A_1\left(\frac{\dot{y}}{U}\right) - A_2\left(\frac{\dot{y}}{U}\right)^2\frac{\dot{y}}{|\dot{y}|} - A_3\left(\frac{\dot{y}}{U}\right)^3 + A_5\left(\frac{\dot{y}}{U}\right)^5 - A_7\left(\frac{\dot{y}}{U}\right)^7 \tag{6.2.15}$$

where the A_i ($i = 1, 2, 3, 5, 7$) are determined by any desired curve-fitting procedure.

Next, since the galloping equation (6.2.9) is now nonlinear, special solution methods must be applied. Ref. [12] employs the method of

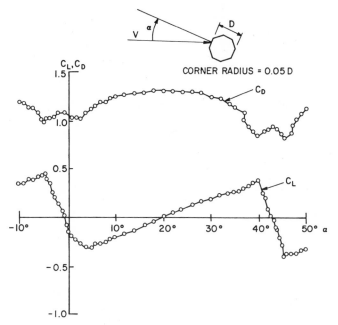

Figure 6.4 Lift and drag coefficients on an octagonal structural section [5].

Kryloff and Bogoliuboff [14], postulating as a first approximation to the response the forms:

$$y = a \cos (\omega t + \varphi) \qquad (6.2.16a)$$

$$\dot{y} = - a\omega \sin (\omega t + \varphi) \qquad (6.2.16b)$$

Where a and φ are considered to be slowly varying function of time.

Full details of the solution method will not be pursued here, but the theory adduced has proven effective in describing the phenomena observed. From analysis of the type suggested Ref. [13] identifies various type-forms of curves $C_F(\alpha)$ and the trends of the corresponding galloping response amplitude a as function of the reduced velocity parameter $U/B\omega_1$ (see Figure 6.5).

All of the above results were obtained assuming laminar incident flow, which often is adequate as a representation of actual conditions. However, the effects of small-scale turbulence upon galloping can, by affecting the average values of C_L and C_D in certain instances, modify the phenomenon. Finally, [13] notes that there exist situations, for example the condition of a large triggering disturbance of the galloping body,

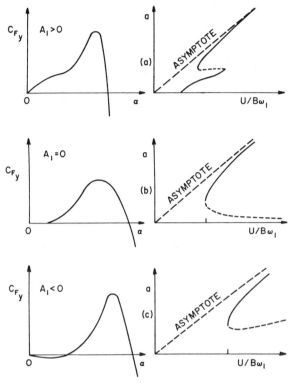

Figure 6.5   Types of across-wind force coefficients and corresponding trends of galloping amplitude [13].

wherein galloping takes place although it can be demonstrated that Den Hartog's criterion (6.2.14) is not satisfied.

In certain cases of elongated sectional dimensions, a more complex form of galloping, involving both across-wind and torsional freedoms, can be demonstrated. Analysis of this borders on the methods of classical flutter analysis, but the necessary aerodynamic information can nonetheless be extracted by average steady-state methods making the analysis 'quasi-steady' in this case also.

In the case of galloping it is not usually necessary to extend the detailed analysis spanwise into the third dimension, though this has been done [15]. Since the phenomenon is usually unwanted, identification of sectional galloping proclivitities and attempts to alleviate them are the main concerns. However, the galloping which occurs on electrical power lines has proven to be elusive and difficult to conquer. The primary reasons for this have been designed into the situation where long, generally unobserved cables are inevitably exposed to the vagaries of the weather. Some anti-galloping dampers, ice-warning and removal devices, the installation of connectors between adjacent cables, and other palliatives have been employed, each with modest success.

Wake galloping

Objects of circular section cannot develop any average lift in a uniform flow. However, if the flow region about them is sheared, that is, contains an across-flow velocity distribution, it is obvious that net pressure differentials over such objects can develop and lead to across-wind and along-wind forces. In the present section the wake galloping phenomenon, which is called 'subspan galloping' in connection with overhead power lines, will be described as a typical instability resulting from sheared flow.

The phenomenon occurs most prominently with bundled-conductor power lines. These are lines in which, instead of a single cable carrying one phase of the current, a group or 'bundle', consisting of two or more parallel cables, is employed. When parallel cables are so used they are held in the 'bundle' configuration by devices called spacers, which are placed perpendicular to the cables periodically along the span between support towers. The spacers divide a cable-bundle span between towers into subspans.

In the situation described, when wind blows across the span of the cable bundle it frequently occurs that one cable of the bundle is located in the aerodynamic wake of another cable. If it happens to lie in certain regions of that wake, particularly where strongly sheared flow is occurring, the downstream cable may exhibit the unstable oscillatory tendency which is called wake galloping or, in power line parlance, 'subspan oscillation'.

Figure 6.6 depicts the geometry of the situation, where the wake of the windward cylinder (cable) lies between the heavy lines. The quasi-steady aerodynamics of the phenomenon have been explored by Cooper [16, 34], who examined the forces of lift and drag acting on a downstream cylinder in various positions. The results observed are depicted qualitatively in Figure 6.7. Aside from expected velocity and drag defects, as indicated, a striking centering tendency of the lift on the downstream cylinder is to be noted, that is, lift is directed downward when the cylinder is above the wake centerline and upward when it is below that line. The physical reasons for this may stem from the averaging, on the

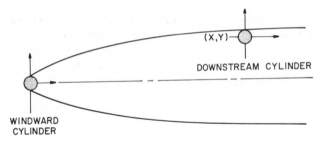

Figure 6.6   Geometry for the wake-galloping phenomenon.

cylinder in the wake, of the inward-directed effects of the alternating vortex flow leaving the windward cylinder.

In any event, when the leeward cylinder is elastically sprung in X- and Y-directions it exhibits oscillatory tendencies when located in certain regions of the wake. Figure 6.8 is an oscillograph trace of the oscillation orbit of a downstream cylinder in this kind of situation.

Figure 6.9 depicts the geometry of the situation for analysis purposes. Equations of motion of the downstream cylinder can be written in the form

$$m\ddot{x} + d_x\dot{x} + k_{xx}x + k_{xy}y = F_x \tag{6.2.17a}$$

$$m\ddot{y} + d_y\dot{y} + k_{yx}x + k_{yy}y = F_y \tag{6.2.17b}$$

where m is its mass per unit span, d_x and d_y are damping constants, $k_{rs}(r, s = x, y)$ are direct and cross-coupling mechanical spring constants, and F_x, F_y are aerodynamic forces. These are given, from theory and from experimental wake explorations of the type suggested above, in the

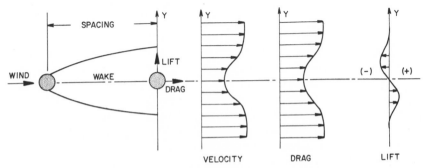

Figure 6.7   Wake properties behind a cylinder [16].

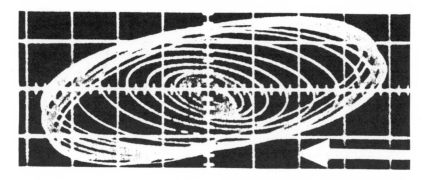

Figure 6.8 Oscillograph trace of cylinder wake galloping [34].

form [17]:

$$F_x = \tfrac{1}{2}\rho U^2 D\left\{\frac{\partial C_x}{\partial x}x + \frac{\partial C_x}{dy}y + C_y\frac{\dot{y}}{U_w} - 2C_x\frac{\dot{x}}{U_w}\right\} \qquad (6.2.18a)$$

$$F_y = \tfrac{1}{2}\rho U^2 D\left\{\frac{\partial C_y}{\partial x}x + \frac{\partial C_y}{\partial y}y - C_y\frac{\dot{y}}{U_w} - 2C_y\frac{\dot{x}}{U_w}\right\} \qquad (6.2.18b)$$

where $\tfrac{1}{2}\rho U^2$ is the free stream dynamic pressure, D is diameter of the wake cylinder, U_w is the wake velocity at point (x, y) in the wake, and C_x, C_y are aerodynamic drag (x) and lift (y) coefficients referred to free-stream velocity. Equations (6.2.18) appear to have been derived originally by Simpson [18].

Solution of the equations of motion is not unlike that of the flutter problem, with the exception that experimental steady-state data, in real form, suffice for the formulation. The equations being linear, solutions for x and y of the form $e^{\lambda t}$ are seen to be stable when $\lambda = \omega_1 + i\omega_2$ is found to have $\omega_1 < 0$. For $\omega_1 = 0$, boundaries in the wake can then be defined which identify the regions of wake galloping susceptibility. Figure 6.10, which depicts a wake region of instability, suggests typical results [16] obtained experimentally and theoretically.

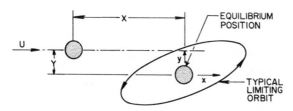

Figure 6.9 Coordinates for the wake galloping problem [16].

293

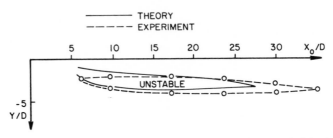

Figure 6.10   Wake galloping: Theoretically and experimentally derived definitions of the wake region of instability [16].

Alleviation of this problem has been experimentally accomplished for power lines by increasing the number of spacers used along bundled spans, thus raising subspan natural frequencies to higher values; and the twisting of the entire bundle throughout the span, destroying the necessary aerodynamic wake coherence for the phenomenon to take place. In practice, however, the isolated occurrence of subspan oscillation has been difficult to detect and arrest, mainly for several economic reasons attendant on operation of long power line systems. Considerable effort has been directed along the lines of various spacer designs (as dampers, line detuners, etc.) with only modest success.

6.3   Vortex shedding

The wind harp of the ancient Greeks consisted of a set of strings which vibrated when exposed to breezes, eliciting various tones. That the oscillations of such strings were mainly across-wind and associated with the shedding of alternating vortices in the wake of the strings was not generally recognized until much later. In 1878 Strouhal [19] published a paper on the eliciting of tones by fluid flow. Since this work the name of Strouhal has been associated with the dimensionless number

$$\frac{n_s D}{U} = S \tag{6.3.1}$$

associated with vortex shedding from various bluff bodies of cross-sectional dimension D. Strouhal noted that the value of S is nearly constant (equal to about 0.2) for flow around a circular section, whereas the frequency of n_s of the associated vortex shedding was directly proportional to the incident laminar flow velocity U.

294

In 1908 Bénard [20] and later von Kármán [21] described the phenomenon of vortex shedding, the latter author offering a tentative theory associated with the organized vortex trail, or wake, behind a fixed object placed in a two-dimensional flow (see also [22]). Equation (6.3.1) has persisted as a basically correct relation for the phenomenon, the value of S being adjusted (from about 0.10 to 0.3) for various geometric shapes traversed by the flow.

The phenomenon of a regular trail of vortices with alternating directions of rotation has proved fascinating to observers, and many striking photographs of it have been made in laboratory air and water flows, and in the field. (See, for example [23]). From an engineering viewpoint, interest in the phenomenon centers on the fact that the shed vortices are accompanied by rhythmic pressure distributions which can induce structural oscillation, particularly at the natural frequency of the structural member in the flow.

The structural member acts as if rigidly fixed, in the main, when the frequency n_s of vortex shedding from it is not close to a natural frequency of the member. On the other hand, the member, usually being elastic, oscillates in resonance when the vortex-induced and natural frequencies coincide. When this occurs—the most important case to be considered during structural design—the structural deflection is found to affect, and in fact for a certain velocity range, to control the vortex shedding. This phenomenon is known as *lock-in*, and the progress of it with increasing flow velocity is suggested in Figure 6.11. When velocity increases, after lock-in occurs, it must be augmented by a considerable amount before the

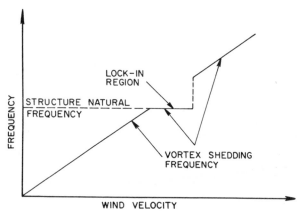

Figure 6.11   Qualitative trend of vortex shedding frequency with wind velocity when lock-in occurs [49].

295

6 Aeroelastic problems of civil engineering structures

natural rhythm of the fluid finally breaks away from that of the structure and returns to adherence to the Strouhal relation (6.3.1).

It has not been possible, to date, to establish purely analytical theories adequate to describe vortex shedding and the lock-in phenomenon. Some experimental element has been found indispensable. For the fixed structural element in air, the across-wind force has predominated, and along-wind effects have been negligible. In water, on the other hand, along-flow effects can be pronounced, giving rise to strong along-flow oscillation (of basic frequency twice the Strouhal frequency). A reasonable approximation to the across-flow force on a fixed body is given by

$$F = \tfrac{1}{2}\rho U^2 D C_{LS} \sin \omega_s t \qquad (6.3.2)$$

where D is the across-flow body dimension, C_{LS} is a lift coefficient, and $\omega_s = 2\pi n_s$, n_s satisfying (6.3.1). (For a circular cylinder at Reynolds number between 40 and 3×10^5 Bishop and Hassan [24] found $C_{LS} \cong 0.6$).

When the body displaces elastically however, the description (6.3.2) is inadequate, and a more realistic model must be sought. One such is the 'lift oscillator' model of Hartlen and Currie [25]. The equation of across-wind motion may be written

$$M[\ddot{y} + 2\zeta\omega_1\dot{y} + \omega_1^2 y] = \tfrac{1}{2}\rho U^2 D C_L \qquad (6.3.3)$$

where C_L is considered to have Strouhal oscillator proclivities. One form of this idea is to require that C_L satisfy the characteristics of a Van der Pol oscillator triggered by the velocity $\dot{y}$ of the across-wind motion of the structure. To this end it is required that C_L satisfy a differential equation of the form

$$\ddot{C}_L + a_1\dot{C}_L + a_2\dot{C}_L^3 + a_3 C_L = a_4\dot{y} \qquad (6.3.4)$$

where $a_1, \ldots a_4$ are constants to be fitted by experiment. Typically, the constant a_3 will be equal to $(2\pi n_s)^2$, and the choice of a_1 and a_2 will be such that low amplitudes of C_L have low or negative damping whereas higher ones will be more strongly damped, resulting in limit cycle oscillations. Extensive developments of this model have been discussed in [26–29]. The usual solution y, in spite of the nonlinear character of the coupled equations (6.3.2) and (6.3.4), is taken as sinusoidal, in close agreement with experimental observations.

A simpler model, used to represent the action in the lock-in range only, is as follows:

$$M[\ddot{y} + 2\zeta\omega_1\dot{y} + \omega_1^2 y] = \tfrac{1}{2}\rho U^2 D \left[Y_1 \frac{\dot{y}}{U} + Y_2 \frac{y}{D} + C_L \sin(\omega_1 t + \varphi) \right]$$

296

where constants Y_1, Y_2, C_L and ψ are fitted to observed experimental results during lock-in.

It should be emphasized that the empirical nature of the models devised requires strong recourse to experiment. The models are not, in themselves, explanatory of the details of the complex observed vortex-shedding and lock-in phenomenon but are merely post-facto fits to the gross features of these phenomena.

The action of actual structures under vortex shedding is widely observed. Tall, unsupported stacks and towers may oscillate across-wind; cables and lines of a great variety exhibit 'Aeolian' vibrations; certain long, unsupported structural members will vibrate in the wind; and the decks of cable-stayed and suspension bridges may undulate under cross-winds of relatively low velocity. The means of prevention or alleviation are several, but basically they are of two types: mechanical or aerodynamic.

Mechanical approaches have included stiffening, or raising the natural frequency, of the structure, increasing the damping, or installing a device such as a tuned-mass damper. Aerodynamic approaches have sought to spoil the coherence of the flow in order to break up the vortex shedding. As examples. Figures 6.12–6.17 illustrate a tuned-mass damper for suppressing the vortex-induced oscillation of a long structural member (bridge deck hanger), the 'Stockbridge' tuned mass damper for suppressing the effects, near a support point, of the aeolian vibration of a power line cable, and several aerodynamic spoiler devices. Refs. [5] and [30–33] discuss some of the latter. Figure 6.18a depicts the results of wind-induced vibration in model tests of various fairings in reducing the vortex-induced response of a cable-stayed bridge deck which was in close proximity to a water surface. Figure 6.18b shows the final choice of fairing on this bridge as installed in the field.

6.4   Flutter

Of the several civil structures exhibiting wind-induced oscillations, the suspension bridge has perhaps been the most notable. This has been a result of the prominent exposure of bridges to wind and the fact that suspension bridges are quite flexible, with generally low natural frequencies. On several occasions bridges of this type have suffered serious damage or even complete destruction under wind. While the Tacoma Narrows disaster of 1940 occasioned the modern experimental and analytical approaches to problems of bridge aeroelasticity, that occurrence

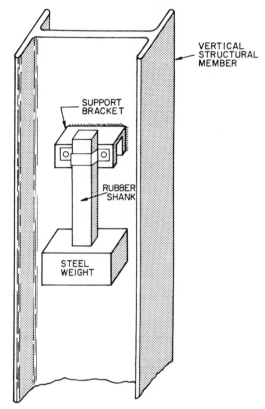

Figure 6.12 *Schematic form of tuned mass damper to suppress vortex-induced oscillations of vertical structural member.*

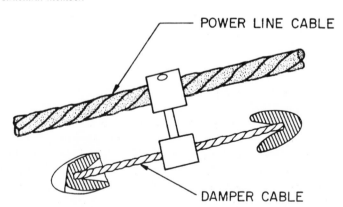

Figure 6.13 *Diagrammatic form of the stockbridge-type power line damper [5].*

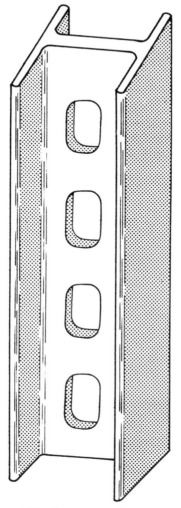

Figure 6.14 Structural member with web holes.

was in fact one of a long line of similar events stretching back 100 years or more into history.

The destructive oscillations of these events are now generally classified as types of flutter, although many appear to be single-degree-of-freedom instabilities in torsion* rather than classical flutter. However, coupled bending-torsion oscillations can and do occur.

---

* These appear to be a type of flutter, akin to stall flutter, associated with partially separated flow during some portion of the oscillation cycle. See Chapter 5.

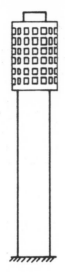

Figure 6.15   Stack with porous shroud [32].

Suspension bridge decks fall broadly into two classifications: those with roadways stiffened by open, latticed trusswork, and those—usually with single deck and of more recent design—having a roadway atop a closed, box-like stiffening structure. Figures 6.19 and 6.2 show examples of the two types. While both types are susceptible to flutter, the more stream-lined box type will tend to have a higher critical flutter speed and may enter into coupled flutter, whereas the truss-stiffened type is very likely to exhibit single-degree torsional flutter. Very bluff deck sections tend strongly to shed vortices as well as enter single-degree flutter. The girder-stiffened original Tacoma Narrows bridge possessed a squat *H*-like section which was extremely unstable aerodynamically, both in its vortex-shedding and its flutter tendencies.

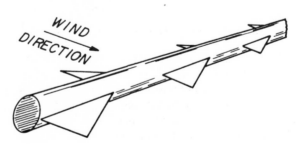

Figure 6.16   Pipeline with fins [50].

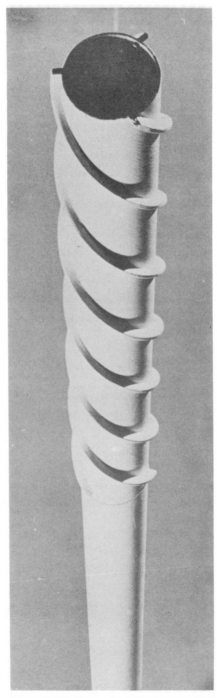

Figure 6.17   Stack with helical strakes [32].

301

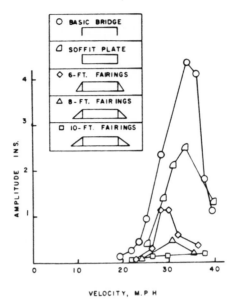

VELOCITY, M.P H

Figure 6.18   (a) Wind tunnel measurements of vertical amplitudes for the Long's Creek bridge—bridge height 15 ft [5].

Figure 6.18   (b) modified Long's Creek bridge [5].

302

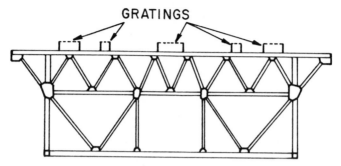

Figure 6.19   Typical cross section of truss-stiffened suspension bridge deck [50].

The collapse of that bridge is often said to be attributable to vortex shedding. In the sense in which this term and the term flutter are presently used, some clarification is perhaps in order. In the usual sense of vortex shedding, a certain rhythm of alternate vortex formation occurs as a result of a fixed, rigidly-supported body being traversed by a smooth flow, with the vortices trailing into its wake. In this case the rhythm is a function only of the geometry of the body and the Reynolds number of the flow. On the other hand, while a fluttering body also produces vortices in its wake, in this case one or more natural structural frequencies influence the rhythm. In fact the shed vortices are the direct result of that ryhthm, which reflects the changing motion (and therefore lift) of the structure itself. Wind speeds at which flow-induced vortex-shedding tends to lock in on a natural structural frequency are usually much lower than flutter speeds; furthermore, increases in vortex-shedding wind speeds tend to break the locked-in condition and leave the structure without much response. However, flutter, once engaged, generally tends to grow more violent as wind speed increases. Thus flutter, while indeed accompanied by a shed vortex wake, has little or nothing to do with a natural vortex rhythm intrinsic to the flow itself.

The study of bridge flutter owes much to the literature on airfoil flutter. The two prove to be enough different in character, however, so that the theoretical aerodynamic formulations for airfoil flutter are not directly transferable to bridges, as will be seen. It is notable, however, that such an attempt was made by Bleich [35].

Further comments on the character of suspended-span bridge flutter will be postponed until an analytical groundwork of the problem has been laid. Let h and α be the vertical and torsional degrees of freedom of a bridge deck, referred to its elastic axis, as illustrated in Figure 6.20. The

303

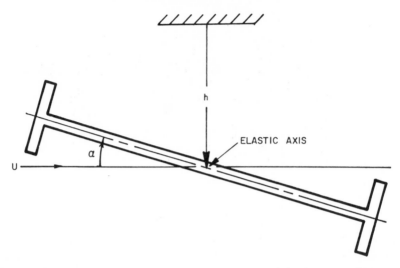

Figure 6.20 Degrees of freedom for analytical description of the divergence, flutter, and buffeting problems.

equations of motion of the bridge deck section are mechanically the same as for the typical airfoil section (recall Section 3.2) namely

$$m\ddot{h} + S_\alpha\ddot{\alpha} + c_h\dot{h} + K_h h = L_h \qquad (6.4.1a)$$

$$S_\alpha\ddot{h} + I\ddot{\alpha} + c_\alpha\dot{\alpha} + K_\alpha\alpha = M_\alpha \qquad (6.4.1b)$$

where m, S_α, I are respectively mass, static unbalance about the elastic axis, and mass moment of inertia, all per unit span; c_h and c_α are damping coefficients, K_h and K_α are sectional stiffnesses, and L_h, M_α are respectively aerodynamic lift and moment at the elastic axis. The majority of bridges are symmetric about their centerlines, and have $S_\alpha \equiv 0$.

Various approaches have been employed—notably in Japan, the United States and France—to obtain experimental results for L_h and M_α. These approaches have centered about two methods: measuring the forces on sinusoidally machine-driven models of bridge deck sections, or inferring the desired results from freely oscillating deck models by system identification techniques. Some results [36] of the U.S. method will be given.

The basic models [36] for the self-excited (flutter) forces and moments on bridge decks were postulated first as linear, thus:

$$L_h = H_1\dot{h} + H_2\dot{\alpha} + H_3 \qquad (6.4.2a)$$

$$M_\alpha = A_1\dot{h} + A_2\dot{\alpha} + A_3\alpha \qquad (6.4.2b)$$

with H_i, A_i ($i = 1, 2, 3$) to be determined, and where possible aerodynamic inertial terms in $\dot{h}$, $\ddot{\alpha}$ were considered negligible. It has become the practice to treat H_i, A_i as real in this context, rather than complex as in the aeronautical context. These coefficients are, as in classical flutter, expected to be functions of the reduced frequency $b\omega/U = k$ where ω is the flutter circular frequency, U the wind velocity, and b the half-chord of the surface under study. Usage in the bridge flutter context has replaced k by $K = B\omega/U$, where B is the full deck width.

It is of course desirable to put the coefficients H_i, A_i into nondimensional forms, which is accomplished in the flutter context by the relations

(a) $H_1^* = \dfrac{mH_1}{\rho B^2 \omega}$ (d) $A_1^* = \dfrac{IA_1}{\rho B^3 \omega}$

(b) $H_2^* = \dfrac{mH_2}{\rho B^3 \omega}$ (e) $A_2^* = \dfrac{IA_2}{\rho B^4 \omega}$ (6.4.3)

(c) $H_3^* = \dfrac{mH_3}{\rho B^3 \omega^2}$ (f) $A_3^* = \dfrac{IA_3}{\rho B^4 \omega^2}$

where ρ is air density and ω is flutter frequency.

With these definitions of nondimensional aerodynamic coefficients H_i^*, A_i^*, the flutter lift and moment may be written

$$L_h = \tfrac{1}{2}\rho U^2 (2B)\left[KH_1^*(K)\frac{\dot{h}}{U} + KH_2^*(K)\frac{B\dot{\alpha}}{U} + K^2 H_3^*(K)\alpha \right] \quad (6.4.4a)$$

$$M = \tfrac{1}{2}\rho U^2 (2B^2)\left[KA_1^*(K)\frac{\dot{h}}{U} + KA_2^*(K)\frac{B\dot{\alpha}}{U} + K^2 A_3^*(K)\alpha \right] \quad (6.4.4b)$$

In the context developed here, factors like $K^2 H_3^*$ are seen to be analogous to derivatives like $dC_L/d\alpha$; hence, loosely speaking, the coefficients H_i^* and A_i^* may be termed 'flutter derivatives' or 'flutter coefficients'. Since in aeronautical practice it is usual to employ the half-chord b in place of B, it is worth noting that replacement of B by b above throughout would result in a comparable set $\tilde{H}_i^*$, $\tilde{A}_i^*$ bearing the following relation to H_i^*, A_i^*:

(a) $\tilde{H}_1^* = 4H_1^*$ (d) $\tilde{A}_1^* = 8A_1^*$
(b) $\tilde{H}_2^* = 8H_2^*$ (e) $\tilde{A}_2^* = 16A_2^*$ (6.4.5)
(c) $\tilde{H}_3^* = 8H_3^*$ (f) $\tilde{A}_3^* = 16A_3^*$

It can then be shown [36] that the set $\tilde{H}_i^*$, $\tilde{A}_i^*$, when written for the classical case of the thin airfoil, has the following values:

(a) $k\tilde{H}_1^*(k) = -2\pi F(k)$

(b) $k\tilde{H}_2^*(k) = -\pi\left[1 + F(k) + \dfrac{2G(k)}{k}\right]$

(c) $k^2\tilde{H}_3^*(k) = -2\pi\left[F(k) - \dfrac{kG(k)}{2}\right]$

(d) $k\tilde{A}_1^*(k) = \pi F(k)$

(e) $k\tilde{A}_2^*(k) = \dfrac{-\pi}{2}\left[1 - F(k) - \dfrac{2G(k)}{k}\right]$

(f) $k^2\tilde{A}_3^*(k) = \pi\left[F(k) - \dfrac{kG(k)}{2}\right]$

(6.4.6)

where $F(k) + iG(k) = C(k)$ is the well-known Theodorsen circulation function (see Section 4.3).

In contrast to the case of the thin airfoil, all values of H_i^* and A_i^* for bridge decks must be obtained from experiment. Methods for accomplishing this are outlined in [36, 37], the second of which discusses obtaining flutter coefficients under turbulent wind. Examples of the type of results obtained are given in Figure 6.21, where coefficients for the airfoil, original Tacoma Narrows bridge, and three truss-stiffened bridge decks are presented as functions of U/NB, where $N = \omega/2\pi$. Figure 6.22 presents analogous results for some box sections, of which 1 and 2 are streamlined to some extent. It is clear from a broad comparison of the thin airfoil results with those of bridge decks that airfoil flutter coefficients are inappropriate for bridge decks.

The numerical solution for the flutter problem, that is, of the problem defined by (6.4.1) and (6.4.4), follows the classical method. A value of K is chosen, and the corresponding values of H_i^* and A_i^* for (6.4.4) are determined from results like those of Figures 6.21 and 6.22. Solutions at the flutter condition depend upon the assumption of responses having the form $h = h_0 e^{i\omega t}$. The determinant of coefficients of the equations (6.4.1) and (6.4.4) is then set equal to zero. Defining the unknown flutter frequency in the form

$$X = \omega/\omega_h \qquad (6.4.7)$$

where $\omega_h^2 \equiv K_{h/m}$ and $\omega_\alpha^2 \equiv K_{\alpha/I}$, one then requires the satisfaction of the following two (real) equations simultaneously:[†]

$$X^4\left(1+\frac{\rho B^4}{I}A_3^*-\frac{\rho B^2}{m}\frac{\rho B^4}{I}A_2^*H_1^*+\frac{\rho B^2}{m}\frac{\rho B^4}{I}A_1^*H_2^*\right)$$

$$+X^3\left(2\zeta_\alpha\frac{\omega_\alpha}{\omega_h}\frac{\rho B^2}{m}H_1^*+2\zeta_h\frac{\rho B^4}{I}A_2^*\right)$$

$$+X^2\left(-\frac{\omega_\alpha^2}{\omega_h^2}-4\zeta_h\zeta_\alpha\frac{\omega_\alpha}{\omega_h}-1-\frac{\rho B^4}{I}A_3^*\right)$$

$$+X\cdot 0$$

$$+\left(\frac{\omega_\alpha}{\omega_h}\right)^2=0 \tag{6.4.8a}$$

$$X^3\left(\frac{\rho B^4}{I}A_2^*+\frac{\rho B^2}{m}H_1^*+\frac{\rho B^2}{m}\frac{\rho B^4}{I}H_1^*A_3^*-\frac{\rho B^2}{m}\frac{\rho B^4}{I}A_1^*H_3^*\right)$$

$$+X^2\left(-2\zeta_\alpha\frac{\omega_\alpha}{\omega_h}-2\zeta_h-2\zeta_h\frac{\rho B^4}{I}A_3^*\right)$$

$$+X\left(-\frac{\rho B^2}{m}H_1^*\frac{\omega_\alpha^2}{\omega_h^2}-\frac{\rho B^4}{I}A_2^*\right)$$

$$+\left(2\zeta_h\frac{\omega_\alpha^2}{\omega_h^2}+2\zeta_\alpha\frac{\omega_\alpha}{\omega_h}\right)=0 \tag{6.4.8b}$$

As one typical method of achieving the final result, the problem is pursued to the above point for each of a number of values of K over a chosen range, and the real solutions ω of each of (6.4.8a) and (6.4.8b) are plotted versus K. The point where the solution curves cross yields the desired real flutter frequency ω_c. This, together with the corresponding K value, K_c, permits determination of the flutter velocity U_c:

$$\frac{B\omega_c}{K_c}=U_c \tag{6.4.9}$$

In the frequently occurring case of single-degree flutter in torsion, the above problem is considerably simplified, since then $h=0$ and a single equation in α remains. This will be further discussed below when the full three-dimensional problem is treated.

[†] These results are written for the common case $S_\alpha \equiv 0$, i.e., a symmetric bridge deck.

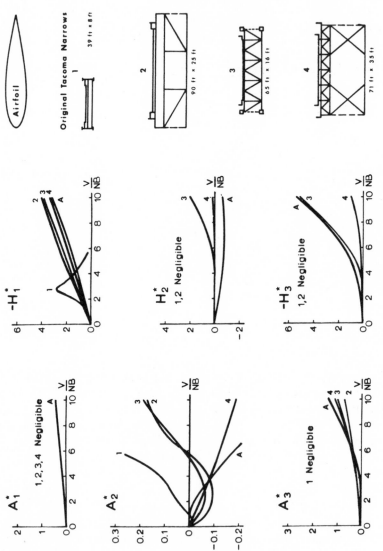

Figure 6.21   Some representative flutter coefficients [36].

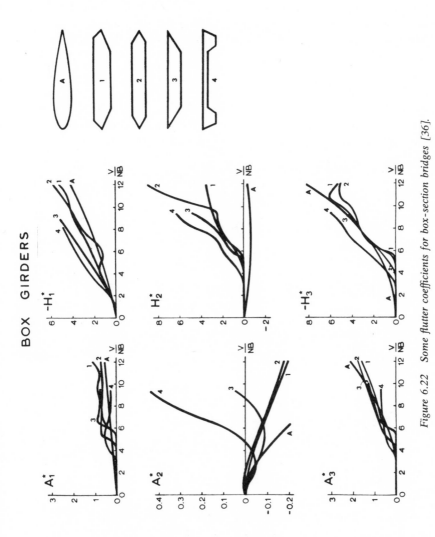

Figure 6.22   Some flutter coefficients for box-section bridges [36].

6 Aeroelastic problems of civil engineering structures

The three-dimensional (full-span) flutter problem for bridges can readily be formulated from the analysis carried to this point. In bridge flutter the lowest modes in bending and torsion suffice for the analysis since they yield the desired lowest critical wind velocity. Letting then:

$$h(x, t) = h(x)p(t) \qquad (6.4.10a)$$

$$\alpha(x, t) = \alpha(x)q(t) \qquad (6.4.10b)$$

represent the deflection at spanwise deck section x where $h(x)$ and $\alpha(x)$ are the desired fundamental modal forms and p, q are generalized coordinates, permits the problem to be re-expressed by the two equations:

$$M_1[\ddot{p} + 2\zeta_h\omega_h\dot{p} + \omega_h^2 p] = \tfrac{1}{2}\rho U^2(2B)$$
$$\times \left[KC_{11}H_1^* \frac{\dot{p}}{U} + KC_{12}H_2^* \frac{B\dot{q}}{U} + K^2 C_{12}H_3^* q \right] \qquad (6.4.11a)$$

$$I_1[\ddot{q} + 2\zeta_\alpha\omega_\alpha\dot{q} + \omega_\alpha^2 q] = \tfrac{1}{2}\rho U^2(2B^2)$$
$$\times \left[KC_{12}A_1^* \frac{\dot{b}}{U} + KC_{22}A_2^* \frac{B\dot{q}}{U} + K^2 C_{22}A_3^* q \right] \qquad (6.4.11b)$$

where subscripts h and α refer to those modes, $\omega_\alpha = 2\pi n_\alpha$, and

(a) $M_1 = \displaystyle\int_0^L m(x)h^2(x)\,\mathrm{d}x$

(b) $I_1 = \displaystyle\int_0^L I(x)\alpha^2(x)\,\mathrm{d}x$

(c) $C_{12} = \displaystyle\int_0^L h(x)\alpha(x)\,\mathrm{d}x$ $\qquad\qquad$ (6.4.12)

(d) $C_{11} = \displaystyle\int_0^L h^2(x)\,\mathrm{d}x$

(e) $C_{22} = \displaystyle\int_0^L \alpha^2(x)\,\mathrm{d}x$

The entire flutter problem then may be carried through as for the two-dimensional case described above by simply replacing H_1^*, H_2^*, H_3^*, A_1^*, A_2^*, A_3^*, respectively, with the generalized flutter coefficients

$$C_{11}H_1^*, \ C_{12}H_2^*, \ C_{12}H_3^*, C_{12}A_1^*, \ C_{22}A_2^*, \ C_{22}A_3^*$$

In the special case of the single-degree problem in torsion, the critical velocity is determined from the condition that the total (mechanical plus aerodynamic) damping must be zero in the mode concerned, the

310

frequency of which is very close to the flutter frequency. This results in the relation

$$A_2^* = \frac{2I_1\zeta_\alpha}{\rho B^4 C_{22}} \tag{6.4.13}$$

If the value of reduced velocity for which (6.4.13) holds is K_c, then equation (6.4.9) yields the flutter velocity, with $\omega_c \cong 2\pi n_\alpha$.

Reduction of flutter tendencies is accomplished aerodynamically by streamlining box sections and by creating vents or open grids in the floors of truss-stiffened decks. However, final checks of flutter proclivities depend upon the results of wind tunnel model tests, either section model results or, occasionally, full dynamically similar bridge models. The latter tend to be much more costly than section model tests. Structural reduction of flutter tendencies is best accomplished by raising torsional stiffness (frequency) to as high a value as possible. Certain sectional forms, notably rectangular and *H*-sections, are quite flutter-prone and should be avoided. Figure 6.23 illustrates three deck section forms with good anti-flutter tendencies. Mechanical dampers have occasionally been suggested for bridges, as have tuned mass dampers, as anti-flutter devices. The latter offer good potential for the purpose but have not to date been exploited to this end.

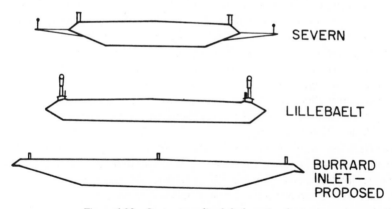

Figure 6.23 *Some streamlined deck section forms.*

6.5 Buffeting theory for line-like (one dimensional) structures

Many nonaeronautical structures are buffeted by the natural wind. Prominent among them are buildings, towers and bridges. The relatively small deflections of most of these structures under wind do not modify the local flow conditions sufficiently to bring self-excited forces into play. However, in the case of such line-like structures as suspension bridges, which are flexible and susceptible to flutter as noted above, self-excited forces can and do arise. These forces, the basic causes of flutter, can also be quite important in buffeting deflections. For this reason the buffeting of suspension bridges provides a representative example of an aeroelastic phenomenon.

This problem, like the divergence and flutter problems, will also be dealt with by the strip theory approach. Therefore, the section lift, drag, and moment are first required. The problem is linearized, and the principle of superposition of the responses to forces from various sources is employed. With these premises the section lift and moment may be expressed as

$$\mathcal{L}(x, t) = L_{se} + L_b \tag{6.5.1a}$$

$$\mathcal{M}(x, t) = M_{se} + M_b \tag{6.5.1b}$$

where the subscripts se and b refer to self-excited and buffeting sources. The self-excited forces are presumed to arise from bridge motion (i.e., values of h and α and their time derivatives), while the buffeting forces are due to time-varying wind components. Since the latter have a random character the bridge response is also random, being a superposition of responses in a number of natural bridge modes. In the cases of bridge buffeting and flutter, the modes in which the bridge responds are found to be practically identical to the mechanical modes of the structure, unmodified in form, and very little in frequency, by the aerodynamic forces. The aerodynamic effects in this context do not appear to be large enough to distort the mechanical modes appreciably in the majority of cases.

Under these conditions, and given the light, often negligible, aerodynamic coupling between modes, approximate buffeting response calculations can be made for each bridge mode individually. This will first be described here, as it permits direct transfer of the analytical forms used for the self-excited forces employed in the flutter problem. In this problem, it will be recalled, oscillation at a single frequency n is postulated. Thus, provisionally neglecting the aerodynamic coupling efficients H_2^*, H_3^*, A_1^*, the following can be written:

$$L_{se} \cong \tfrac{1}{2}\rho \bar{U}^2 (2B) K H_1^*(K) \tag{6.5.2a}$$

$$M_{se} \cong \tfrac{1}{2}\rho \bar{U}^2 (2B^2)[K A_2^*(K) + K^2 A_3^*(K)] \tag{6.5.2b}$$

where $K = B\omega/U$, $\omega = 2\pi n$, n being the oscillation frequency. Most often $n \cong n_i$ where n_i is the natural structural frequency of the ith mode, that is, the particular mode for which the buffeting study is being made. $\bar{U}$ is the mean wind velocity across the bridge.

On the other hand, expressions for the time-dependent buffeting forces, which are functions of the wind gust components and the geometry of the structure, can be specified as follows. Let the three-dimensional components of the wind be $\bar{U} + u(t)$, $v(t)$, $w(t)$, the gust components u, v, w being time-dependent. If C_D, C_L, C_M are respectively the steady drag, lift, and moment coefficients per unit bridge span (based on deck width B and referred to the deck section mass center), and if it is tentatively assumed that these quantities are independent of frequency (i.e., of the frequencies of the Fourier components of u, v, w) then the total drag, lift and moment acting on unit span of the bridge deck are given by*

$$\frac{D(t)}{\frac{1}{2}\rho\bar{U}^2 B} = C_D(\alpha_0)\frac{A}{B}\left[1 + 2\frac{u(x,t)}{\bar{U}}\right] \tag{6.5.3a}$$

$$\frac{-L(t)}{\frac{1}{2}\rho\bar{U}^2 B} = C_L(\alpha_0)\left[1 + 2\frac{u(x,t)}{\bar{U}}\right] + \left[\frac{dC_L}{d\alpha}\bigg|_{\alpha=\alpha_0} + \frac{A}{B}C_D(\alpha_0)\right]\frac{w(x,t)}{\bar{U}} \tag{6.5.3b}$$

$$\frac{M(t)}{\frac{1}{2}\rho\bar{U}^2 B^2} = \left[C_M(\alpha_0) + C_D(\alpha_0)\frac{Ar}{B^2}\right]\left[1 + 2\frac{u(x,t)}{\bar{U}}\right] + \frac{dC_m}{d\alpha}\bigg|_{\alpha=\alpha_0}\frac{w(x,t)}{\bar{U}} \tag{6.5.3c}$$

where A is across-wind projected area (per unit span) normal to $\bar{U}$, r is distance from the deck section mass to the elastic (or effective rotation) axis, and α_0 is the mean angle of attack under the wind velocity $\bar{U}$. (Note that the quantity $1 + 2(u/\bar{U})$ has been obtained by squaring $1 + (u/\bar{U})$ and neglecting $(u/\bar{U})^2$.) The value of α_0 can be obtained to sufficient accuracy by a static analysis of the bridge under non-time-varying wind loads, i.e., with u, v, w equal to zero. Measured values of C_D, C_L, C_M are assumed available as functions of angle of attack α.

Analysis of the buffeting of torsion modes will be discussed first. Let the total twist $\alpha(x, t)$ at spanwise section x be given in terms of torsion

---

*Equations (6.5.3) do not contain any *aerodynamic admittance* factors in their present form. If included, they would be factors correcting D, L, M for their dependence on the frequency content of u, v, w. Such factors most commonly are included when the spectra of D, L, M are employed. Refs. [38–40] discuss aerodynamic admittance functions. They will not be included in what follows.

modes $\alpha_i(x)$ by

$$\alpha(x, t) = \sum_i \alpha_i(x)p_i(t) \tag{6.5.4}$$

where p_i is a generalized coordinate. The deck section (assumed symmetrical here for simplicity) responds at spanwise section x according to the equation

$$I(x)\ddot{\alpha}(x, t) + c_\alpha(x)\dot{\alpha}(x, t) + k_\alpha(x)\alpha(x, t) = \mathcal{M}(x, t) \tag{6.5.5}$$

where $I(x)$, $c_\alpha(x)$, $k(x)$ are respectively section mass moment of inertia, damping coefficient, and stiffness, and $\mathcal{M}$ is total aerodynamic moment on the section. For the single mode $\alpha_i(x)$, of natural frequency n_{α_i}, this equation becomes

$$I_i[\ddot{p}_i(t) + 2\zeta_{\alpha i}(2\pi\eta_{\alpha i})\dot{p}_i(t) + (2\pi\eta_{\alpha i})^2 p_i(t)] = M_{\alpha_i}(t) \tag{6.5.6}$$

where

$$I_i = \int_0^L I(x)\alpha_i^2(x)\,\mathrm{d}x \tag{6.5.7}$$

L being the bridge span, $\zeta_{\alpha i}$ the damping ratio in mode α_i, and M_{α_i} being the generalized force

$$\mathcal{M}_{\alpha_i} = \int_0^L \mathcal{M}(x, t)\alpha_i(x)\,\mathrm{d}x \tag{6.5.8}$$

It is now noted that, by equation (6.5.1b), the moment $\mathcal{M}_{\alpha_i}$ contains contributions of both self-excited and buffeting origin. The self-excited contributions, stemming from equations (6.5.2), are terms in $\dot{\alpha}$ and α that may be brought to the left side of equation (6.5.6) and combined with their mechanical counterparts. This leads to the equation

$$I_i[\ddot{p}_i(t) + 2\tilde{\zeta}_{\alpha_i}(2\pi\tilde{n}_{\alpha_i})p_i(t) + (2\pi\tilde{n}_{\alpha_i})^2 p_i(t)] = \int_0^L \mathcal{M}_b(x, t)\alpha_i(x)\,\mathrm{d}x \tag{6.5.9}$$

where $\tilde{\zeta}_{\alpha_i}$ and $\tilde{n}_{\alpha_i}$ are modified damping and frequency defined as follows:

$$\tilde{n}_{\alpha_i}^2 = n_{\alpha_i}^2 - n^2\left[\frac{\rho B^4 L}{I_i} A_3^*(K)\right] \tag{6.5.10}$$

$$\tilde{\zeta}_{\alpha_i} = \frac{1}{\tilde{n}_{\alpha_i}}\left[\zeta_{\alpha_i}n_{\alpha_i} - \frac{\rho B^4 Ln}{2I_i} A_2^*(K)\right] \tag{6.5.11}$$

The parameter $K = B(2\pi n)/\bar{U}$ is the reduced frequency, n being the oscillation frequency. In bridge torsional buffeting problems, $n \cong n_{\alpha i}$ will usually remain a good approximation.

Equation (6.5.9) is now the standard form of a single-degree linear oscillator with random excitation. It can be shown (see, for example [23], [41])* that the spectrum of the response $\alpha(x, t)$, namely $S_\alpha(x, n)$, may, under these conditions, be approximated by

$$S_\alpha(x, n) \cong \sum_{i=1}^{N} \frac{\alpha_i^2(x) \displaystyle\int_0^L \int_0^L \alpha_i(x_1)\alpha_i(x_2) S_{M_1 M_2}^C(n)\, dx_1\, dx_2}{16\pi^4 \tilde{n}_{\alpha_i}^4 I_i^2 \left\{ \left[1 - \left(\dfrac{n}{\tilde{n}_{\alpha_i}}\right)^2 \right]^2 + 4\tilde{\zeta}_{\alpha_i}^2 \left(\dfrac{n}{\tilde{n}_{\alpha_i}}\right)^2 \right\}} \tag{6.5.12}$$

the sum being over any desired number N of modes, and where $S_{M_1 M_2}^C(n)$ is the cross co-spectrum† of buffeting moments M_1 and M_2 per unit span acting respectively at x_1, x_2.

The form of (6.5.12) requires that a representation for $S_{M_1 M_2}^C(n)$ be developed, using equation (6.5.3c). This brings in the cross-spectra of the wind components u and w at points x_1, x_2 on the span.

The result is

$$S_{M_1 M_2}^C(n) = [\tfrac{1}{2}\rho\bar{U}^2 B^2]^2 \left\{ 4 C_{ME}[\alpha_0(x_1)] C_{ME}[\alpha_0(x_2)] \frac{S_{u_1 u_2}^C(n)}{\bar{U}^2} \right.$$

$$+ 2 C_{ME}[\alpha_0(x_1)] C_M'[\alpha_0(x_2)] \frac{S_{u_1 w_2}^C(n)}{\bar{U}^2}$$

$$+ 2 C_{ME}[\alpha_0(x_1)] C_M'[\alpha_0(x_1)] \frac{S_{u_2 w_1}^C(n)}{\bar{U}^2}$$

$$\left. + C_M'[\alpha_0(x_1)] C_M'[\alpha_0(x_2)] \frac{S_{w_1 w_2}^C(n)}{\bar{U}^2} \right\} \tag{6.5.13}$$

where terms like S_{uw}^C represent wind component spectra, and where the

---

* Also see Appendix I, 'A Primer for Structural Response to Random Pressure Fluctuations'.

† The co-spectrum, rather than the full complex cross-spectrum, appears here because it can be argued for problems in isotropic homogeneous turbulence that the corresponding quadrature spectrum of wind components is null. In the natural wind the quadrature spectrum, while not exactly null, can still be considered negligible. Thus this effect carries over to moments derived linearly from wind forces as in equation (6.5.3c).

6 Aeroelastic problems of civil engineering structures

abbreviations

$$C_{ME}[\alpha_0(x)] \equiv C_M[\alpha_0(x)] + C_D[\alpha_0(x)]\frac{Ar}{B^2} \qquad (6.5.14)$$

$$\left.\frac{dC_M}{d\alpha}\right|_{\alpha=\alpha_0(x)} = C'_M[\alpha_0(x)] \qquad (6.5.15)$$

have been used.

The root mean square of the fluctuating torsional response (beyond the value $\alpha_0(x)$) at section x is obtained from (6.5.12) by the relation

$$\overline{\alpha^2(x)}^{\frac{1}{2}} = \int_0^\infty S_\alpha(x, n)\, dn \qquad (6.5.16)$$

Completely analogous results to those above are obtained for the purely bending modes $h_i(x)$ of the bridge.

The results are as follows:

Response spectrum:

$$S_h(x, n) \cong \sum_i \frac{h_i^2(x)\int_0^L\int_0^L h_i(x_1)h_i(x_2)S^C_{L_1L_2}(n)\, dx_1\, dx_2}{16\pi^4 n_{h_i}^4 M_i^2\left\{\left[1-\left(\dfrac{n}{n_{h_i}}\right)^2\right]^2 + 4\bar{\zeta}_{n_i}^2\left(\dfrac{n}{n_i}\right)^2\right\}} \qquad (6.5.17)$$

Generalized mass:

$$M_i = \int_0^L \mathcal{M}(x)h_i^2(x)\, dx \qquad (6.5.18)$$

Overall damping ratio:

$$\tilde{\zeta}_{h_i} = \zeta_{h_i} - \frac{\rho B^2 Ln}{2M_i n_{h_i}} H_1^*(K) \qquad (6.5.19)$$

where subscripts h_i refer to the ith bending mode.

Cross co-spectrum:

$$\begin{aligned}
S^C_{L_1L_2}(n) = [\tfrac{1}{2}\rho\bar{U}^2 B]^2\Big\{ &4C_L[\alpha_0(x_1)]C_L[\alpha_0(x_2)]\frac{S^C_{u_1u_2}(n)}{\bar{U}^2} \\
&+ 2C_L[\alpha_0(x_1)]C'_{LE}[\alpha_0(x_2)]\frac{S^C_{u_1w_2}(n)}{\bar{U}^2} \\
&+ 2C_L[\alpha_0(x_2)]C'_{LE}[\alpha_0(x_1)]\frac{S^C_{u_2w_1}(n)}{\bar{U}^2} \\
&+ C'_{LE}[\alpha_0(x_1)]C'_{LE}[\alpha_0(x_2)]\frac{S^C_{w_1w_2}(n)}{\bar{U}^2}\Big\}
\end{aligned} \qquad (6.5.20)$$

316

where

$$C'_{LE}[\alpha_0(x)] = \frac{dC_L}{d\alpha}\bigg|_{\alpha=\alpha_0(x)} + \frac{A}{B}C_D[\alpha_0(x)] \tag{6.5.21}$$

The problem, as described in somewhat condensed form up to this point, is phrased in terms which are most adaptable to spectral, or frequency-domain, analysis. In particular, the self-excited forces are given in terms of a frequency parameter K and the buffeting loads are presented ultimately in their spectral form. It is of some further interest to comment on an alternative manner in which the self-excited forces can be phrased if completely arbitrary time-dependent motions h and α occur.

The analogous self-excited force problem for the thin airfoil depends essentially upon the indicial lift function known as the Wagner function (see Section 4.3). This function, which is the key around which all self-excited forces and moments can be formulated for arbitrary airfoil motion, must be modified and generalized in the case of the suspension bridge deck. Under continuing assumption of the linear superposition principle, the lift and moment per unit span for arbitrary motions h and α are given as follows [42]

$$L_{se} = \tfrac{1}{2}\rho\bar{U}^2(2B)\frac{dC_L}{d\alpha}\int_0^s \Phi_L(s-\sigma)\frac{d\alpha(s)}{ds}d\sigma \tag{6.5.21a}$$

$$M_{se} = \tfrac{1}{2}\rho\bar{U}^2(2B)\frac{dC_M}{d\alpha}\int_0^s \Phi_M(s-\sigma)\frac{d\alpha(s)}{ds}d\sigma \tag{6.5.21b}$$

where $s = \bar{U}t/B$ is a dimensionless time or distance, and in which there appear two new indicial functions $\Phi_L(s)$ and $\Phi_M(s)$ for lift and moment, respectively.

These functions cannot be evaluated from first theoretical principles for the case of arbitrary, bluff shapes of the variety of bridge deck sections. They must therefore be evolved from experimental data. Refs. [42] and [43] discuss the formation of Φ_L and Φ_M out of data provided by the coefficients H_i^*, A_i^*, which are themselves valid only for purely sinusoidal response. This is accomplished, in brief, by equating L_{se} (or M_{se}) in the form (6.5.2) to its counterpart in the form (6.5.21) when α is sinusoidal in s. Then presuming a modified exponential expression (for example for Φ_M) of the type:

$$\Phi_M(s) = c_1 + c_2 e^{d_1 s} + c_3 e^{d_2 s} \tag{6.5.22}$$

the equated results provide enough conditions to determine c_1, c_2, c_3, d_1, d_2.

317

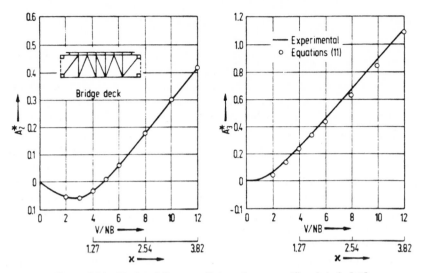

Figure 6.24   Torsional flutter coefficients for truss-stiffened deck [42].

It may be of interest here to offer a single example, i.e., the form developed for $\Phi_M(s)$ for a given deck section whose coefficients A_2^* and A_3^* are already experimentally obtained. This is presented in Figures 6.24 and 6.25, the latter including also the classical Wagner function[†] for comparison. The unusual (but for this situation more characteristic) shape of the curve of $\Phi_M(s)$ makes it abundantly clear that quite different self-excited aerodynamic moments are active in this instance as compared to the thin airfoil case. In fact, either the form of the evolution of A_2^* (which changes sign with increasing velocity parameter U/nB) or that of $\Phi_M(s)$, with its striking initial rise and asymptotic behavior toward steady state *from above*, characterizes a system capable of single-degree flutter in torsion—a phenomenon impossible with the centrally balanced thin air-foil. A study of suspension bridge buffeting using the indicial function approach has been reported in [44].

The above rather brief commentary will serve to cover those facets of the buffeting theory that are presented here. *Along-wind* buffeting has not been described here in detail, but it develops along similar lines to those described earlier, employing (6.5.3a) and proceeding without self-excited terms. The general buffeting problem of line-like structures in the absence of self-excited forces has been dealt with by Davenport [40, 51]. An outline of modal buffeting calculation is given in the next section.

[†] Recall Section 4.3.

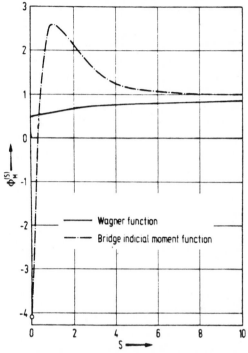

Figure 6.25   Indicial moment function for bridge deck of Figure 6.24.

6.6   Practical formulas for bridge buffeting response

A summary of results is given here to facilitate the practical application of the foregoing theory to the buffeting of self-excited, line-like structures. The essentials of the method used consist of simplifications, through various approximations, of the integral expressions (6.5.12) and (6.5.17) of the previous section, which are then integrated over all frequencies to yield mean square results, as presented in equations (6.6.1) and (6.6.2) below. In the process, a number of results from [45], [40], and [46] are brought in, notably formulas for wind spectra [46] and an assumed lateral correlation of random wind effects of the form $e^{-C_0 \Delta x/L}$ where Δx is separation distance between points of interest for correlation purposes and C_0 is a constant derived from meteorological experience. Of necessity, considerable condensation is employed in the following presentation. Fuller details may be found in [23] and [45].

Consider a bridge (or similar structure) with the following characteristics: I = span, B = deck width, $\alpha_r(x)$, $h_r(x)$ = rth natural modes in

319

torsion and bending, respectively, x = coordinate along the span, $n_{\alpha r}$ and n_{hr} = rth natural frequencies of vibration in torsion and bending, $\zeta_{\alpha r}$ and ζ_{hr} = mechanical damping in rth torsion and bending modes, $m(x)$ = mass of bridge deck per unit length, $I(x)$ = mass moment of inertia per unit length about effective structural rotation axis (elastic axis), r = distance from center of mass to elastic axis of bridge deck, A = net area per unit span of bridge deck projected on a vertical plane normal to the mean wind, z = height of bridge deck above ground (water) level. Let $\alpha(x)$ denote the fluctuation of the twist angle about the mean position α_0, and let $h(x)$ denote the vertical deflection. The mean square values of $\alpha(x)$ and $h(x)/B$ can be written approximately as

$$\sigma_\alpha^2(x) \cong \sum_{r=1}^N \frac{\pi \alpha_r^2(x)}{4\zeta_{\alpha r}\tilde{K}_{\alpha r}^3}\left(\frac{\rho B^4 L}{I_r}\right)^2\left(\int_0^L \alpha_r^2(x)\frac{dx}{L}\right)\frac{2(c_\alpha-1)}{c_\alpha^2}\frac{S_{MU}(\tilde{n}_{\alpha r})}{\bar{U}^2} \quad (6.6.1)$$

$$\sigma_{h/B}^2(x) \cong \sum_{r=1}^N \frac{\pi h_r^2(x)}{4\zeta_{hr}K_{hr}^3}\left(\frac{\rho B^2 L}{M_r}\right)^2\left(\int_0^L h_r^2(x)\,dx/L\right)\frac{2(c_h-1)}{c_h^2}\frac{S_{LU}(\tilde{n}_{hr})}{\bar{U}^2}$$
$$(6.6.2)$$

where N = number of vibration modes in torsion or bending, $\bar{U}$ = mean wind speed and ρ = air density (≈ 1.25 kg/m^3). In equation (6.6.1)

$$\tilde{K}_{\alpha r} = K_{\alpha r}\left[1 - \frac{I_{ar}}{I_r}A_3^*\left(\frac{U}{n_{\alpha r}B}\right)\right]^{\frac{1}{2}} \quad (6.6.3)$$

$$K_{\alpha r} = \frac{B(2\pi n_{\alpha r})}{\bar{U}} \quad (6.6.4)$$

$$I_{ar} = \int_0^L \rho B^4 \alpha_r^2(x)\,dx \quad (6.6.5)$$

(aerodynamic moment of inertia in rth mode)

$$I_r = \int_0^L I(x)\alpha_r^2(x)\,dx \quad (6.6.6)$$

(generalized moment of inertia in rth mode)

$$\zeta_{\alpha r} = \frac{1}{\tilde{K}_{\alpha r}}\left[2\zeta_{\alpha r}K_{\alpha r} - \frac{I_{ar}}{I_r}\tilde{K}_{\alpha r}A_2^*\left(\frac{U}{\tilde{n}_{\alpha r}B}\right)\right] \quad (6.6.7)$$

$$\tilde{n}_{\alpha r} = \frac{U\tilde{K}_{\alpha r}}{2\pi B} \quad (6.6.8)$$

$$c_\alpha = \frac{7\tilde{n}_{\alpha r}L}{\bar{U}} \quad (6.6.9)$$

$$S_{MU}(\tilde{n}_{\alpha r}) \simeq C_{ME}^2 S_u(\tilde{n}_{\alpha r}) + \tfrac{1}{4} C_{M_0}'^2 S_w(\tilde{n}_{\alpha r}) \tag{6.6.10}$$

$$C_{ME} = C_M(\alpha_0) + \frac{Ar}{B} C_D(\alpha_0) \tag{6.6.11}$$

$$S_u(\tilde{n}_{\alpha r}) = \frac{u_*^2}{\tilde{n}_{\alpha r}} \frac{200 \tilde{f}_{\alpha r}}{(1 + 50 \tilde{f}_{\alpha r})^{5/3}} \tag{6.6.12}$$

(spectrum of longitudinal velocity fluctuations, see [23, 46])

$$S_w(\tilde{n}_{\alpha r}) = \frac{u_*^2}{\tilde{n}_{\alpha r}} \left[\frac{3.36 f_r}{1 + 10 \tilde{f}_{\alpha r}^{5/3}} \right] \tag{6.6.13}$$

(spectrum of vertical velocity fluctuations, see [23, 47])

$$f_{\alpha r} = \frac{\tilde{n}_{\alpha r} z}{\bar{U}} \quad (z = \text{bridge height}) \tag{6.6.14}$$

$$u_* = \frac{\bar{U}(z_{\text{ref}})}{2.5 \ln \left(\dfrac{z_{\text{ref}}}{z_0} \right)} \tag{6.6.15}$$

The quantities A_2^*, A_3^*, $C_M(\alpha_0)$, $C_D(\alpha_0)$, $C_{M_0}' = \dfrac{dC_M}{d\alpha}\bigg|_{\alpha=\alpha_0}$ (equations (6.6.3), (6.6.7), (6.6.10), (6.6.11)) are aerodynamic properties of the bridge deck sections which must be determined by experiment. In equation (6.6.15), $z_{\text{ref}} =$ any reference height at which a mean wind velocity $\bar{U}(z_{\text{ref}})$ is specified (for example, the bridge height above ground or water); $z_0 =$ roughness length† (characterizing the terrain roughness).

Items pertinent to vertical mode response calculation are given below:

$$K_{hr} = \frac{B(2\pi n_{hr})}{\bar{U}} \tag{6.6.16}$$

$$\tilde{\zeta}_{hr} = \zeta_{hr} - \frac{M_{ar}}{2M_r} H_1^* \left(\frac{\bar{U}}{n_{hr} B} \right) \tag{6.6.17}$$

$$M_{ar} = \int_0^L \rho B^2 h_r^2(x) \, dx \tag{6.6.18}$$

---

† z_0 ranges from 0.0003 cm (calmest sea surface) to 60 cm at the centers of largest cities. $z_0 = 4$ cm for highgrass, 10 cm for palmetto, 35 cm inside towns. Ref. [23] lists fuller details on values of z_0.

(aerodynamic mass in rth mode)

$$M_r = \int_0^L m(x)h_r^2(x)\,\mathrm{d}x \qquad (6.6.19)$$

(generalized mass in rth mode)

$$c_h = \frac{7 n_{hr} L}{\bar{U}} \qquad (6.6.20)$$

$$S_{LU}(n_{hr}) \cong C_L^2(\alpha_0)S_u(n_{hr}) + \tfrac{1}{4}C_{LE}'^2 S_w(n_{hr}) \qquad (6.6.21)$$

$$C_{LE}' = C_{L_0}' + \frac{A}{B} C_D(\alpha_0) \qquad (6.6.22)$$

$$S_u(n_{hr}) = \frac{u_*^2}{n_{hr}} \left[\frac{200 f_{hr}}{(1 + 50 f_{hr})^{5/3}} \right] \qquad (6.6.23)$$

$$f_{hr} = \frac{n_{hr} z}{\bar{U}} \qquad (6.6.24)$$

$$S_w(n_{hr}) = \frac{u_*^2}{n_{hr}} \left[\frac{3.36 f_{hr}}{1 + 10 f_{hr}^{5/3}} \right] \qquad (6.6.25)$$

The quantities H_1^*, $C_L(\alpha_0)$, $C_D(\alpha_0)$, $C_{L_0}' = \dfrac{\mathrm{d}C_L}{\mathrm{d}\alpha}\bigg|_{\alpha=\alpha_0}$ (equations (6.6.17), (6.6.21), (6.6.22)) are experimentally determined aerodynamic properties of the bridge. For example u_* in equation (6.6.23), see equation (6.6.15).

The somewhat lengthy set of formulas appurtenant to (6.6.1) and (6.6.2) provided above appear to contain the essential parameters to which the buffeting phenomenon is sensitive. Hence, emphasis upon the roles of these parameters can act as a guide to buffeting proclivities of a given design. For example, a bridge with an aerodynamic damping coefficient A_2^* that exhibits reversal in sign with advancing wind velocity can become critical in both flutter and buffeting responses. Bridges stiff in torsion are desirable against both phenomena. Wind vertical components appear to play a somewhat greater role in bridge buffeting than do horizontal components. Hence wide bridges tend to attract greater buffeting forces; on the other hand, basic flutter stability appears to be enhanced by greater bridge width. Thus, for example, a wide bridge may be buffeted in an annoying though not necessarily dangerous manner. It will be noted also that explicit forms for gust spectra of the natural wind (horizontal and vertical components) are introduced into the calculations implied in this section. The spectra used are the latest available, based upon empirical curve fits to meteorological data. Ref. [23] treats atmospheric spectra somewhat more fully.

References for Chapter 6

[1] Wardlaw, R. L., 'Static Force Measurements of Six Deck Sections for the proposed New Burrard Inlet Crossing', *Report LTR-LA-53, NAE*, National Research Council, Ottawa, Ont., Canada (June 1970).

[2] Den Hartog, J. P., 'Transmission Line Vibration Due to Sleet', *Transactions, AIEE* (1932) pp. 1074–1076.

[3] Richards, D. J. W., 'Aerodynamic Properties of the Severn Crossing Conductor', *Wind Effects on Buildings and Structures*, National Physical Laboratory, Teddington, U.K., 1965, Vol. II, pp. 687–765.

[4] Den Hartog, J. P., 'Mechanical Vibrations', McGraw-Hill, New York, 4th Ed., 1956.

[5] Scanlan, R. H. and Wardlaw, R. L., 'Reduction of Flow-Induced Structural Vibrations', *Section 2 of: Isolation of Mechanical Vibration, Impact, and Noise*, AMD Vol. 1, 1973 American Society of Mechanical Engineers, New York.

[6] Parkinson, G. V. and Brooks, N. P. H., 'On the Aeroelastic Instability of Bluff Cylinders', *Trans. ASME Jnl. Appl. Mech.*, Vol. 83 (1961) pp. 252–258.

[7] Parkinson, G. V. and Smith, J. D., 'An Aeroelastic Oscillator with Two Stable Limit Cycles', *Trans. ASME, Jnl. Appl. Mech.*, Vol. 84 (1962) pp. 444–445.

[8] Parkinson, G. V., 'Aeroelastic Galloping in One Degree of Freedom', *Proc. Sympos. on Wind Effects on Bldgs. and Struct.*, Teddington, England, pp. 581–609, 1963.

[9] Parkinson, G. V. and Smith, J. D., 'The Square Prism as an Aeroelastic Nonlinear Oscillator', *Quart. Jnl. of Mech. and Appl. Math.*, Oxford Press, London, Vol. XVII, Pt. 2 (1964) pp. 225–239.

[10] Parkinson, G. V. and Santosham, T. V., 'Cylinders of Rectangular Section as Aeroelastic Nonlinear Oscillators', *Paper Vibrations Conf.*, ASME, Boston, Mass. (March 1967).

[11] Parkinson, G. V., 'Mathematical Models of Flow-Induced Vibrations of Bluff Bodies', in *Flow-induced Structural Vibrations*, (Ed. Naudascher) Springer-Verlag (1974) pp. 81–127.

[12] Novak, M., 'Aeroelastic Galloping of Prismatic Bodies', *Jnl. EMD, Proc. ASCE*, Vol. 9, No. EM1 (Feb. 1969) pp. 115–142.

[13] Novak, M., 'Galloping Oscillations of Prismatic Structures', *Jnl. EMD, Proc. ASCE* (Feb. 1972) Vol. 98, No. EM1 (Feb. 1972) pp. 27–46.

[14] Kryloff, N. and Bogoliuboff, N., 'Introduction to Nonlinear Mechanics', *(Annals of Math. Studies, No. 11) Trans. S. Lefschetz*, Princeton Univ. Press, Princeton, N.J., 1947.

[15] Richardson, A. S., Martuccelli, J. R. and Price, W. S., 'Research Study on Galloping of Electric Power Transmission Lines', *Paper 7, Proceedings, Symp. on Wind Effects on Buildings and Structures*, Vol. II, Teddington, U.K., 1965.

[16] Wardlaw, R. L., Cooper, K. R. and Scanlan, R. H., 'Observations on the Problem of Subspan Oscillation of Bundled Power Conductors', *DME/NAE Quarterly Bulletin*, 1973(1), National Research Council (April 1973) Ottawa, Canada.

[17] Scanlan, R. H., 'A Wind Tunnel Investigation into the Aerodynamic Stability of Bundled Power Line Conductors for Hydro-Quebec', *Part VI, Report LTR-LA-121 NAE*, National Research Council, Ottawa, Canada (Sept. 1972).

[18] Simpson, A., 'On the Flutter of a Smooth Cylinder in a Wake', *The Aeronautical Quarterly* (Feb. 1971) pp. 25–41.

[19] Strouhal, V., 'Ueber eine besondere Art der Tonerregung', *Annalen der Physik, Leipzig*, Vol. 5, 1878.

[20] Bénard, H., *Formation de centres de giration à l'arrière d'un obstacle en mouvement*, Comptes Rendus de l'Académie des Sciences, Paris, Vol. 146, 1908.

[21] von Kármán, Th., *Ueber den Mechanismus des Widerstandes den ein bewegter Koerper in einer Fluessigkeit erfaehrt*, Nachricht der Koeniglichen Gesellschaft der Wissenschaft, Goettingen, 1912.

[22] Wille, R., 'Karman Vortex Streets', *Advances in Apilied Mechanics, Vol. VI*, Academic Press, New York, 1960, pp. 273–287.

[23] Simiu, E. and Scanlan, R. H., *Wind Effects on Structures (An Introduction to Wind Engineering)* Wiley, N.Y. (in press 1977–1978).

[24] Bishop, R. E. D. and Hassan, A. Y., 'The Lift and Drag Forces on a Circular Cylinder Oscillating in a Flowing Fluid', *Proc. Royal Society (London), Series A*, Vol. 277 (1964) pp. 51–74.

[25] Hartlen, R. T. and Currie, I. G., 'A Lift-Oscillator Model For Vortex-Induced Vibrations', *Proc. ASCE, Jnl. Engr. Mech.*, Vol. 96, pp. 577–591, 1970.

[26] Griffen, O. M., Skop, R. A. and Koopmann, G. H., 'The Vortex-Excited Resonant Vibrations of Circular Cylinders', *J. Sound Vibration*, Vol. 31 (1973) pp. 235–249.

[27] Skop, R. A. and Griffin, O. M., 'A Model for the Vortex-Excited Response of Bluff Cylinders', *J. Sound Vibration*, Vol. 27 (1973) pp. 225–233.

[28] Skop, R. A. and Griffin, O. M., 'On a Theory for the Vortex-Excited Oscillations of Flexible Cylindrical Structures', *J. Sound Vibration*, Vol. 41 (1975) pp. 263–274.

[29] Griffin, O. M., Skop, R. A. and Ramberg, S. E., 'Modeling of the Vortex-Induced Oscillations of Cables and Bluff Structures', *Paper, Spring Mtg., Soc. for Exper. Stress Analysis*, Silver Spring, Md. (May 1976).

[30] Scruton, C., 'Note on a device for the suppression of the vortex-excited oscillations of flexible structures of circular or near circular section, with special reference to its application to tall stacks', *NPL Aero Report* 1012, National Physical Laboratory, U.K., 1963.

[31] Walshe, D. E. and Wooton, L. R., 'Preventing wind-induced oscillations of structures of circular section', *Proceedings, The Institution of Civil Engineers*, Vol. 47 (Sept., 1970) pp. 1–24.

[32] Wooton, L. R. and Scruton, C., 'Aerodynamic Stability', *Modern design of wind-sensitive structures*, Seminar, Construction Research and Information Association, U.K., 1970.

[33] Wardlaw, R. L. and Cooper, K. R., 'Mechanisms and Alleviation of Wind-Induced Structural Vibrations', *Proceedings, Second Symposium on Applications of Solid Mechanics*, McMaster Univ., Hamilton, Ont., Canada (June 1974) pp. 369–399.

[34] Cooper, K. R., 'A Wind Tunnel Investigation of Twin-Bundled Power Conductors', *Report LTR-LA-96 NAE*, National Research Council, Ottawa, Canada (May 1972).

[35] Bleich, F., 'Dynamic Instability of Truss-Stiffened Suspension Bridges Under Wind Action', *Proc. ASCE*, Vol. 75, No. 3 (March 1949) pp. 413–416, and Vol. 75 No. 6 (June, 1949) pp. 855–865.

[36] Scanlan, R. H. and Tomko, J. J., 'Airfoil and Bridge Deck Flutter Derivatives', *Jnl. Engrg. Mech. Div.*, ASCE, Vol. 97, No. EM6 (Dec. 1971) pp. 1717–1737.

[37] Lin, W.-H., *Forced and Self-Excited Responses of a Bluff Structure in a Turbulent Wind*, Doctoral Dissertation, Dept. of Civil Engrg., Princeton University (Dec. 1976).

References for Chapter 6

[38] Sears, W. R., 'Some Aspects of Non-Stationary Airfoil Theory and its Practical Application', *Jnl Aeron. Sci.*, Vol. 8 (1941) pp. 104–108.
[39] Liepmann, H. W., 'On the Application of Statistical Concepts to the Buffeting Problem', *Jnl. Aeron. Sci.*, Vol. 19, No 12 (Dec. 1952) pp. 793–800, 822.
[40] Davenport, A. G., 'The Response of Slender, Line-Like Structures to a Gusty Wind', *Proc. Instn. of Civil Engineers, London* (1962) Paper 6610, pp. 389–407.
[41] Robson, J. D., *Random Vibration*, Elsevier, N.Y. 1964.
[42] Scanlan, R. H. and Budlong, K. S., 'Flutter and Aerodynamic Response Considerations for Bluff Objects in a Smooth Flow', *Flow-Induced Structural Vibrations* (Ed. Naudascher) Springer-Verlag, N.Y. (1974) pp. 339–354.
[43] Scanlan, R. H., Béliveau, J.-G. and Budlong, K. S., 'Indicial Aerodynamic Functions for Bridge Decks', *Proc. ASCE, Jnl. Eng. Mech. Div.* (August 1974) pp. 657–672.
[44] Béliveau, J.-G., Vaicaitis, R. and Shinozuka, M., 'Motion of a Suspension Bridge Subject to Wind Loads'. (In Press, 1977) *Proc. ASCE, Jnl. of the Struct. Div.*
[45] Scanlan, R. H. and Gade, R. H., 'Motion of Suspended Bridge Spans under Gusty Wind' (in Press, 1977). *Proc.-ASCE, Jnl. of the Struct. Div.*
[46] Simiu, E., 'Wind Spectra and Dynamic Alongwind Response'. *Jnl. Struct. Div. ASCE*, Vol. ST 9 (Sept. 1974) pp. 1897–1910.
[47] Lumley, J. L. and Panofsky, H. A., *The Structure of Atmospheric Turbulence*, Wiley, N.Y. 1964.
[48] Fung, Y. C., *The Theory of Aeroelasticity*. Wiley, N.Y. (1955) p. 74. (Also reprinted by Dover).
[49] Baird, R. C., 'Wind-Induced Vibration of a Pipe-Line Suspension Bridge and its Cure'. *Trans. Am. Soc. of Mech. Engrs.*, Vol. 77 (August 1955) pp. 797–804.
[50] Okubo, T., Okauchi, I. and Murakami, E., 'Aerodynamic Response of a Large-Scale Bridge Model Against Natural Wind' *Reliability Approach in Structural Engineering.* Maruzen Co. Ltd. Tokyo (1975) pp. 315–328.
[51] Davenport, A. G., 'The Action of Wind on Suspension Bridges'. *International Symposium on Suspension Bridges*, National Civil Engineering Laboratory, Lisbon, Portugal (Nov. 1966) pp. 79–100.

7

Aeroelastic problems of Rotorcraft

In this chapter we will examine a number of aeroelastic phenomena associated with helicopters and other rotor or propeller driven aircraft. Certain areas have been selected for treatment to illustrate some significant stability problems which are associated with the design of helicopters. The approach to be followed employs simplified modelling of various problems such that physical insight into the nature of the phenomena can be obtained. In general a complete and precise formulation of many of the problem areas discussed is highly complex and the reader is referred to the literature for these more detailed formulations.

A basic introduction to the mechanics and aerodynamics of helicopters may be found in [1] and [2]. Extensive reviews of helicopter aeroelasticity may be found in [3] and [4]. Ref. [4] provides an excellent discussion of the considerations necessary in modelling helicopter aeroelasticity and illustrates the complexity of a general formulation as well as the care required to obtain a complete and precise analytical model.

Helicopter rotors in use may be broadly classified in three types, semi-articulated or see-saw, fully-articulated, and hingeless. This classification is based on the manner in which the blades are mechanically connected to the rotor hub. The see-saw rotor (Bell) is typically a two-bladed rotor with the blades connected together and attached to the shaft by a pin which allows the two-blade assembly to rotate such that tips of the blades may freely move up and down with respect to the plane of rotation (flapping motion). In the fully-articulated rotor, each blade is individually attached to the hub through two perpendicular hinges allowing rigid motion of the blade in two directions, out of the plane of rotation (flapping motion) and in the plane of rotation (lag motion) as typified by rotors on Sikorsky, Boeing-Vertol and Hughes Helicopters. The third type, a more recent development, is the hingeless rotor (Lockheed, MBB, Westland) in which the rotor blade is a cantilever beam.

Being a long thin member, elastic deformations of the hingeless blade are significant in the analysis of the dynamics of the vehicle. Bending out of the plane of rotation is referred to as flap bending and in-plane as lag bending. These three rotor configurations are shown schematically in Figure 7.1. Rotation of the blade about its long axis is controlled by a pitch change mechanism suitably connected to the pilot's stick. For further details see [1] for articulated rotors and [5] for hingeless rotors. Other variations in rotor hub geometry are found such as the gimballed rotor described in [6]. We will concentrate our discussion on the aeroelastic behavior of fully-articulated and hingeless rotors. However, it is important to realize that for the aeroelastic analysis of rotors the precise details of the hub and blade geometry must be carefully modelled.

Problems in helicopter aeroelasticity may be classified by the degrees-of-freedom which are significantly coupled. Typically, the dynamics of a single blade are of interest although coupling among blades can be present through the elasticity of the blade pitch control system or the aerodynamic wake [7]. The degrees-of-freedom of a single blade include rigid body motion in the case of the articulated system as well as elastic motion. Elastic motions of interest include bending in two directions and twisting or torsion. These elastic deformations are coupled in general. In addition to individual blade aeroelastic problems, the blade degrees-of-freedom can couple with the rigid body degrees-of-freedom of the fuselage in flight as well as the elastic deformations of the fuselage [8–10] or with the fuselage/landing gear system on the ground [10]. In fact a complete aeroelastic model of the helicopter typically involves a dynamic model with a large number of degrees-of-freedom. We do not propose to examine these very complex models. but rather will consider simple formulations of certain significant stability problems which will give some insight into the importance of aeroelasticity in helicopter design. Avoiding resonances is also of considerable significance, but is not discussed here. First, aeroelastic problems associated with an individual blade are described and then those associated with blade/body coupling are examined.

7.1   Blade dynamics

Classical flutter and divergence of a rotor blade involving coupling of flap bending and torsion have not been particularly significant due to the fact that, in the past, rotor blades have been designed with their elastic axis, aerodynamic center, and center of mass coincident at the quarter chord.

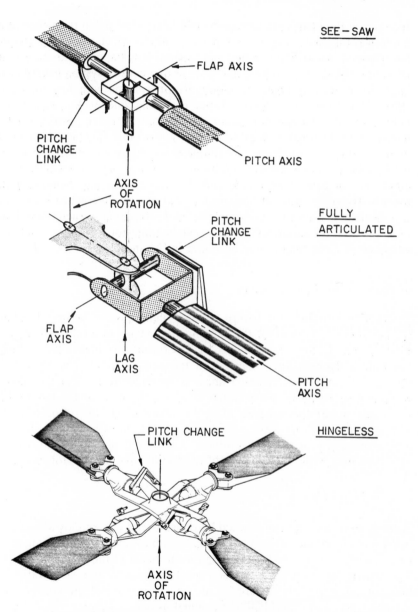

SEE – SAW

FLAP AXIS

PITCH
CHANGE
LINK

PITCH AXIS

AXIS
OF
ROTATION

FULLY
ARTICULATED

PITCH
CHANGE
LINK

FLAP
AXIS

LAG
AXIS

PITCH
AXIS

PITCH CHANGE
LINK

HINGELESS

AXIS
OF
ROTATION

Figure 7.1 *Various rotor hub configurations.*

In addition the blades are torsionally stiff (a typical torsion frequency of a modern rotor blade is about 5–8 per revolution) which minimizes coupling between elastic flap bending and torsion. It is important to note that torsional stiffness is used here in the sense that control system flexibility is included as well as blade flexibility. Rotor systems with low torsional stiffness [11] have experienced flutter problems, and on hingeless rotor helicopters, the blade section center of mass and elastic axis position can be moved from the quarter chord to provide a favorable effect on the overall flight stability [12] which may mean that these classical problem areas will have to be reviewed more carefully in the future. Sweep has also been employed on rotor blades [13] and this will couple flap bending with torsion. However, we will not consider flutter and divergence here, but will instead concentrate on problems more frequently encountered in practice. Further discussion of classical bending-torsion flutter and divergence of rotor blades may be found in [3] and [14].

Articulated, rigid blade motion

In order to introduce the nature of rotor blade motion we first develop the equations of motion for the flapping and lagging of a fully articulated blade assuming that the blade is rigid. Consider a single blade which has only a flapping hinge located on the axis of rotation as shown in Figure 7.2. The blade flapping angle is denoted by β_S and the blade rotational speed by Ω. We proceed to derive the equation of motion of the blade about the flapping axis. We assume that the rotor is in a hovering state with no translational velocity. It is most convenient to use a Newtonian approach to this problem. Since the flapping pin is at rest in space we may write the equation of motion for the blade as follows [15]

$$\dot{\bar{H}}_P + \bar{\Omega}_B \times \bar{H}_P = \int_0^R \bar{r} \times d\bar{F}_A \qquad (7.1.1)$$

A blade-body axis system denoted by the subscript $_B$ is employed and $\bar{H}_P$ is the moment of momentum of the blade with respect to the flapping pin. $d\bar{F}_A$ is the aerodynamic force acting on the blade at the radial station $\bar{r}$. The gravity force on the blade is neglected owing to the comparatively high rotational velocity. Figure 7.2 also shows the coordinate system and variables involved. The blade is modelled as a very slender rod and the body axes are principal axes such that the inertia characteristics of the blade are

$$I_B \cong I_y \cong I_z; \qquad I_x \cong 0$$

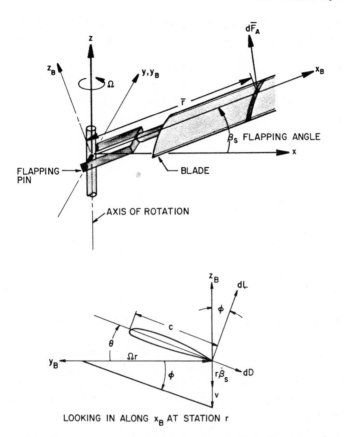

Figure 7.2   Coordinate systems and aerodynamics for blade flapping analysis.

Therefore

$$\bar{H}_P = (I_B q_B)\bar{j}_B + (I_B r_B)\bar{k}_B \tag{7.1.2}$$

where

$$\bar{\Omega}_B = p_B \bar{i}_B + q_B \bar{j}_B + r_B \bar{k}_B \tag{7.1.3}$$

The equation of motion, (7.1.1), becomes

$$I_B[\dot{q}_B - p_B r_B]\bar{j}_B + I_B[\dot{r}_B + p_B q_B]\bar{k}_B = \int_0^R \bar{r} \times \mathrm{d}\bar{F}_A \tag{7.1.4}$$

Now we must express the angular body rates in terms of the variables of interest in the problem, Ω the angular velocity, and β_S the flap angle.

The angular velocity must be resolved into the blade axis system by rotation through β_S, and then the flapping velocity $\dot\beta_S$ added.

$$\begin{Bmatrix} p_B \\ q_B \\ r_B \end{Bmatrix} = \begin{bmatrix} \cos\beta_S & 0 & \sin\beta_S \\ 0 & 1 & 0 \\ -\sin\beta_S & 0 & \cos\beta_S \end{bmatrix} \begin{Bmatrix} 0 \\ 0 \\ \Omega \end{Bmatrix} + \begin{Bmatrix} 0 \\ -\dot\beta_S \\ 0 \end{Bmatrix}$$

That is

$$p_B = \Omega \sin\beta_S$$

$$q_B = -\dot\beta_S \tag{7.1.5}$$

$$r_B = \Omega \cos\beta_S$$

Substitution of (7.1.5) into (7.1.4) gives

$$I_B[-\ddot\beta_S - \Omega^2 \cos\beta_S \sin\beta_S]\bar{j}_B + I_B[-2\Omega \sin\beta_S\dot\beta_S]\bar{k}_B = \int_0^R \bar{r} \times \mathrm{d}\bar{F}_A \tag{7.1.6}$$

The first term on the left hand side is the angular acceleration of the blade about the y_B axis and the second term is the angular acceleration of the blade about the z_B axis. The second term can be considered an inertial torque about the z_B axis, i.e., in the lag direction, which arises as a result of out-of-plane (flapping) motion of the blade. The aerodynamic force on the blade element comprises the lift and drag and is formulated using strip theory (usually called blade element theory) [1, 2]. Also see the discussion in Section 3.4. Three-dimensional effects are obtained by including the induced velocity which for our purposes may be calculated by momentum theory [1]. Thus from Figure 7.2

$$\mathrm{d}\bar{F}_A = \mathrm{d}L\bar{k}_B + (-\mathrm{d}D - \phi\,\mathrm{d}L)\bar{j}_B \tag{7.1.7}$$

where the inflow angle ϕ is assumed to be small and is made up of the effect of induced velocity (downwash) and the induced angle due to flapping velocity. Therefore

$$\mathrm{d}L = \tfrac{1}{2}\rho(\Omega r)^2 c\,\mathrm{d}ra(\theta - \phi)$$

$$\mathrm{d}D = \tfrac{1}{2}\rho(\Omega r)^2 c\,\mathrm{d}r\delta$$

$$\phi = \frac{r\dot\beta_S + v}{\Omega r}$$

Define

$$x \equiv \frac{r}{R} \qquad \lambda \equiv -\frac{v}{\Omega R} \qquad \gamma \equiv \frac{\rho a c R^4}{I_B}, \qquad \text{the Lock number} \tag{7.1.8}$$

The blade chord, c, and pitch angle, θ, are taken to be independent of x, for simplicity, although rotor blades are usually twisted. The blade section drag coefficient is denoted by δ and is also assumed to be independent of the radial station. Thus

$$dL = \frac{I_B \gamma \Omega^2}{R} \left[\theta - \frac{\dot{\beta}_S}{\Omega} + \frac{\lambda}{x} \right] x^2 \, dx$$

$$dD = \frac{I_B \gamma \Omega^2}{R} \left(\frac{\delta}{a} \right) x^2 \, dx \tag{7.1.9}$$

and

$$\bar{r} = x R \bar{i}_B$$

The total rotor thrust is found by integrating the lift along the radius, averaging over one revolution, and multiplying by the number of blades to give [1]

$$\frac{2 C_T}{a \sigma} = \frac{\theta}{3} + \frac{\lambda}{2} \tag{7.1.10}$$

where

$$\sigma = \frac{bc}{\pi R}$$

and b is the number of blades. The thrust coefficient is

$$C_T = \frac{T}{\rho \pi R^2 (\Omega R)^2}$$

Momentum theory results in the following expression for the induced velocity

$$\lambda = - \sqrt{\frac{C_T}{2}}$$

so that the integral

$$\int_0^R \bar{r} \times d\bar{F}_A = - \frac{I_B \gamma \Omega^2}{8} \left[\theta + \frac{4\lambda}{3} - \frac{\dot{\beta}_S}{\Omega} \right] \bar{j}_B$$
$$+ \frac{I_B \gamma \Omega^2}{8} \left[- \frac{\delta}{a} + \frac{\dot{\beta}_S}{\Omega} \left(\frac{\dot{\beta}_S}{\Omega} - \theta \right) + \frac{4}{3} \left(\theta - 2 \frac{\dot{\beta}_S}{\Omega} \right) \lambda + 2 \lambda^2 \right] \bar{k}_B \tag{7.1.11}$$

The $\bar{j}_B$ components contribute to the flapping equation of motion which may be expressed from equations (7.1.6) and (7.1.11) as

$$\ddot{\beta}_S + \frac{\gamma \Omega}{8} \dot{\beta}_S + \Omega^2 \cos \beta_S \sin \beta_S = \frac{\gamma \Omega^2}{8} \left[\theta + \frac{4\lambda}{3} \right] \tag{7.1.12}$$

The $\bar{k}_B$ component of equation (7.1.11) is the aerodynamic torque about the z_B axis or in the lag direction. There is a steady component and a component proportional to flapping velocity. This aerodynamic component proportional to flapping velocity is small compared to the inertial component (equation (7.1.6)) proportional to flapping velocity. The inertial component may be considered as a reason why a lag hinge is desirable to relieve the bending moment at the root of the blade.

If we assume that the flapping motion is small as is typical of rotor blade motion then the flapping equation becomes linear.

$$\ddot{\beta}_S + \frac{\gamma\Omega}{8}\dot{\beta}_S + \Omega^2\beta_S = \frac{\gamma\Omega^2}{8}\left[\theta + \frac{4\lambda}{3}\right] \qquad (7.1.13)$$

The linearized blade flapping equation may be recognized as a second order system with a natural frequency equal to the rotor angular velocity and a damping ratio equation to $\gamma/16$ which arises from the aerodynamic moment about the flapping pin. This motion is well damped as γ is between 5 and 15 for typical rotor blades. This system must, of course, be well damped since the aerodynamic inputs characteristically occur in forward flight at Ω and thus the blade flapping motion is forced at resonance.

The spring or displacement term can be interpreted as arising from the centrifugal force [1]. This same stiffening effect will appear in the flexible blade analysis and will increase the natural frequency as rotational speed is increased.

If the more general case of flapping in forward flight is considered then the equation of motion for flapping (7.1.12) will contain periodic coefficients which can lead to instabilities [16]. However, the flight speed at which such instabilities occur is well beyond the performance range of conventional helicopters.

Now we include the lag degree-of-freedom to obtain a complete description of rigid motion of a fully-articulated rotor blade. The complete development of this two-degree-of-freedom problem is quite lengthy and will not be reproduced here [17].

Following the approach given above, assuming that the flap angle and lag angle are small and that the lag hinge and flap hinge are coincident and located a small distance e (hinge offset) from the axis of rotation as shown in Figure 7.3, and further accounting for the effect of lag velocity on the aerodynamic forces acting on the blade, the lift is given by

$$dL = \tfrac{1}{2}\rho[(\Omega + \dot{\zeta}_s)r]^2 c\,dra\left[\theta - \frac{(r\dot{\beta}_S + v)}{(\Omega + \dot{\zeta}_s)r}\right] \qquad (7.1.14)$$

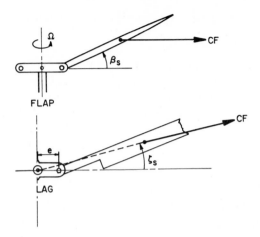

Figure 7.3   Direction of centrifugal force for flap and lag motion.

where the effect of the small distance e on the aerodynamics is neglected. The lag angle is defined as positive in the direction of rotor rotation. Care must be taken in formulating the inertial terms since we have noted above that a term like $\dot{\beta}_S \sin \beta_S$ is of significance in the equations of motion, and thus the small angle assumption must not be made until after the expressions for the acceleration have been obtained. Rotating by the flap angle first and then by the lag angle, the angular rates in the blade body axis system are given by

$$
\begin{Bmatrix} p_B \\ q_B \\ r_B \end{Bmatrix} = \begin{bmatrix} \cos \zeta_S & \sin \zeta_S & 0 \\ -\sin \zeta_S & \cos \zeta_S & 0 \\ 0 & 0 & 1 \end{bmatrix} \begin{bmatrix} \cos \beta_S & 0 & \sin \beta_S \\ 0 & 1 & 0 \\ -\sin \beta_S & 0 & \cos \beta_S \end{bmatrix} \begin{Bmatrix} 0 \\ 0 \\ \Omega \end{Bmatrix}
$$
$$
+ \begin{bmatrix} \cos \zeta_S & \sin \zeta_S & 0 \\ -\sin \zeta_S & \cos \zeta_S & 0 \\ 0 & 0 & 1 \end{bmatrix} \begin{Bmatrix} 0 \\ -\dot{\beta}_S \\ 0 \end{Bmatrix} + \begin{Bmatrix} 0 \\ 0 \\ \dot{\zeta}_S \end{Bmatrix} \tag{7.1.15}
$$

We must also account for the fact that the hinge point of the blade is no longer at rest but is accelerating [15]. Since the hinge point is located at a distance e from the axis of rotation, the equation of motion, (7.1.1), must be modified to read

$$
\dot{\bar{H}}_P + \bar{\Omega}_B \times \bar{H}_P = \int_0^R \bar{r} \times d\bar{F}_A + \bar{E} \times M_B \bar{a}_P \tag{7.1.16}
$$

335

where $\bar{a}_P$ is the acceleration of the hinge point

$$\bar{a}_P = \bar{\Omega} \times (\bar{\Omega} \times \bar{E}) + \dot{\bar{\Omega}} \times \bar{E} \tag{7.1.17}$$

$\bar{E}$ is the offset distance and M_B is the blade mass.

Accounting for all of these factors, and assuming that the flapping and lagging motion amplitudes are small, the equations of motion for this two-degree-of-freedom system may be expressed [17, 18] as

$$-\ddot{\beta}_S - \Omega^2\left(1 + \frac{3}{2}\bar{e}\right)\beta_S - 2\beta_S\dot{\zeta}_S\Omega = -\frac{\gamma\Omega^2}{8}\left[\theta + \frac{4}{3}\lambda - \frac{\dot{\beta}_S}{\Omega} + \left(2\theta + \frac{4}{3}\lambda\right)\frac{\dot{\zeta}_S}{\Omega}\right]$$

$$-2\beta_S\dot{\beta}_S\Omega + \ddot{\zeta}_S + \frac{3}{2}\bar{e}\Omega^2\zeta_S = \frac{\gamma\Omega^2}{8}\left[-\left(\theta + \frac{8}{3}\lambda\right)\frac{\dot{\beta}_S}{\Omega} - \left(2\frac{\delta}{a} - \frac{4}{3}\lambda\theta\right)\frac{\dot{\zeta}_S}{\Omega}\right.$$

$$\left. -\frac{\delta}{a} + \frac{4}{3}\lambda\theta + 2\lambda^2\right] \tag{7.1.18}$$

where

$$\bar{e} = \frac{e}{R}.$$

It has been assumed that the blade has a uniform mass distribution. These results can be displayed more conveniently by nondimensionalizing time by rotor angular velocity Ω and also expressing the variables as the sum of a constant equilibrium part and a perturbation

$$\beta_S = \beta_0 + \beta$$
$$\zeta_S = \zeta_0 + \zeta \tag{7.1.19}$$

Retaining only linear terms, the equilibrium equations are

$$\beta_0 = \frac{\gamma}{8(1 + \frac{3}{2}\bar{e})}\left[\theta + \frac{4\lambda}{3}\right]$$

$$\zeta_0 = \frac{\gamma}{12\bar{e}}\left[-\frac{\delta}{a} + \frac{4}{3}\lambda\theta + 2\lambda^2\right] = -\frac{1}{3}\frac{\gamma}{\bar{e}}\left(\frac{2C_q}{a\sigma}\right) \tag{7.1.20}$$

The steady value of the flapping, β_0, is referred to as the coning angle. The steady value of the lag angle, ζ_0, is proportional to the rotor torque coefficient, C_q [1].

The perturbation equations are

$$\ddot{\beta} + \frac{\gamma}{8}\dot{\beta} + \left(1 + \frac{3}{2}\bar{e}\right)\beta + \left[2\beta_0 - \frac{\gamma}{8}\left(2\theta + \frac{4}{3}\lambda\right)\right]\dot{\zeta} = 0$$

$$\left[-2\beta_0 + \frac{\gamma}{8}\left(\theta + \frac{8}{3}\lambda\right)\right]\dot{\beta} + \ddot{\zeta} + \frac{\gamma}{8}\left(2\frac{\delta}{a} - \frac{4}{3}\lambda\theta\right)\dot{\zeta} + \frac{3}{2}\bar{e}\zeta = 0 \tag{7.1.21}$$

These equations describe the coupled flap-lag motion of a rotor blade. A number of features can be noticed. The effect of the blade angular velocity on the lag frequency is much weaker than on flap frequency. The uncoupled natural frequency in flap expressed as a fraction of the blade angular velocity is

$$\frac{\omega_\beta}{\Omega} = \sqrt{1 + \tfrac{3}{2}\bar{e}} \qquad (7.1.22)$$

and the uncoupled frequency in lag is

$$\frac{\omega_\zeta}{\Omega} = \sqrt{\tfrac{3}{2}\bar{e}} \qquad (7.1.23)$$

For a typical hinge offset of $\bar{e} = 0.05$, the rigid flap frequency is

$$\frac{\omega_\beta}{\Omega} = 1.04$$

and the rigid lag frequency is

$$\frac{\omega_\zeta}{\Omega} = 0.27$$

The flap natural frequency is thus somewhat higher than the rotational speed and the lag frequency is roughly one-quarter of the rotational speed. This difference is due to the weaker effect of the restoring moment due to centrifugal force in the lag direction as indicated in Figure 7.3.

The uncoupled lag damping arises primarily from the blade drag and is equal to

$$D_L \cong 2\frac{\delta}{a}\left(\frac{\gamma}{8}\right) \qquad (7.1.24)$$

The lift curve slope of the blade, a, is the order of 6 per radian and the drag coefficient, δ, is the order of 0.015 giving a physical lag damping which is 0.005 times the flap damping or characteristically negligible. The damping ratio of the uncoupled lag motion for a Lock number of 8 is

$$\zeta_L = 0.009$$

This low value of aerodynamic damping indicates that structural damping will be of significance in estimating the lag damping. Any coupling between these equations which reduces the lag damping tends to result in an instability. Equations (7.1.21) can be rewritten

$$\ddot{\beta} + \frac{\gamma}{8}\dot{\beta} + \left(1 + \frac{3}{2}\bar{e}\right)\beta + \left(\beta_0 - \frac{\gamma}{8}\theta\right)\dot{\zeta} = 0$$

$$-\frac{\gamma}{8}\theta\dot{\beta} + \ddot{\zeta} + \frac{\gamma}{8}\left[2\frac{\delta}{a}\right]\dot{\zeta} + \frac{3}{2}\bar{e}\zeta = 0$$

(7.1.25)

where the equilibrium relationship for β_0 has been introduced (7.1.20) with the effect of hinge offset on coning neglected. It can be shown that the coupling present in this two-degree-of-freedom system arising from inertial and aerodynamic forces will not lead to an instability. However, minor features of the hub geometry can lead to instability. The equilibrium lag angle is proportional to rotor torque (equation (7.1.20)) and consequently varies over a wide range from high power flight to autorotation as a result of the weak centrifugal stiffening. Thus the simple pitch link geometry shown in Figure 7.4 will produce a pitch change with lag

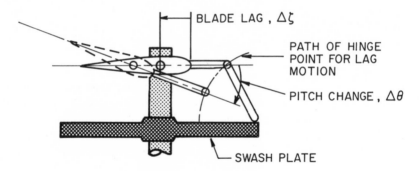

BLADE LAG , $\Delta\zeta$

PATH OF HINGE POINT FOR LAG MOTION

PITCH CHANGE , $\Delta\theta$

SWASH PLATE

Figure 7.4 *Pitch-lag coupling due to pitch link geometry. Articulated rotor.*

depending upon the equilibrium lag angle. The blade pitch angle variation with lag angle can be expressed as

$$\Delta\theta = \theta_\zeta\zeta$$

This expression is inserted into equations (7.1.18). Retaining only the linear terms, the perturbation equations are

$$\ddot{\beta} + \frac{\gamma}{8}\dot{\beta} + \left(1 + \frac{3}{2}\bar{e}\right)\beta + \left(\beta_0 - \frac{\gamma}{8}\theta\right)\dot{\zeta} - \frac{\gamma}{8}\theta_\zeta\zeta = 0$$

$$-\frac{\gamma}{8}\theta\dot{\beta} + \ddot{\zeta} + \frac{\gamma}{8}\left[2\frac{\delta}{a}\right]\dot{\zeta} + \frac{3}{2}\bar{e}\zeta - \frac{\gamma}{6}\lambda\theta_\zeta\zeta = 0$$

(7.1.26)

We can now sketch a root locus for the effect of θ_ζ on the dynamics of this system. Expressing the equations of motion in operational notation, the root locus equation for variations in θ_ζ is

$$\frac{-\theta_\zeta \dfrac{\gamma\lambda}{6}\left[s^2+\dfrac{3}{4}\dfrac{\beta_0}{\lambda}s+1+\dfrac{3}{2}\bar{e}\right]}{\left[s^2+\dfrac{\gamma}{8}s+\left(1+\dfrac{3}{2}\bar{e}\right)\right]\left[s^2+\dfrac{\gamma}{8}\left[2\dfrac{\delta}{a}\right]s+\dfrac{3}{2}\bar{e}\right]+\dfrac{\gamma}{8}\theta\left(\beta_0-\dfrac{\gamma}{8}\theta\right)s^2}=-1$$

$$(7.1.27)$$

The root locus shown in Figure 7.5 illustrates the effect of this geometric coupling, indicating that the critical case where instability occurs corresponds to the 180° locus (θ_ζ is positive). Recall that λ is negative. Thus, if forward lag produces an increase in pitch, an instability is likely to occur. The effect is also proportional to thrust coefficient indicating that the instability is more likely to occur as the thrust is increased [18, 19]. Increasing thrust also increases the steady-state lag angle, hence increasing the geometric coupling for the geometry shown. In general, this instability tends to be of a rather mild nature, however it can act to reduce blade fatigue life. Mechanical dampers are

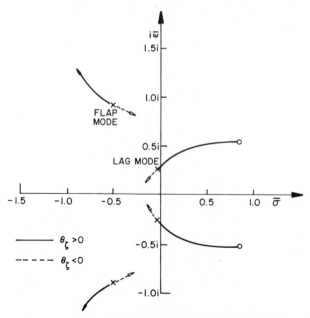

Figure 7.5   *Effect of pitch-lag coupling on flap-lag stability.*

often installed about the lag axes for reasons to be discussed and these also provide additional lag damping and thus alleviate the instability.

This example serves to illustrate that great care must be taken in the geometric design of the articulated rotor hub to avoid undesirable couplings and possible instabilities. We now turn to the elastic hingeless blade.

Elastic motion of hingeless blades

The dynamics of a single hingeless blade will now be examined. Again we will use a simplified analysis which yields the essential features of the dynamic motion and the reader is referred to the literature for a more detailed approach. In general, the flap and lag elastic deformations referred to a shaft axis system are coupled as a result of the fact that the blade will have some average pitch angle, and therefore the principal elastic axes of the blade will be inclined with respect to the shaft. In fact, the term flexible blade as used here includes the hub as well as the blade itself. Hub is used to refer to the portion of the blade structure inboard of the radial location where the pitch change takes place. The rotation of the blade principal elastic axes with blade pitch will depend upon the relative stiffness of the hub and the blade. It can be seen physically that if the hub is soft in comparison to the blade then the principal axes of this flexible system tend to remain fixed as the pitch of the blade is changed. However, if the hub is stiff and the blade is soft the principal elastic axes rotate in a 1 : 1 relationship with blade pitch. An additional source of elastic coupling between flap and lag deflections arises from the built-in blade twist. This flap-lag coupling tends to be small for a helicopter blade where the twist over the length of the blade is of the order of 8° washout, but is significant for blades such as a prop-rotor [6] where the twist is 30°. A third source of elastic coupling between flap and lag arises from inclusion of torsion as a degree-of-freedom. For the typical rotor blade with a high torsional frequency, the effect of torsional flexibility on flap-lag coupling can be obtained through a quasistatic approximation to the torsional motion. That is, for a first order estimate the torsional inertia and damping can be neglected, and the coupling effects of torsional flexibility expressed in terms of a geometric coupling similar in form to the hub geometry effects described in connection with the fully-articulated rotor. A detailed analysis of the flap-lag-torsion motion of a hingeless rotor blade may be found in [20] and [21], and the complete equations of motion for elastic bending and torsion of rotor blades may be found in [22].

We now proceed to examine the flap-lag motion of a hingeless rotor blade from a simplified viewpoint.

If it is assumed that the rotor blade is untwisted, has zero pitch, and is torsionally rigid, the natural frequencies of the rotating blade can be expressed in terms of its mode shapes, ϕ, and derivatives with respect to radial distance ϕ' and ϕ'' as [23–25]

$$\omega_\beta^2 = \frac{\int_0^R EI_\beta(\phi_\beta'')^2\,dr + \Omega^2 \int_0^R (\phi_\beta')^2 \left(\int_r^R mn\,dn\right) dr}{\int_0^R m\phi_\beta^2\,dr}$$

$$\omega_\zeta^2 = \frac{\int_0^R EI_\zeta(\phi_\zeta'')^2\,dr + \Omega^2 \left\{\int_0^R (\phi_\zeta')^2 \left(\int_r^R mn\,dn\right) dr - \int_0^R m\phi_\zeta^2\,dr\right\}}{\int_0^R m\phi_\zeta^2\,dr}$$

(7.1.28)

m is the running mass of the blade and EI is the stiffness. The first term in each of these expressions gives the nonrotating natural frequency and the second term gives the effect of centrifugal stiffening due to rotation. The coefficient of the square of the angular velocity Ω in the expression for flapping frequency is usually referred to as the Southwell coefficient. Note that the effect of the centrifugal stiffening is considerably weaker in the lag direction than in the flap direction as would be expected from previous discussion of the articulated rotor.

Denoting the Southwell coefficient by K_S

$$K_S = \frac{\int_0^R (\phi_\beta')^2 \left(\int_r^R mn\,dn\right) dr}{\int_0^R m\phi_\beta^2\,dr}$$

(7.1.29)

and the nonrotating frequencies by

$$\omega_{\beta_0}^2 = \frac{\int_0^R EI_\beta(\phi_\beta'')^2\,dr}{\int_0^R m\phi_\beta^2\,dr}$$

$$\omega_{\zeta_0}^2 = \frac{\int_0^R EI_\zeta(\phi_\zeta'')^2\,dr}{\int_0^R m\phi_\zeta^2\,dr}$$

(7.1.30)

341

the rotating frequencies can be written as

$$\omega_\beta^2 = \omega_{\beta_0}^2 + K_S \Omega^2$$
$$\omega_\zeta^2 = \omega_{\zeta_0}^2 + (K_S - 1)\Omega^2$$

(7.1.31)

if the flap and lag mode shapes are assumed to be the same. It is interesting to note that if the mode shape is assumed to be that of a rigid articulated blade with hinge offset $\bar{e}$, i.e.,

$$\phi = 0 \qquad 0 < x < \bar{e}$$
$$\phi = (x - \bar{e}) \qquad \bar{e} < x < 1$$

(7.1.32)

for a uniform mass distribution and small $\bar{e}$, then from (7.1.29)

$$K_S \cong 1 + \tfrac{3}{2}\bar{e}$$

(7.1.33)

Thus the natural frequencies are from (7.1.30), (7.1.31) and (7.1.33)

$$\omega_\beta^2 = \Omega^2(1 + \tfrac{3}{2}\bar{e})$$
$$\omega_\zeta^2 = \Omega^2(\tfrac{3}{2}\bar{e})$$

(7.1.34)

reducing to the results for the rigid blade. For typical blade mass and stiffness distributions the Southwell coefficient is of the order of 1.2 [24].

A simplified model for the elastic rotor blade follows. The elastic blade is modelled as a rigid blade with hinge offset $\bar{e}$ and two orthogonal springs (K_β and K_ζ) located at the hinge to represent the flap and lag stiffness characteristics. The natural frequencies for this model of the blade are

$$\omega_\beta^2 = \frac{K_\beta}{I_B} + (1 + \tfrac{3}{2}\bar{e})\Omega^2$$
$$\omega_\zeta^2 = \frac{K_\zeta}{I_B} + (\tfrac{3}{2}\bar{e})\Omega^2$$

The spring constants K_β and K_ζ can be chosen to match the nonrotating frequencies of the actual elastic blade and the offset is chosen to match the Southwell coefficient and in this way the dependence of frequency on rotor angular velocity is matched. Owing to the fact that the Southwell coefficient is close to one, i.e., the equivalent offset, $\bar{e}$, is small, in many investigations the dependence of the Southwell coefficient on $\bar{e}$ is neglected [17] giving

$$\omega_\beta^2 = \frac{K_\beta}{I_B} + \Omega^2$$
$$\omega_\zeta^2 = \frac{K_\zeta}{I_B}$$

(7.1.36)

Thus, with this approximation there is no centrifugal stiffening in the lag direction. We will use this approximation in the analysis which follows. Recall that these frequencies are assumed to be uncoupled and therefore are defined with respect to the blade axes. Thus, they will appear coupled in a shaft oriented axis system. In order to include the effect of hub flexibility in the analysis the hub (the portion of the blade system which does not rotate with pitch) is modelled by a second pair of orthogonal springs which are oriented parallel and perpendicular to the shaft and do not rotate when the blade pitch is changed [17]. These spring constants are denoted K_{β_H} and K_{ζ_H}. The springs representing blade stiffness (K_{β_B} and K_{ζ_B}) are also located at the root since offset has been neglected. However, this pair of springs rotate with the blade as pitch is changed. Figure 7.6 shows the geometry.

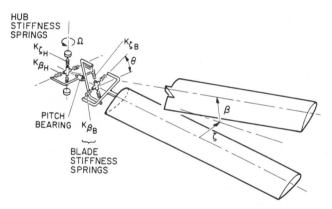

Figure 7.6 *Spring model for elastic blade and hub.*

This model for the hub and blade gives rise to elastic coupling between flap and lag motion. Essentially a mode shape $\phi = x$ is being employed to describe the elastic deflection of the blade in both directions such that the aerodynamic and inertial coupling terms developed for the articulated blade model (equation (7.1.21) apply directly to this approximate model of the hingeless blade. The equations of motion for flap-lag dynamics are therefore

$$\ddot{\beta} + \frac{\gamma}{8}\dot{\beta} + p^2\beta - \left\{\frac{\gamma}{8}\left(2\theta + \frac{4}{3}\lambda\right) - 2\beta_0\right\}\dot{\zeta} + z^2\zeta = 0$$

$$\left[-2\beta_0 + \frac{\gamma}{8}\left(\theta + \frac{8}{3}\lambda\right)\right]\dot{\beta} + z^2\beta + \ddot{\zeta} + \frac{\gamma}{8}\left(2\frac{\delta}{a} - \frac{4}{3}\lambda\theta\right)\dot{\zeta} + q^2\zeta = 0$$

(7.1.37)

343

where the difference between these equations of motion and those presented for the articulated blade (7.1.21) arise from the terms p, q, and z. p and q are the ratios of the uncoupled natural frequencies, i.e., those at zero pitch, to the rotor rpm and z is the elastic coupling effect. For the spring model described above these terms can be expressed as [17]

$$p^2 = 1 + \frac{1}{\Delta}(\bar{\omega}_\beta^2 + R(\bar{\omega}_\zeta^2 - \bar{\omega}_\beta^2)\sin^2\theta)$$

$$q^2 = \frac{1}{\Delta}(\bar{\omega}_\zeta^2 - R(\bar{\omega}_\zeta^2 - \bar{\omega}_\beta^2)\sin^2\theta)$$

$$z^2 = \frac{R}{2\Delta}(\bar{\omega}_\zeta^2 - \bar{\omega}_\beta^2)\sin^2\theta$$

$$\Delta = 1 + R(1-R)\frac{(\bar{\omega}_\zeta^2 - \bar{\omega}_\beta^2)^2}{\bar{\omega}_\zeta^2\bar{\omega}_\beta^2}\sin 2\theta \qquad (7.1.38)$$

$$\bar{\omega}_\beta^2 = \frac{K_\beta}{I_B\Omega^2} \qquad \bar{\omega}_\zeta^2 = \frac{K_\zeta}{I_B\Omega^2}$$

$$K_\beta = \frac{K_{\beta_B}K_{\beta_H}}{K_{\beta_B} + K_{\beta_H}} \qquad K_\zeta = \frac{K_{\zeta_B}K_{\zeta_H}}{K_{\zeta_B} + K_{\zeta_H}} \qquad R = \frac{\bar{\omega}_\zeta^2\dfrac{K_\beta}{K_{\beta_B}} - \bar{\omega}_\beta^2\dfrac{K_\zeta}{K_{\zeta_B}}}{\bar{\omega}_\zeta^2 - \bar{\omega}_\beta^2}$$

R is referred to as the elastic coupling parameter. The physical significance of this parameter can be understood by examining the relationship between the rotation of the principal axes of the blade-hub system, η, and the blade pitch angle, θ, [26]

$$\tan 2\eta = \frac{R\sin 2\theta}{R\cos 2\theta + (1-R)} \qquad (7.1.39)$$

It can be seen from this expression that if $R = 0$ the principal axes remain fixed as blade pitch is changed and consequently there is no elastic coupling. The flap and lag natural frequencies are

$$p^2 = \bar{\omega}_\beta^2 + 1$$
$$q^2 = \bar{\omega}_\zeta^2$$

where $\bar{\omega}_\beta^2$ and $\bar{\omega}_\zeta^2$ are the dimensionless nonrotating frequencies. This is the case in which the hub is flexible and the blade is rigid. At the other limit $R = 1$, equation (7.1.39) indicates that the principal axes rotate in a $1:1$ relationship with the blade pitch ($\eta = \theta$). In this case elastic coupling is present, and expressions for the natural frequencies (7.1.38) simply represent the fact that as the blade is rotated through 90° pitch the nonrotating frequencies must interchange. In addition to the case $R = 0$ where the elastic coupling between flap and lag vanishes, another interesting case exists in which no elastic coupling is present. This is the case referred to as matched stiffness, i.e., when the nonrotating frequencies of the blade are equal in both directions ($\bar{\omega}_\zeta = \bar{\omega}_\beta$). Various advantages accrue from this particular design choice as will be discussed below.

In principle the designer has at his disposal the selection of the nonrotating frequencies of the blade. Consider some of the options in this regard. For simplicity only the behavior of the rotor at zero pitch is examined. One choice is the matter of the hub stiffness relative to the blade stiffness which has an important impact on the flap-lag behavior of the rotor through the parameter R as will be discussed below. The flap frequency is largely chosen on the basis of the desired helicopter stability and control characteristics [5, 12]. Since the rotor blade is in general a long slender member, the flap frequency will tend to be relatively near to the rotor rpm. Typical ratios of flap frequency to blade angular velocity for hingeless rotor helicopters are of the order of $p = 1.05-1.15$ [25] although at least one helicopter has flown with a flap frequency ratio of 1.4 [27]. The second major design decision is the choice of the lag frequency. Characteristically, the nonrotating lag frequency will tend to be considerably higher than the flap frequency owing to the larger dimensions of the blade and hub in the chordwise direction compared to the flapwise direction. As mentioned above, lag hinges are provided on articulated rotors to relieve lag stresses arising from flapping. Owing to the fact that the flap frequency is only slightly larger than once per revolution on a typical hingeless blade there will be considerable flap bending of the rotor blades. In fact, the amplitude of the vertical displacement of the blade tip on a hingeless blade will be quite similar to the flapping amplitude of the fully articulated rotor. The relationship between amplitude of tip motion of the hingeless blade and the flapping amplitude of the articulated blade is given by [28]

$$|\beta_H| = \frac{|\beta_A|}{\left\{ 1 + \left(\frac{8}{\gamma}(p^2 - 1) \right)^2 \right\}^{\frac{1}{2}}}$$

Therefore, the inplane forces due to flap bending will cause the significant root stresses on a hingeless rotor. The dependence of these stresses on the selection of lag frequency can be seen by assuming that the

345

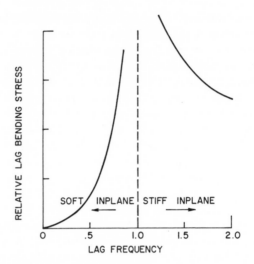

Figure 7.7   Dependance of lag bending stress on lag frequency.

flap and lag bending are loosely coupled ($z = 0$). The lag bending amplitude arising from sinusoidal flap bending at one per rev can be expressed from the equations (7.1.37), neglecting the lag damping, as

$$\left|\frac{\zeta}{\beta}\right| = \frac{\left[2\beta_0 - \frac{\gamma}{8}\left(\theta + \frac{8}{3}\lambda\right)\right]}{(q^2 - 1)} \tag{7.1.40}$$

The lag bending moment at the blade root, $K_\zeta\zeta$, thus varies as $q^2/(q^2 - 1)$ as shown in Figure 7.7. It can be seen that if the lag frequency is selected above one per rev, large root bending stresses occur. The bending moment is reduced by choosing a lag frequency well below one per rev. A lag frequency below one per rev incidentally would be characteristic of a matched stiffness blade. For example, if

$$p^2 = 1.2 = \bar{\omega}_\beta^2 + 1$$

and

$$\bar{\omega}_\zeta^2 = \bar{\omega}_\beta^2$$

then

$$\bar{\omega}_\zeta = 0.45$$

Rotor blades are usually characterized by their lag frequency as soft inplane ($\bar{\omega}_\zeta < 1$) or stiff ($\bar{\omega}_\zeta > 1$). Thus rotor blade lag stresses can be reduced by choosing a soft inplane blade design and it should be kept in mind in the discussion that follows that there is a significant variation in the root bending stress with lag frequency. In the following, the influence of lag frequency on the dynamics of a hingeless blade is examined. Also it may be noted at this point that in contrast to the articulated rotor in which mechanical motion exists in lag at the root such that mechanical lag dampers can be fitted, this is usually difficult with the hingeless rotor if the need should arise as a result of a stability problem. However, hingeless rotor helicopters have been equipped with lag dampers [10, 29].

Note also that if the lag frequency is selected such that at the operating condition of the rotor it is less than one per rev, then resonance in the lag mode will be encountered as the rotor is run up to operating speed.

Flap-lag stability characteristics as predicted by the equations of motion given by equations (7.1.37) are now examined. First consider the case in which the hub is considerably more flexible than the blade ($R = 0$). In Figure 7.8 the stability boundaries given by equations (7.1.37) are shown as a function of flap and lag frequency and blade pitch angle for a typical rotor blade. This figure was obtained by determining the conditions under which Routh's discriminant equals zero. It can be seen that an approximately elliptical region of instability occurs which increases in extent as

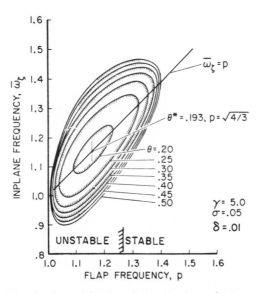

Figure 7.8   Flap lag stability boundaries. $R = 0$, no elastic coupling [17].

347

blade pitch is increased. It is centered around a lag bending frequency of 1.15 and flap frequency of 1.15 indicating that in this particular case, flap-lag instability is likely to be a problem for stiff inplane rotors. A root locus is shown in Figure 7.9 for increasing blade pitch with $R = 0$. It can be seen that the stiff inplane blade $(1.1 < \bar{\omega}_\zeta < 1.2)$ is destabilized with increasing pitch; however, the instability is rather mild even at very large pitch angles. Figure 7.10 shows the effect of various ratios of hub stiffness

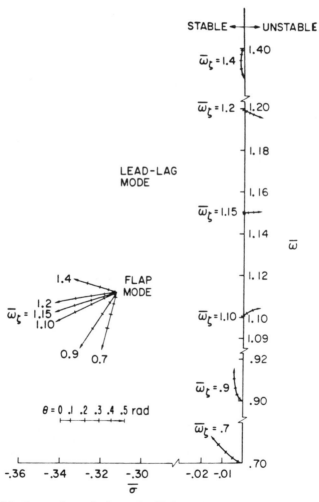

Figure 7.9   Locus of roots for increasing blade pitch. $R = 0$, no elastic coupling [17].

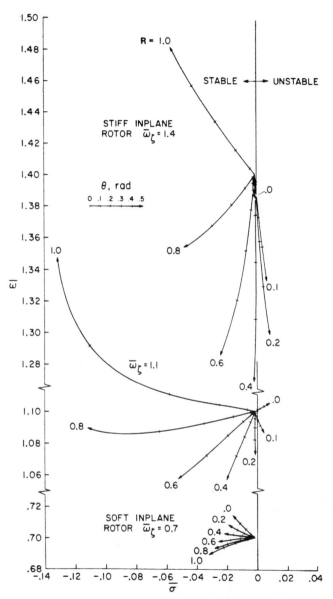

Figure 7.10   Locus of roots for increasing blade pitch with various levels of elastic coupling [17].

to blade stiffness (different values of R) indicating the importance of careful modelling of the blade and hub in the study of flap-lag stability. This theory has been correlated with experiment in [30] and [31]. At large pitch angles where the blade encounters stall, wider ranges of instability occur as shown in [30]. This increase in the region of instability is primarily a result of the loss in flap damping owing to reduction in blade lift curve slope, a.

Various other configuration details have an impact on the flap-lag stability such as precone (the inclination of the blade feathering or pitch change axis with respect to a plane perpendicular to the hub). Precone is usually employed to relieve the root bending stresses that arise from the steady flap bending moment due to average blade lift. The blade may also have droop and sweep [21] (the inclinations of the blade axis with respect to the pitch change axis in the flap and lag directions respectively) which will also have an impact on the flap-lag stability. The presence of kinematic pitch-lag coupling will have important effects on hingeless blade stability which depend strongly on the lag stiffness and the elastic coupling parameter R [17].

If torsional flexibility is included, elastic coupling between pitch, lag and flap will exist. This can be most readily understood by extending the simple spring model of blade flexibility to include a torsion spring. Consider a blade hub system as shown in Figure 7.11 with a flap angle β

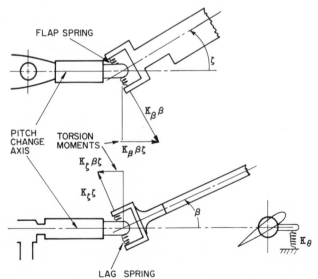

Figure 7.11   Simplified blade model for flap-lag-torsion coupling.

and a lag angle ζ. Owing to the root spring orientation there will be torques exerted about the torsion axis which depend on the respective stiffnesses in the two directions. Representing the torsional stiffness of the blade and control system by K_θ, the equation for torsional equilibrium is (neglecting torsional inertia and damping)

$$K_\theta \theta = (K_\beta - K_\zeta)\beta\zeta \tag{7.1.41}$$

Linearizing about the blade equilibrium position, β_0, ζ_0,

$$\Delta\theta = \frac{1}{K_\theta}[(K_\beta - K_\zeta)\beta_0\,\Delta\zeta + (K_\beta - K_\zeta)\zeta_0\,\Delta\beta] \tag{7.1.42}$$

That is, torsional flexibility results in both pitch-lag coupling

$$\theta_\zeta = \left(\frac{K_\beta - K_\zeta}{K_\theta}\right)\beta_0 \tag{7.1.43}$$

and pitch-flap coupling

$$\theta_\beta = \left(\frac{K_\beta - K_\zeta}{K_\theta}\right)\zeta_0 \tag{7.1.44}$$

These couplings depend upon the relative stiffness of the blade in the flap and lag directions and the equilibrium values of the flap deflection and the lag deflection. A matched stiffness blade ($K_\beta = K_\zeta$) eliminates these couplings and is perhaps the primary reason for interest in a matched stiffness blade. For typical blade frequencies, K_ζ is larger than K_β and therefore θ_ζ tends to be negative and θ_β positive. θ_β is equivalent to what is usually referred to as a δ_3 hinge on an articulated blade. In powered flight ζ_0 is negative (7.1.20), and the sign of the effect is equivalent to negative δ_3 [32]. This pitch change arising from flapping is statically destabilizing in the sense that an upward flapping produces an increase in pitch. If this term becomes sufficiently large, flapping divergence can occur. In autorotation, this coupling would change sign as the equilibrium lag angle is positive. The characteristically negative value of pitch-lag coupling θ_ζ tends to produce a stabilizing effect in most cases as may be seen from the articulated rotor example. Negative values of θ_ζ can be destabilizing for a stiff inplane rotor with small values of R [17]. Precone, that is rotation of the pitch change axis in the flap direction, has a significant effect on the pitch-lag coupling. The coning angle β_0, in equation (7.1.43) refers only to the elastic deflection of the blade. Consequently, with perfect precone, that is, when the precone angle is equal to the equilibrium steady flap angle given by equation (7.1.20), the elastic deflection is zero and the pitch-lag coupling is zero. For over precone i.e., if the rotor is operated well below its design thrust, β_0 is negative and a destabilizing pitch-lag coupling occurs. It should be noted that hub flexibility will also have an important impact on these kinematic couplings since it will determine the deflection of the pitch change axis. A precise formulation of flap-lag-torsion coupling as well as further discussion of its influence on blade stability can be found in [21].

Ormiston has explored the gains in flap-lag stability which may be achieved by various combinations of flap-lag and pitch-lag coupling in [33].

In summary, a soft-inplane rotor blade tends to be less susceptible to isolated blade instabilities while the stiff-inplane blade tends to exhibit instabilities along with a considerably more complex behavior with changes in parameters. Figure 7.12 contrasts the effect of pitch-lag coupling on these two rotor blade types illustrating the complexity of the stability boundaries for the stiff inplane case in contrast to the soft inplane case which is quite similar to the articulated rotor.

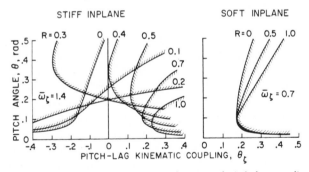

Figure 7.12 Flap-lag stability boundaries as a function of pitch-lag coupling and elastic coupling [17].

7.2 Stall flutter

A single degree-of-freedom instability encountered by helicopter blades which also occurs in gas turbines is referred to as stall flutter. The reader should consult Chapter 5 for a discussion of stall flutter on a nonrotating airfoil. Stall flutter is primarily associated with high speed flight and maneuvering of a helicopter and arises from the fact that stalling of the rotor blade is encountered at various locations on the rotor disc. For a rotor blade, stall flutter does not constitute a destructive instability, but rather produces a limit cycle behavior owing to the varying aerodynamic conditions encountered by the blade as it rotates in forward flight.

Consider the aerodynamic conditions existing on a rotor blade in high speed flight. On the advancing side of the rotor disc, the dynamic pressure at a blade section depends on the sum of the translation velocity of the helicopter and the rotational velocity, while on the retreating side of the disc it depends on the difference between these two velocities. Consequently, if the rolling moment produced by the rotor is equal to zero, as required for equilibrium flight, the angle-of-attack of the blade is considerably smaller on the advancing side than on the retreating side. A typical angle-of-attack distribution at an advance ratio of 0.33 (140 kts) is shown in Figure 7.13. This resulting distribution is produced by a combination of flapping or flap bending motion and the pilot's control input.

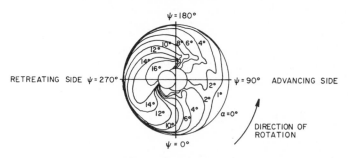

Figure 7.13   *Angle of attack distribution of helicopter rotor at 140 knots (advance ratio =* 0.33) [34]

Note that on the advancing side the angle-of-attack is small and varies comparatively slowly with azimuth angle. On the retreating side the angle-of-attack is large and changes rapidly with azimuth angle. Consequently, prediction of the airload on the blade requires a model for the aerodynamics of the blade element which includes unsteady effects both in the potential flow region as well as in the stalled region. The source of the stall flutter instability is related to the unsteady aerodynamic characteristics of an airfoil under stalled conditions. Since the stalled region is only encountered by the blade over a portion of the rotor disc, however, if an instability occurs as a result of the aerodynamics at stall, it will not give rise to continuing unstable motion since a short time later the blade element will be at a low angle-of-attack, well below stall.

Owing to the complexity of the flow field around a stalled airfoil, we must have recourse to experimental data in order to determine the unsteady aerodynamic characteristics of an airfoil oscillating at high angle-of-attack. Experimental data have become available in recent years on typical helicopter airfoil sections [34–36], which make it possible to characterize the aerodynamics of an airfoil oscillating about stall. In addition a number of investigations have been conducted which give insight into the nature and complexity of the aerodynamic flow field under stalled conditions [37–39].

For a simplified treatment of stall flutter, it is assumed that the blade motion can be adequately described by a model involving only the blade torsional degree-of-freedom. The influence of flapping or heave motion of the section is neglected such that $\theta = \alpha$. The equation of motion for this single degree-of-freedom system is therefore

$$\ddot{\alpha} + \omega_\theta^2 \alpha = \left(\frac{\rho(\Omega R)^2 c^2}{2 I_\theta}\right) C_M(\dot{\alpha}) \tag{7.2.1}$$

where aerodynamic strip theory analysis is employed. Since the aerodynamic damping is a complex function of the angular velocity it is convenient to express equation (7.2.1) as an energy equation by multiplying by $d\alpha$ and integrating over one cycle to obtain

$$\Delta\left\{\frac{\dot{\alpha}^2}{2} + \omega_\theta^2 \, \frac{\alpha^2}{2}\right\} = \frac{\rho(\Omega R)^2 c^2}{2 I_\theta} \oint C_M(\dot{\alpha}) \, d\alpha \tag{7.2.2}$$

The left hand side of equation (7.2.2) expresses the change in energy over one cycle which is produced by the dependence of aerodynamic pitching moment on angle-of-attack rate as given by the right hand side. Figure 7.14 shows the time history of the pitching moment and normal force coefficients as a function of angle-of-attack for an airfoil oscillating at a reduced frequency typical of one per rev motion at three mean angles-of-attack. The arrows on the figure denote the direction of change of C_N and C_M. Note the large hysteresis loop which occurs in the normal force in the dynamic case when the mean angle-of-attack is near stall. This effect was first noted by Farren [40]. In the potential flow region the effects are rather small and are predicted by Theodorsen's method (see Chapter 4). Proper representation of the unsteady lift behavior does have an important bearing on the prediction of rotor performance, but will not be discussed further [41]. The pitching moment characteristics are of primary interest here.

The pitching moment is well behaved in the potential flow region, and well above stall, resulting in small elliptically shaped loops over one

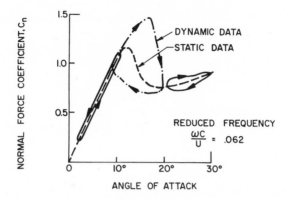

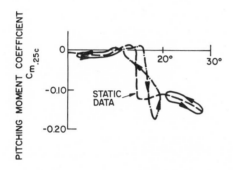

Figure 7.14   Typical oscillating airfoil data [35].

cycle. In the vicinity of lift stall two interesting effects occur, the average pitching moment increases markedly in a nose down sense, a phenomenon that is referred to as moment stall [35] and the moment time history looks like a figure eight. The change in energy over one cycle given by equation (7.2.2) is proportional to

$$\oint C_M(\dot\alpha)\,d\alpha$$

The value of this integral is equal to the area enclosed by the loop, and its sign is given by the direction in which the loop is traversed. If the loop is traversed in a counter clockwise direction, then this integral will have a negative value indicating that energy is being removed from the structure or that there is positive damping. Thus the low and high angle-of-attack traces indicate positive damping. Around the angle-of-attack at which moment stall occurs statically however, the figure-eight-like behavior

355

indicates that there is essentially no net dissipation of energy over a cycle or possibly that energy is being fed into the structure (the integral on the right hand side of equation (7.2.2) is positive). This pitching moment characteristic gives rise to the phenomenon referred to as stall flutter. To actually encounter stall flutter, this behavior must occur over some appreciable span of the blade [42]. Ref. [42] also discusses the importance of the rate of change of angle-of-attack with time on the dynamics of this process and concludes that delay in the development of dynamic stall depending upon $\dot{\alpha}$ is responsible for stall occurring over a significant radius of the blade with consequent effects on the rotor loads and vibrations. Of course, the rotor blade only encounters this instability over a small azimuth range and consequently the complete motion is essentially a limit cycle. The loss in damping at stall coupled with the marked change in the average pitching moment gives rise to large torsional motion with perhaps 2 or 3 cycles of the torsion excited before being damped by the low angle-of-attack aerodynamics. A typical time history of blade torsional motion and angle-of-attack when stall flutter is encountered is shown in Figure 7.15. The dominant effect of the occurrence of

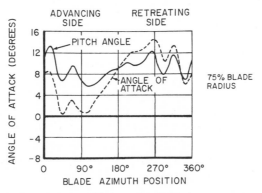

Figure 7.15   Typical time history of blade motion for blade encountering stall flutter [42].

stall flutter on a helicopter is to give rise to a marked increase in the vibratory loads in the blade pitch control system [42]. Ref. [42] discusses approximate methods for incorporating unsteady stall aerodynamics into the rotor blade equations of motion. The most significant assumption in the analysis of rotor blade stall flutter relates to the applicability of two-dimensional data on airfoils oscillating sinusoidally to a highly three-dimensional flow field in which the motions are nonsinusoidal.

Further understanding of the aerodynamics of stall may make it possible to design airfoil sections which would minimize the occurrence of stall flutter and the associated control loads. Blade section design, however, has many constraints owing to the wide range in aerodynamic conditions encountered in one revolution and the aerodynamic phenomena described appear to be characteristic of airfoils oscillating about static stall angle-of-attack.

7.3 Blade motion/body coupling

As the last major topic, an aeroelastic instability of helicopters associated with coupling of blade motion and body motion is examined. This problem is of considerable significance in articulated and hingeless rotor helicopter design, and was first encountered on autogiros. It differs from the problems previously discussed in the sense that it is a destructive instability much like classical bending-torsion flutter rather than being a mild instability which ultimately becomes a small amplitude limit cycle. In fact, this violent instability was at first attributed to rotor blade flutter until a theory was developed during the period 1942 to 1947 showing it to be a new phenomenon. The instability is called *ground resonance* and was first analyzed and explained by Coleman [43] who modelled the essential features of the instability for articulated rotor helicopters. The name ground resonance is somewhat confusing since, in fact, the dynamic system of the helicopter and blades is unstable. The instability occurs at a particular rotor angular velocity and therefore it appears in some sense like a resonance, but it is not. Further the ground enters the problem owing to the mechanical support provided the helicopter fuselage by the landing gear. The particularly interesting result obtained by Coleman is that the instability can be predicted neglecting the rotor aerodynamics, that is, ground resonance is purely a mechanical instability, the energy source being the rotor angular velocity. In the discussion Coleman's development is followed. Then there is qualitative discussion of the more complex formulation of this problem as applied to hingeless rotors. For an articulated rotor, the aerodynamics tend to be unimportant and only the lag degree-of-freedom needs to be included. For hingeless rotors, the flapping degree-of-freedom is important in the analysis and aerodynamic forces play a significant role [10]. The addition of the flapping degrees-of-freedom leads to a similar instability in flight referred to as air resonance.

Following Coleman's analysis we consider a simplified model of a helicopter resting on the ground. The degrees-of-freedom assumed are: pitch and roll of the rotor shaft or pylon which arise from the landing gear oleo strut flexibility and the lag degree-of-freedom of each rotor blade. Ref. [44] presents a generalized approach including additional support degrees-of-freedom and Ref. [45] presents results of an experimental investigation of the phenomenon. Discussion is restricted to the case in which the rotor has three or more rotor blades and thus has polar symmetry. The two-bladed rotor is a somewhat more complex problem and a few remarks on this special case are made at the end of this section. A four-bladed rotor system is used as the example since the approach is most easily visualized in this case. The generalization to three or more blades is described at the end of this section.

Consider then the helicopter shown in Figure 7.16. The system has six degrees-of-freedom, the lag motion of each of the four rotor blades and the two pylon deflections. Each rotor blade is modelled as an articulated blade with hinge offset $\bar{e}$. A spring is included at the root since a centering spring may be employed about the lag hinge.

A coordinate system is chosen which is fixed in space in order to allow the simplest mathematical treatment of asymmetric stiffness and inertia characteristics associated with pitch and roll motion of the fuselage on the landing gear. If a rotating coordinate system is employed then the differential equations describing the dynamics would involve periodic coefficients with attendant problems in unravelling the solution. In fact, it is this difference in the form of the equations of motion in fixed and rotating coordinate systems which gives rise to difficulties in analyzing the two-bladed rotor system with asymmetric pylon characteristics. The two-bladed rotor lacks polar symmetry and, therefore, a fixed coordinate system approach will give rise to periodic coefficients from the rotor while a rotating coordinate system analysis will give rise to periodic coefficients from the asymmetric pylon characteristics. Thus periodic coefficients can not be eliminated in the two-bladed case unless the pylon frequencies are equal. For a rotor with three or more blades the use of a fixed coordinate system allows treatment of asymmetric pylon characteristics without encountering the problem of solving equations with periodic coefficients.

First we consider the equations of motion describing blade lag dynamics in a fixed coordinate system to illustrate the influence of coordinate system motion. All of our previous examples have used a coordinate system rotating with the blade. Simplification of this problem can be effected by defining new coordinates to describe the rotor lag motion. These new coordinates are linear combination of the lag motion of the individual blades. They usually are referred to as multi-blade

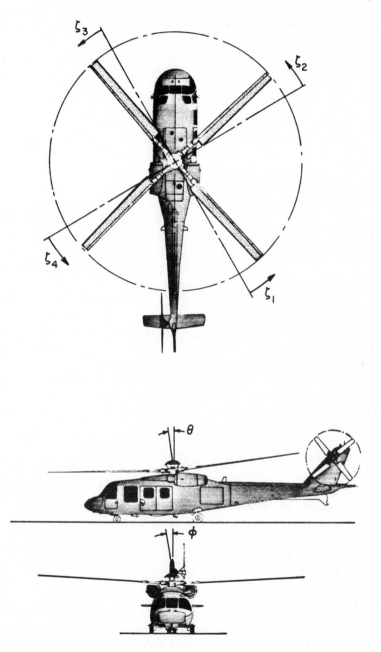

Figure 7.16   Mechanical degrees of freedom for ground resonance analysis.

359

coordinates [46] and are defined for a four-bladed rotor as

$$\gamma_0 = \frac{\zeta_1 + \zeta_2 + \zeta_3 + \zeta_4}{4}$$

$$\gamma_1 = \frac{\zeta_1 - \zeta_3}{2}$$

$$\gamma_2 = \frac{\zeta_2 - \zeta_4}{2}$$

(7.3.1)

$$\gamma_3 = \frac{(\zeta_1 + \zeta_3) - (\zeta_2 + \zeta_4)}{4}$$

The new coordinates γ_1 and γ_2 describe the motion of the center of mass of the rotor system with respect to the axis of rotation and thus are responsible for coupling of rotor motion to pylon motion, while γ_0 and γ_3 describe motions of the rotor in which the center of mass of the rotor system remains on the axis of rotation. If $\gamma_1 = \gamma_2 = 0$, then motions corresponding to γ_0 and γ_3 are such that opposite blades move as though rigidly attached together with a vertical pin at the root. These motion variables, γ_0 and γ_3 will be uncoupled for the dynamic problem of interest and consequently the system is reduced to four degrees-of-freedom by introducing these coordinates as will be shown.

Now the equations of motion for γ_1 and γ_2 are developed in a moving coordinate system and then transformed to a stationary coordinate system. With the hub fixed, the lag motion of each blade without aerodynamics is, as shown earlier,

$$\ddot{\zeta}_i + \Omega^2(\bar{\omega}_\zeta^2)\zeta_i = 0 \qquad i = (1, 2, 3, 4)$$

(7.3.2)

The natural frequency, $\bar{\omega}_\zeta^2$ arises from a mechanical spring on the hinge and the offset or centrifugal stiffening effect and is given by equation (7.1.35). The equations of motion for γ_1 and γ_2 are from (7.3.1) and (7.3.2)

$$\ddot{\gamma}_1 + \Omega^2(\bar{\omega}_\zeta^2)\gamma_1 = 0$$

$$\ddot{\gamma}_2 + \Omega^2(\bar{\omega}_\zeta^2)\gamma_2 = 0$$

(7.3.3)

These equations may be thought of as describing the motion of the center of mass of the rotor system in two directions with respect to the coordinate system rotating at the rotor angular velocity Ω. Resolving to a

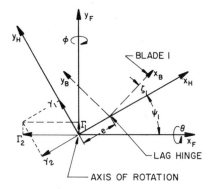

Figure 7.17 Axis systems and coordinates for ground resonance analysis.

fixed coordinate system as shown in Figure 7.17,

$$\gamma_1 = \Gamma_1 \cos \Omega t + \Gamma_2 \sin \Omega t$$
$$\gamma_2 = -\Gamma_1 \sin \Omega t + \Gamma_2 \cos \Omega t$$

(7.3.4)

Differentiating and substituting (7.3.4) into (7.3.3) we obtain the equations

$$[\ddot{\Gamma}_1 + 2\Omega\dot{\Gamma}_2 + \{\Omega^2(\bar{\omega}_\zeta^2 - 1)\}\Gamma_1] \cos \Omega t$$
$$+ [\ddot{\Gamma}_2 - 2\Omega\dot{\Gamma}_1 + \{\Omega^2(\bar{\omega}_\zeta^2 - 1)\}\Gamma_2] \sin \Omega t = 0$$
$$[\ddot{\Gamma}_1 + 2\Omega\dot{\Gamma}_2 + \{\Omega^2(\bar{\omega}_\zeta^2 - 1)\}\Gamma_1] \sin \Omega t$$
$$- [\ddot{\Gamma}_2 - 2\Omega\dot{\Gamma}_1 + \{\Omega^2(\bar{\omega}_\zeta^2 - 1)\}\Gamma_2] \cos \Omega t = 0$$

(7.3.5)

The second equation appears similar to the first with the coefficients of the sin and cos terms reversed. While the variables have been transformed to a fixed system, the equations of motion are still referred to a moving frame which accounts for the presence of the cos and sin terms. To complete the transformation multiply the first equation by cos and add to the second multiplied by sin to obtain one fixed axis equation. Multiplying the first equation by sin and subtracting from the second multiplied by cos yields the second equation. The two equations of motion are

$$\ddot{\Gamma}_1 + 2\Omega\dot{\Gamma}_2 + \{\Omega^2(\bar{\omega}_\zeta^2 - 1)\}\Gamma_1 = 0$$
$$\ddot{\Gamma}_2 - 2\Omega\dot{\Gamma}_1 + \{\Omega^2(\bar{\omega}_\zeta^2 - 1)\}\Gamma_2 = 0$$

(7.3.6)

361

These are the equations of motion for the new lag coordinates (or CM motion) in the nonrotating coordinate system. Note that the variables are coupled due to the effects of rotation. The characteristic equation for the dynamics of the lag motion is now obtained from equations (7.3.6) as

$$\{s^2 + \Omega^2(\bar{\omega}_\zeta^2 - 1)\}^2 + 4\Omega^2 s^2 = 0 \tag{7.3.7}$$

The roots of this characteristic equation are

$$
\begin{aligned}
s_{1,2} &= \pm i\Omega(\bar{\omega}_\zeta + 1) \\
s_{3,4} &= \pm i\Omega(\bar{\omega}_\zeta - 1)
\end{aligned}
\tag{7.3.8}
$$

Thus, the coordinate transformation has resulted in natural frequencies in the fixed coordinate system which are equal to the natural frequencies in the rotating system given by equations (7.3.3) ($\bar{\omega}_\zeta \Omega$) plus or minus the rotational speed (the angular velocity of the coordinate system). This is a basic characteristic of natural frequencies when calculated in rotating and fixed coordinate systems which must be kept in mind in analyzing rotating systems. At this point we consider one other aspect of the dynamics of this type of system which is helpful in visualizing the motion. Consider the eigenvectors describing the amplitude and phase of the two variables in transient motion. These ratios are obtained from the equations of motion and the characteristic roots.

$$\left|\frac{\Gamma_1}{\Gamma_2}\right| = \frac{-2\Omega s}{s^2 + \Omega^2(\bar{\omega}_\zeta^2 - 1)}\Bigg|_{s_{1,2,3,4}}$$

Therefore

$$
\begin{aligned}
\left|\frac{\Gamma_1}{\Gamma_2}\right| &= \pm i\Bigg|_{s_{1,2}} \\
\left|\frac{\Gamma_1}{\Gamma_2}\right| &= \mp i\Bigg|_{s_{3,4}}
\end{aligned}
\tag{7.3.9}
$$

The upper sign corresponds to the upper sign in the roots (7.3.8).

In either of these characteristic motions, Γ_1 and Γ_2 are of equal amplitude and Γ_1 either leads or lags Γ_2 by 90°. Thus the transient motion of the rotor system center of mass is a circular motion. This symmetry which occurs in many rotating systems permits an elegant formulation using complex coordinates [28, 43]. The two variables Γ_1 and Γ_2 can be combined into one single complex variable, as will be discussed below. Further since the transient motion is circular, these modes are referred to as whirling modes and the whirling may be described as advancing or regressing depending upon whether the mode of motion corresponds to transient motion in the direction of rotor rotation or against the direction of rotation. Consider the root

$$s_1 = +i\Omega(\bar{\omega}_\zeta + 1)$$

corresponding to a counter clockwise rotation of the variables Γ_1 and Γ_2 in Figure 7.17. From the eigenvectors (7.3.9) we see that Γ_1 leads Γ_2 by 90°. Thus Γ_1 reaches a maximum and then Γ_2 reaches a maximum and so the oscillation proceeds in the direction of rotation and is an advancing mode. Similarly $s_2 = -\Omega(\bar{\omega}_\zeta + 1)$, corresponds to the two vectors rotating in a clockwise direction, but now Γ_1 lags Γ_2 and so this is also an advancing mode. Hence, the mode with frequency $(\bar{\omega}_\zeta + 1)\Omega$ is an advancing mode. Following a similar argument for the mode $(\bar{\omega}_\zeta - 1)\Omega$ we find that it is a regressing mode, when $\bar{\omega}_\zeta$ is greater than 1. One must be careful of this terminology, since in a rotating coordinate system modes are also described as advancing and regressing modes, but because of the change in coordinate system angular velocity, modes may be regressing in this system and advancing in the stationary system. From a geometric point of view there are two whirling modes corresponding to the four characteristic roots. Use of complex coordinates helps to visualize the direction of rotation of the modes simply [28].

Thus, the transient motion of the center of mass of the rotor system may be described in terms of two circular or whirling modes. When viewed in the fixed frame of reference one is an advancing whirl (in the same direction as the rotation of the rotor) and one is a regressing whirl at low rotor angular velocity ($\omega_\zeta > \Omega$) and a slow advancing whirl at larger angular velocities ($\omega_\zeta < \Omega$). Recall that the frequencies as seen in the rotating frame are simply equal to $\pm\omega_\zeta$ and one is an advancing mode while the other is always a regressing mode.

Now the effect of the pylon motion is added. It is assumed that the pylon is sufficiently long and the angular deflections are sufficiently small such that the hub motion lies in a horizontal plane. The equations of motion are developed using a Newtonian approach. First a single blade is considered and then the effects of the other blades are added. It is most convenient to derive the equations with the pylon or fuselage motion referred to a fixed axis system and the lag angle referred to a moving axis system and then transform the lag angle to a fixed coordinate system. This will illustrate the manner in which periodic coefficients enter the equations. The equations of motion for the blade and fuselage system may be

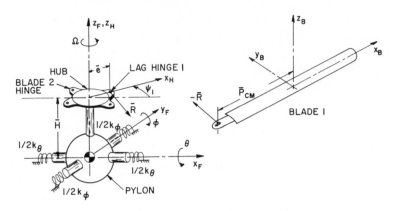

Figure 7.18   *Free body diagram for ground resonance analysis.*

written from the free body diagram shown in Figure 7.18, as

$$\dot{\bar{H}}_b + \bar{\Omega}_B \times \bar{H}_B = \bar{P}_{CM} \times \bar{R}$$

$$M_B \bar{a}_{CM} = -\bar{R} \qquad (7.3.10)$$

$$\dot{\bar{H}}_F = \bar{H} \times \bar{R}$$

where $\bar{R}$ is the reaction force at the hinge, $\bar{H}_B$ is the moment of momentum of the blade about its center of mass and $\bar{H}_F$ is the moment of momentum of the fuselage about its center of mass which is assumed fixed in space. $\bar{\Omega}_B$ is the angular velocity of the blade, $\bar{P}_{CM}$ is the distance from the hinge to the blade center of mass and $\bar{H}$ is the height of the rotor hub above the *CM*. The acceleration of the blade center of mass in terms of the acceleration at the hinge point is

$$\bar{a}_{CM} = \bar{a}_E + \bar{\Omega}_B \times (\bar{\Omega}_B \times \bar{P}_{CM}) + \dot{\bar{\Omega}}_B \times \bar{P}_{CM} \qquad (7.3.11)$$

and the acceleration of the hinge in terms of the acceleration of the hub, $\bar{a}_0$, and the rotational velocity of the hub is

$$\bar{a}_E = \bar{a}_0 + \bar{\Omega} \times (\bar{\Omega} \times \bar{e}) \qquad (7.3.12)$$

The angular velocity of the hub is assumed constant. Three sets of unit vectors are defined. The subscript $_B$ refers to the set of unit vectors fixed to the blade, the subscript $_H$ to the set fixed in the hub and the subscript $_F$ refers to a set fixed in space. The lag angle is assumed to be small, so that the relationships among these unit vectors for blade number 1 (Figure

364

7.17) are

$$\begin{Bmatrix} \bar{i}_F \\ \bar{j}_F \\ \bar{k}_F \end{Bmatrix} = \begin{bmatrix} \cos\psi_1 & -\sin\psi_1 & 0 \\ \sin\psi_1 & \cos\psi_1 & 0 \\ 0 & 0 & 1 \end{bmatrix} \begin{Bmatrix} \bar{i}_H \\ \bar{j}_H \\ \bar{k}_H \end{Bmatrix} \qquad (7.3.13)$$

$$\begin{Bmatrix} \bar{i}_H \\ \bar{j}_H \\ \bar{k}_H \end{Bmatrix} = \begin{bmatrix} 1 & -\zeta_1 & 0 \\ \zeta_1 & 1 & 0 \\ 0 & 0 & 1 \end{bmatrix} \begin{Bmatrix} \bar{i}_B \\ \bar{j}_B \\ \bar{k}_B \end{Bmatrix} \qquad (7.3.14)$$

The various quantities involved in the equations of motion are

$$\bar{H}_B = I_{CM}(\Omega + \dot{\zeta}_1)\bar{k}_B$$
$$\bar{H}_F = I_x\dot{\theta}\bar{i}_F + I_y\dot{\phi}\bar{j}_F$$
$$\bar{\Omega} = \Omega\bar{k}_H$$
$$\bar{\Omega}_B = (\Omega + \dot{\zeta}_1)\bar{k}_B$$
$$\bar{H} = h\bar{k}_F \qquad\qquad (7.3.15)$$
$$\bar{e} = e\bar{i}_H$$
$$\bar{P}_{CM} = r_{CM}\bar{i}_B$$
$$\bar{a}_0 = \ddot{x}_H\bar{i}_F + \ddot{y}_H\bar{j}_F = h\ddot{\phi}\bar{i}_F - h\ddot{\theta}\bar{j}_F$$

Substituting equations (7.3.11)–(7.3.15) into the equations of motion (7.3.10) for the blade and body motion, noting that $\bar{R} = -M_B\bar{a}_{CM}$, we obtain one blade equation of motion and two body equations of motion

$$\ddot{\zeta}_1 + \frac{er_{CM}M_B}{I_B}\Omega^2\zeta_1 = \frac{M_Br_{CM}h}{I_B}(\ddot{\theta}\cos\psi_1 + \ddot{\phi}\sin\psi_1)$$

$$(I_y + M_Bh^2)\ddot{\phi} = M_Bh[e\Omega^2 + r_{CM}(\Omega + \dot{\zeta}_1)^2]\cos\psi_1$$
$$+ M_Br_{CM}h(\ddot{\zeta}_1 - \Omega^2\zeta_1)\sin\psi_1 \qquad (7.3.16)$$
$$(I_x + M_Bh^2)\ddot{\theta} = -M_Bh[e\Omega^2 + r_{CM}(\Omega + \dot{\zeta}_1)^2]\sin\psi_1$$
$$+ M_Br_{CM}h(\ddot{\zeta}_1 - \Omega^2\zeta_1)\cos\psi_1$$

The subscript 1 has been added to note that only one blade has been considered.

The equations of motion for the other three blades are identical to blade one with the azimuth angle suitably shifted, i.e., the equation of motion of blade 2 in terms of the azimuth angle of blade 1 is

$$\ddot{\zeta}_2 + \frac{er_{CM}M_B}{I_B}\Omega^2\zeta_2 = \frac{M_Br_{CM}h}{I_B}\left(\ddot{\theta}\cos\left(\psi_1 + \frac{\pi}{2}\right) + \ddot{\phi}\sin\left(\psi_1 + \frac{\pi}{2}\right)\right) \qquad (7.3.17)$$

The equations of motion for the new coordinates, γ_0, γ_1, etc., are formulated by linear combinations of the blade equations and are

$$\ddot{\gamma}_0 + \frac{er_{CM}M_B}{I_B}\Omega^2\gamma_0 = 0$$

$$\ddot{\gamma}_1 + \frac{er_{CM}M_B}{I_B}\Omega^2\gamma_1 = \frac{M_Br_{CM}h}{I_B}(\ddot{\theta}\cos\psi_1 + \ddot{\phi}\sin\psi_1)$$

$$\ddot{\gamma}_2 + \frac{er_{CM}M_B}{I_B}\Omega^2\gamma_2 = \frac{M_Br_{CM}h}{I_B}(-\ddot{\theta}\sin\psi_1 + \ddot{\phi}\cos\psi_1) \qquad (7.3.18)$$

$$\ddot{\gamma}_3 + \frac{er_{CM}M_B}{I_B}\Omega^2\gamma_3 = 0$$

We thus see as discussed earlier that γ_0 and γ_3 are not coupled to the hub motion and thus do not need to be considered further. Note that the equations of motion for γ_1 and γ_2 have periodic coefficients since $\psi_1 = \Omega t$. The influence of the other three blades must be added to the fuselage equations. The first equation becomes

$$(I_y + 4M_Bh^2)\ddot{\phi} = \sum_{i=1}^{4} M_Bh(e\Omega^2 + r_{CM}(\Omega + \dot{\zeta}_i)^2)\cos\left(\psi_1 + \frac{(i-1)\pi}{2}\right)$$

$$+ M_Br_{CM}h(\ddot{\zeta}_i - \Omega^2\zeta_i)\sin\left(\psi_1 + \frac{(i-1)\pi}{2}\right) \qquad (7.3.19)$$

and a similar form is obtained for the other fuselage equation. Using trigonometric identities and the definitions of the multi-blade coordinates (7.3.1) the two fuselage equations become, retaining only linear terms,

$$(I_y + 4M_Bh^2)\ddot{\phi} = 2M_Br_{CM}h\{(\ddot{\gamma}_1 - 2\Omega\dot{\gamma}_2 - \Omega^2\gamma_1)\sin\psi_1$$

$$+ (\ddot{\gamma}_2 + 2\Omega\dot{\gamma}_1 - \Omega^2\gamma_2)\cos\psi_1\}$$

$$(I_x + 4M_Bh^2)\ddot{\theta} = 2M_Br_{CM}h\{(\ddot{\gamma}_1 - 2\Omega\dot{\gamma}_2 - \Omega^2\gamma_1)\cos\psi_1 \qquad (7.3.20)$$

$$- (\ddot{\gamma}_2 + 2\Omega\dot{\gamma}_1 - \Omega^2\gamma_1)\sin\psi_1\}$$

Again we see that the coordinates γ_0 and γ_3 do not appear.

These equations involve periodic coefficients. The periodic coefficients are a consequence of defining the lag motion in a rotating system and the fuselage motion in a fixed system as noted earlier. The periodic coefficients can be eliminated by transforming the lag motion to fixed coordinates as described above. This transformation involves the relationships given by equations (7.3.4). A centering spring about the lag hinge is incorporated in the lag equations such that the lag frequency is given by

$$\omega_\zeta^2 = \frac{K_\zeta}{I_B} + \frac{er_{CM}M_B}{I_B}\Omega^2$$

This is equivalent to equation (7.1.35) without the assumption of a uniform blade mass distribution. Employing equations (7.3.4) the blade equations (7.3.18) become

$$\ddot{\Gamma}_1 + (\omega_\zeta^2 - \Omega^2)\Gamma_1 + 2\Omega\dot{\Gamma}_2 = \frac{M_B r_{CM} h}{I_B}\ddot{\theta}$$

$$-2\Omega\dot{\Gamma}_1 + \ddot{\Gamma}_2 + (\omega_\zeta^2 - \Omega^2)\Gamma_2 = \frac{M_B r_{CM} h}{I_B}\ddot{\phi}$$

(7.3.21)

The fuselage equations, including the effects of the supporting springs k_ϕ and k_θ, are

$$(I_y + 4M_B h^2)\ddot{\phi} + k_\phi\phi = 2M_B r_{CM} h\ddot{\Gamma}_2$$

$$(I_x + 4M_B h^2)\ddot{\theta} + k_\theta\theta = 2M_B r_{CM} h\ddot{\Gamma}_1$$

(7.3.22)

Equations (7.3.22) can be placed in the form given in [43] by converting the pylon rotations θ and ϕ to linear hub translations x and y. From Figure 7.17, dropping the subscripts $_F$ on x and y,

$$x = h\phi, \qquad \bar{x} = \frac{x}{R}$$

$$y = -h\theta, \qquad \bar{y} = \frac{y}{R}$$

A uniform mass blade is assumed such that

$$I_B = M_B\frac{R^2}{3}$$

Define effective fuselage mass and spring constants by

$$M_{Fx} = \frac{I_y}{h^2} \qquad k_x = \frac{k_\phi}{h^2}$$

$$M_{Fy} = \frac{I_x}{h^2} \qquad k_y = \frac{k_\theta}{h^2}$$

These definitions eliminate the parameter h from equations (7.3.21) and

367

(7.3.22). Equations (7.3.21) and (7.3.22) become

$$\ddot{\Gamma}_1 + (\omega_\zeta^2 - \Omega^2)\Gamma_1 + 2\Omega\dot{\Gamma}_2 = -\tfrac{3}{2}\ddot{y}$$
$$-2\Omega\dot{\Gamma}_1 + \ddot{\Gamma}_2 + (\omega_\zeta^2 - \Omega^2)\Gamma_2 = \tfrac{3}{2}\ddot{x}$$
$$(M_{Fx} + 4M_B)\ddot{x} + k_x\bar{x} = M_B\ddot{\Gamma}_2 \tag{7.3.23}$$
$$(M_{Fy} + 4M_B)\ddot{y} + k_y\bar{y} = -M_B\ddot{\Gamma}_1$$

Note that the periodic coefficients will not be eliminated if we attempt to transform the body motion into rotating coordinates except in the special case where fuselage inertias and springs are identical about both axes. The procedure followed above for four blades will produce identical results for three or more blades. A generalized description of this procedure may be found in [6].

If the rotor has two blades, the only way to eliminate the periodic coefficients is to convert the pylon motion to rotating coordinates. Only in the special case of equal inertia and stiffness will the periodic coefficients be eliminated [43]. In general, if there is polar symmetry in one frame of reference and a lack of symmetry in the other frame, expressing the equation of motion in this latter frame will eliminate the necessity of dealing with periodic coefficients. With a two bladed rotor, the rotor lacks symmetry. If the support system also does not have polar symmetry, the periodic coefficients cannot be eliminated and Floquet theory must be employed to analyze the stability of the system. The simplest case (equal support stiffness) of the two-bladed rotor is analyzed in the rotating system and the three or more bladed rotor in the fixed frame. Recall from the previous discussion that this will give quite a different picture of the variation of system natural frequencies with rpm. For simplicity only the multibladed rotor with pylon symmetry is discussed which may be treated in either reference frame. The pylon characteristics are assumed to be

$$M_{Fx} = M_{Fy} = M_F$$
$$k_x = k_y = k_F$$

The important parameter governing the coupling between the blade motion and the fuselage motion is the ratio of the total blade mass to the total system mass defined by μ,

$$\mu = \frac{4M_B}{M_F + 4M_B}$$

It is convenient to nondimensionalize the time by the support frequency

368

with the blade mass concentrated at the hub

$$\omega_F^2 = \frac{k_F}{M_F + 4M_B}$$

since rotor angular velocity is considered to be the variable parameter. The frequencies nondimensionalized in this fashion are denoted by $\hat{\omega}_\zeta$ and $\hat{\Omega}$. Introducing these definitions, equations (7.3.23) become,

$$\ddot{\Gamma}_1 + (\hat{\omega}_\zeta^2 - \hat{\Omega}^2)\Gamma_1 + 2\hat{\Omega}\dot{\Gamma}_2 + \tfrac{3}{2}\ddot{\bar{y}} = 0$$

$$-2\hat{\Omega}\dot{\Gamma}_1 + \ddot{\Gamma}_2 + (\hat{\omega}_\zeta^2 - \hat{\Omega}^2)\Gamma_2 - \tfrac{3}{2}\ddot{\bar{x}} = 0$$

$$\frac{\mu}{4}\ddot{\Gamma}_1 + \ddot{\bar{y}} + \bar{y} = 0 \qquad\qquad (7.3.24)$$

$$-\frac{\mu}{4}\ddot{\Gamma}_2 + \ddot{\bar{x}} + \bar{x} = 0$$

The stability of the system defined by equations (7.3.24) is examined as a function of the various physical parameters of the problem. First, consider the limiting case in which the blade mass is zero ($\mu = 0$). This eliminates the coupling between the fuselage motion and the blade motion. The natural frequencies of the system are composed of the uncoupled blade dynamics and the fuselage dynamics. The roots of the characteristic equation are therefore

$$\pm i(\hat{\Omega} + \hat{\omega}_\zeta), \qquad \pm i(\hat{\Omega} - \hat{\omega}_\zeta), \qquad \pm i, \qquad \pm i$$

The latter two pairs correspond to the fuselage motion and the former to the blade motion. The modes of motion are whirling or circular modes owing to polar symmetry. Figure 7.19 shows the whirling modes, i.e., only the four frequencies with signs that correspond to the direction of whirling, positive being an advancing mode. The frequencies are shown as a function of rotor angular velocity. In the numerical example shown, a centering spring (K_ζ) is included such that $\hat{\omega}_\zeta = 0.3$ when the rotor is not rotating ($\hat{\Omega} = 0$) and the hinge offset $\bar{e} = 0.05$. ($\hat{\Omega} - \hat{\omega}_\zeta$) is a regressing mode when negative, ($\hat{\Omega} + \hat{\omega}_\zeta$) is an advancing mode. The fuselage modes ($\pm i$) are advancing and regressing modes respectively. These four whirling modes constitute the dynamics of the system in the limiting case of no hub mass. For comparison purposes, if the system is analyzed in the rotating frame, the result will be equivalent to subtracting the angular velocity $\hat{\Omega}$ from the frequencies shown in Figure 7.19 resulting in the diagram shown in Figure 7.20. Thus, the appearance of the figure depends upon the coordinate system. For two bladed rotors one is likely

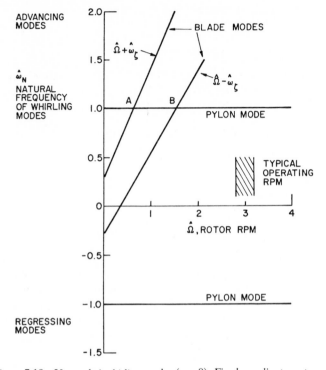

Figure 7.19   Uncoupled whirling modes ($\mu = 0$). Fixed coordinate system.

to see a graph similar to Figure 7.20 while for multibladed rotor analysis
one usually sees the fixed coordinate plot shown in Figure 7.19. It may be
noted that at two rotor angular velocities ($\hat{\Omega} = 0.65$ and $\hat{\Omega} = 1.51$) the
frequency of one of the blade modes is equal to a pylon mode. It would
be expected that the coupling effects due to blade mass are most significant in these regions.

Now the influence of the mass ratio μ on the dynamics of the system
is examined. It would be most convenient if root locus techniques could
be used. This is not possible directly with equations (7.3.24) since μ will
not appear linearly in the characteristic equation. Through introduction of
complex coordinates, root locus techniques can be employed [28].

Define

$$\bar{z} = \bar{x} + i\bar{y}$$

$$\delta = \Gamma_2 - i\Gamma_1$$

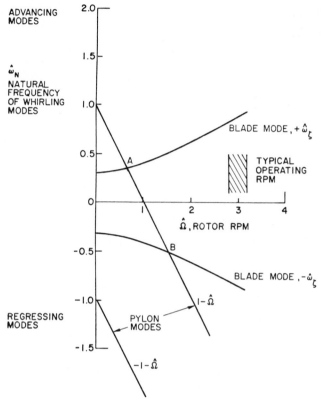

Figure 7.20   Uncoupled whirling modes ($\mu = 0$). Rotating coordinate system.

This coordinate change reduces the four equations (7.3.24) to two equations owing to the symmetry properties of these equations, i.e.,

$$\ddot{\delta} - 2i\hat{\Omega}\dot{\delta} + (\hat{\omega}_\zeta^2 - \hat{\Omega}^2)\delta - \tfrac{3}{2}\ddot{z} = 0 \qquad (7.3.25)$$
$$-\frac{\mu}{4}\ddot{\delta} + \ddot{z} + \bar{z} = 0$$

We now have a fourth order system in place of an eighth order system (Equations 7.3.24), and the roots of this system are the whirling modes only, i.e.,

$$i(\hat{\Omega} + \hat{\omega}_\zeta)$$
$$i(\hat{\Omega} - \hat{\omega}_\zeta)$$
$$\pm i$$

371

The characteristic equation of this system can now be written as

$$(s^2+1)(s^2-2i\hat{\Omega}s+(\hat{\omega}_\zeta^2-\hat{\Omega}^2))-\tfrac{3}{8}\mu s^4=0$$

Now μ appears as a linear parameter and the characteristic equation can be written as

$$\frac{\tfrac{3}{8}\mu s^4}{(s^2+1)(s^2-2i\hat{\Omega}s+(\hat{\omega}_\zeta^2-\hat{\Omega}^2))}=1 \qquad (7.3.26)$$

Equation (7.3.26) is in root locus form with μ as the variable parameter and a zero degree locus is indicated. The usual root locus rules apply to equations with complex coefficients as well as to those with real coefficients. Figure 7.21 shows root loci for increasing μ for two values of $\hat{\Omega}$ (0.2 and 1.3) and indicates that the influence of μ on the dynamics is quite different depending upon the sign of $(\hat{\Omega}-\hat{\omega}_\zeta)$. When $(\hat{\Omega}-\hat{\omega}_\zeta)$ is negative it can be seen from Figure 7.21 that the coupling effect of

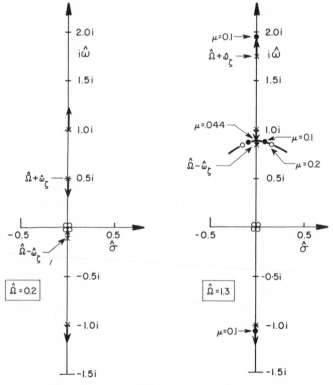

Figure 7.21   *Root locus for increasing blade mass ratio (μ) for two operating conditions.*

increasing μ is to separate the system frequencies. However, when $(\hat{\Omega} - \hat{\omega}_\zeta)$ is positive, the two intermediate frequencies come together and if μ is sufficiently large instability occurs. The most critical case occurs at intersection B of Figure 7.19 i.e., when the regressing mode frequency $(\hat{\Omega} - \hat{\omega}_\zeta) = 1$ such that the two intermediate frequencies are equal. At this operating condition any value of μ causes instability. Intersection A is not critical because of the large separation of the two intermediate frequencies $(\hat{\Omega} + \hat{\omega}_\zeta$ and $\hat{\Omega} - \hat{\omega}_\zeta)$. In the typical case, for an articulated rotor $\hat{\omega}_\zeta$ is the order of one and $\hat{\Omega}$ at operating condition $(\hat{\Omega}_{OP})$ is the order of three so that intersection B occurs below operating rpm. That is

$$\hat{\Omega}_{CR} = 1 + \hat{\omega}_\zeta \cong 2$$

Thus to completely eliminate the possibility of this instability which is called ground resonance, one must have $\hat{\Omega}_{CR} > \hat{\Omega}_{OP} = 3$, and a very large offset is required since $\hat{\omega}_\zeta > 2$ and therefore, $\bar{\omega}_\zeta = 0.67$ which corresponds to a hinge offset of 0.3 (7.1.35) without a centering spring. Note that this ratio $\hat{\Omega}_{OP}$ is largely determined by considerations other than rotor stability, such as the rotor operating rpm and the shock absorbing character of the landing gear. Since this large hinge offset is not practical, a centering spring may be employed to increase $\bar{\omega}_\zeta$; however this will increase the root bending moment, the reason the lag hinge was installed.

   The unstable region extends below and above this intersection to an extent depending upon the mass ratio as well as the other geometric parameters. Various criteria can be found in the literature as to the size of the unstable region as a function of mass ratio. A typical graph of the frequencies as well as extent of the unstable region as a function of μ is shown in Figure 7.22 taken from [10]. This result applies to the case where the pylon has only one degree-of-freedom in contrast to the example here where the pylon has two degrees-of-freedom.

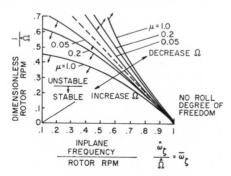

Figure 7.22   Extent of unstable region for various mass ratios [10].

Now we examine the influence of mechanical damping on the stability of the system. Damping in the rotating (lag damping) and nonrotating (pylon damping) parts of the system is considered separately. Consider first the influence of damping on the pylon. This would lead to terms $C_F\dot\phi$ and $C_F\dot\theta$ in equations (7.3.22). Adding this in complex form to the second of equations (7.2.25) and expressing the characteristic equation in root locus form

$$\frac{\hat{C}_F s[s^2 - 2i\hat\Omega s + (\hat\omega_\zeta^2 - \hat\Omega^2)]}{(s^2+1)(s^2 - 2i\hat\Omega s + (\hat\omega_\zeta^2 - \hat\Omega^2)) - \frac{3}{8}\mu s^4} = -1 \qquad (7.3.27)$$

where

$$\hat{C}_F = \frac{C_F \omega_F}{k_F h^2}$$

This root locus has two zeros at the uncoupled lag modes. Figure 7.23 shows the influence of increasing damping for two cases. In the first, μ is small so that the basic system is neutrally stable. Adding only fixed axis damping destabilizes the system. In the second case where μ is large enough such that the basic system is unstable no amount of damping will stabilize the system.

Now consider adding damping to the lag motion of the blades. It must be noted that this damping will be in the rotating coordinate system (about the blade hinge) and so to directly add damping terms to the equations of motion, the rotating frame equations must be used. The damping then appears as $C_R\dot\gamma_1$ and $C_R\dot\gamma_2$. If the transformations are followed, this will ultimately result (in the rotating frame with complex notation) in the damping appearing in the first of equations (7.3.25) as

$$\hat{C}_R(\dot\delta - i\hat\Omega\delta)$$

The $i\hat\Omega$ term appears because rotating coordinate system damping is expressed with respect to a fixed frame. Adding this damping to the first of the two equations and expressing the characteristic equation in root locus form as

$$\frac{\hat{C}_R(s - i\hat\Omega)(s^2+1)}{(s^2+1)(s^2 - 2i\hat\Omega s + (\hat\omega_\zeta^2 - \hat\Omega^2)) - \frac{3}{8}\mu s^4} = -1 \qquad (7.3.28)$$

the root locus shown in Figure 7.24 is obtained. Again it is interesting to note that adding damping only in the rotating frame results in destabilizing one of the fuselage modes when the system is initially neutrally stable (small μ). For large μ the situation is similar to the fixed axis damping case. These rather surprising effects of damping in a rotating

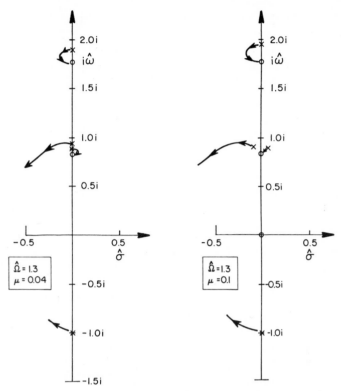

Figure 7.23   Root locus for increasing pylon (fixed axis) damping on ground resonance stability.

system indicate that damping must be handled with considerable care. Owing to the order of the system it is rather difficult to obtain physical insight into the source of these effects. A *combination* of damping in the rotating frame (blades) and stationary frame (pylon) is required to stabilize the system, although as can be seen from the root locus sketches there will always be one zero near to the fuselage, blade lag mode making it difficult to provide a large amount of damping in one of the modes. There would, of course, generally be damping in the pylon especially if it is the landing gear flexibility which provides the pylon frequencies. Particularly on articulated rotors, blade lag dampers are added since as noted this region of instability must be run through in increasing the rotor speed to operating rpm. Ref. [43] presents boundaries showing the damping required to eliminate the instability region for articulated rotor helicopters.

375

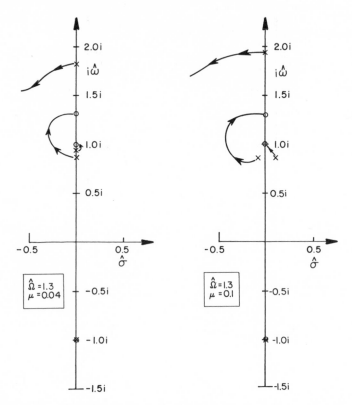

Figure 7.24   Root locus for increasing lag (rotating axis) damping on ground resonance stability.

The treatment of more general problems, including the blade flapping degree-of-freedom and discussion of its importance in the hingeless rotor case, may be found in [8], [9], and [10]. The two-bladed rotor is treated in [43].

Flapping motion of articulated rotors with small hinge offset does not produce appreciable hub moments and consequently there is only weak coupling between the flapping motion and the pylon motion. The hingeless rotor, however, produces large hub moments and consequently the flapping motion is coupled into the pylon, lag dynamics [10]. Figure 7.25 shows the influence of the flapping frequency on the stability boundaries for $\mu = 0.1$.

There are now three frequencies involved in the problem, the pylon frequency, ω_F, the lag frequency, $(\hat{\Omega} - \hat{\omega}_\zeta)$, and the flap frequency p $(p - 1$ in the stationary frame). In addition to the destructive instability which

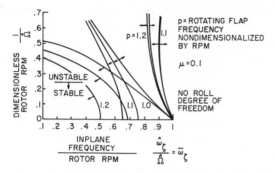

Figure 7.25   *Influence of flap frequency on ground resonance stability boundaries* [10].

occurs when the coupled pylon frequency is equal to the lag frequency $(\hat{\Omega} - \hat{\omega}_\zeta)$, a mild instability occurs when the coupled flap frequency is equal to the lag frequency $(\hat{\Omega} - \hat{\omega}_\zeta)$ as shown by Figure 7.26. Note that pylon and flap frequencies are significantly changed by the coupling. The ground resonance problem now becomes quite complex and difficult to generalize. The reader is referred to [10] for further detail. It may be noted that an increasingly detailed model of the rotor blades must be employed. Since both flap and lag degrees-of-freedom are involved it is

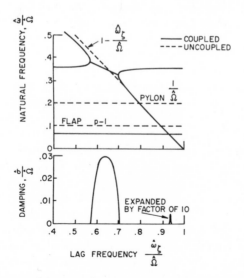

Figure 7.26   *Frequency and damping of rotor-body system. Flap, lag and pylon pitch degrees of freedom. No aerodynamics* [10].

377

important to model the coupling between these motions which occurs as a result of hub and blade geometry. Aerodynamic forces and structural damping are also significant.

Air resonance refers to the form this dynamic problem takes with the landing gear restraint absent, that is, with the vehicle in the air. Coupling of flapping motion, body motion and lag motion is involved. Fuselage inertia and damping characteristics can have a significant impact on the stability. The air resonance problem is clearly asymmetrical and characteristically the roll axis is more critical owing to its low inertia and small aerodynamic damping [8].

Since the primary source of damping in this physical system arises from flap bending, it is possible that the nature of the flight control system can have an impact on air resonance stability as shown in [47]. Essentially an attitude feedback from the body to cyclic pitch tends to maintain the rotor in a horizontal plane thus effectively removing the aerodynamic damping from the flapping/body dynamics.

There are also other indications that the flight control system feedbacks have an impact on rotor system stability [48] however, this problem does not appear to be well understood.

Another problem associated with propeller and prop/rotor driven aircraft which involves a blade motion-support coupling is whirl flutter which has been experienced on conventional aircraft [49] as well as on V/STOL aircraft [6]. This instability in the case of the conventional aircraft can be explained by considering only the wing as flexible, i.e., the propeller blades may be assumed to be rigid. For the tilt prop/rotor aircraft where blade flexibility is important, the primary source of the instability is the same as in the rigid propeller case. It is a result of the aerodynamic characteristics of propellers and prop rotors at high inflows typical of cruising flight. It can be shown that the source of the whirl flutter instability is primarily associated with the fact that an angle-of-attack change on a propeller produces a yawing moment and a sideslip angle produces a pitching moment. Further the magnitude of this moment change grows with the square of the tangent of the inflow angle [6] and results in a rapid onset of the instability.

For the prop/rotor a complex model with a large number of degrees-of-freedom is required to predict the dynamics of the system accurately [6]. The whirl flutter instability can occur on articulated rotors as well as hingeless rotors although for somewhat different physical reasons [50]. Here inplane force dependence on angular rate produces unstable damping moments acting on the support. The hingeless rotor which produces significant hub moments is similar to the rigid propeller. Young [51] has shown by a simplified analysis that under certain circumstances the occurrence of this instability can be minimized by suitable selection of the flapping frequency. Ref. [6] contains an excellent discussion of these various problem areas.

A typical predicted variation of damping with flight speed for a tilt-prop-rotor aircraft is shown in Figure 7.27. As mentioned earlier, it is important to insure in modelling this dynamic system that the structural details of the hub, blade and pitch control system are precisely modelled. Ref. [52] indicates the impact that relatively small modelling details can have on the flutter speed, as well as describing in detail the modelling requirements for prop-rotor whirl flutter.

Two-bladed rotors

Aeroelastic analysis of two-bladed rotors requires special considerations since the two blades are connected together. The reader is referred to [53] and [54] for the analysis of two-bladed rotors.

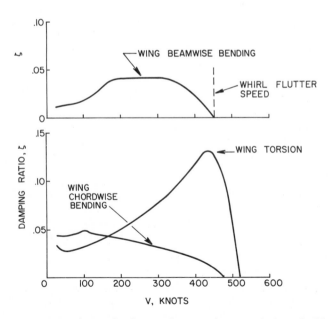

Figure 7.27   Damping of wing bending modes as a function of airspeed. Tilt prop/rotor aircraft with gimballed rotor [6].

References for Chapter 7

[1] Gessow, A. and Myers, G. C., Jr. *Aerodynamics of the Helicopter*, The Macmillan Company, New York, 1952.

[2] Bramwell, A. R. S., *Helicopter Dynamics*, John Wiley & Sons, New York, 1976.

[3] Loewy, R. G., 'Review of Rotary Wing V/STOL Dynamics and Aeroelastic Problems', *Journal of American Helicopter Society*, Vol. 14, No. 3 (July 1969).

[4] Friedmann, P., 'Recent Developments in Rotary Wing Aeroelasticity', Paper No. 11, *Second European Rotorcraft and Powered Lift Aircraft Forum* (September 20–22, 1977).

[5] Hohenemser, K. H., 'Hingeless Rotorcraft Flight Dynamics', *Agardograph 197*, 1974.

[6] Johnson, W., 'Dynamics of Tilting Proprotor Aircraft in Cruise Flight', *NASA TN D-7677* (May 1974).

[7] Spierings, P. A. M., *Analytical Study of the Aerodynamic Coupling of Helicopter Rotor Blades in Hover Flight*. Ph.D. Thesis, Princeton University Department of Aerospace and Mechanical Sciences Report 1163-T, (May 1974).

[8] Burkham, J. E. and Miao, W. L., 'Exploration of Aeroelastic Stability Boundaries with a Soft-in-Plane Hingeless-Rotor Model', *Journal of the American Helicopter Society*, Vol. 17, No. 4 (October 1972).

[9] Donham, R. E., *et. al.*, 'Ground and Air Resonance Characteristics of a Soft In-Plane Rigid-Rotor System', *Journal of the American Helicopter Society*, Vol. 14, No. 4 (October 1969).

[10] Ormiston, R. A., 'Aeromechanical Stability of Soft Inplane Hingeless Rotor Helicopters', Paper No. 25, *Third European Rotorcraft and Powered Lift Aircraft Forum*, Aix-en-Provence, France (September 7–9, 1977).

[11] Goland, L. and Perlmutter, A. A., 'A Comparison of the Calculated and Observed Flutter Characteristics of a Helicopter Rotor Blade', *Journal of the Aeronautical Sciences*, Vol. 24, No. 4 (April 1957).

[12] Reichert, G. and Huber, H., 'Influence of Elastic Coupling Effects on the Handling Qualities of a Hingeless Rotor Helicopter', *Agard Conference Proceedings No. 121 Advanced Rotorcraft* (February 1973).

[13] Lentine, F. P. *et al.*, 'Research in Manuverability of the XH-51A Compound Helicopter', *USAAVLABS TR 68–23* (June 1968).

[14] Ham, N. D., 'Helicopter Blade Flutter', *AGARD Report 607* (January, 1973).

[15] Shames, I. H., *Engineering Mechanics, Vol. II: Dynamics*, Prentice-Hall, Inc., Englewood Cliffs, N.J., 1958.

[16] Peters, D. A. and Hohenemser, K. H., 'Application of the Floquet Transition Matrix to Problems of Lifting Rotor Stability', *Journal of the American Helicopter Society*, Vol. 16, No. 2 (April 1971).

[17] Ormiston, R. A. and Hodges, D. H., 'Linear Flap-Lag Dynamics of Hingeless Helicopter Rotor Blades in Hover', *Journal of the American Helicopter Society*, Vol. 17, No. 2 (April 1972).

[18] Chou, P. C., 'Pitch Lag Instability of Helicopter Rotors', *Institute of Aeronautical Sciences Preprint 805*, 1958.

[19] Bennett, R. M. and Curtiss, H. C., Jr., 'An Experimental Investigation of Helicopter Stability Characteristics Near Hovering Flight Using a Dynamically Similar Model', *Princeton University Department of Aeronautical Engineering Report 517* (July 1960).

[20] Hansford, R. E. and Simons, I. A., 'Torsion-Flap-Lag Coupling on Helicopter Rotor Blades', *Journal of the American Helicopter Society*, Vol. 18, No. 4 (October 1973).

[21] Hodges, D. H. and Ormiston, R. A., 'Stability of Elastic Bending and Torsion of Uniform Cantilever Rotor Blades in Hover with Variable Structural Coupling', *NASA TN D*-8192 (April 1976).

[22] Hodges, D. H. and Dowell, E. H., 'Non-Linear Equations of Motion for the Elastic Bending and Torsion of Twisted Nonuniform Rotor Blades', *NASA TN D-7818*, (July 1974).

[23] Flax, A. H. and Goland, L., 'Dynamic Effects in Rotor Blade Bending', *Journal of the Aeronautical Sciences*, Vol. 18, No. 12 (December 1951).

[24] Yntema, R. T., 'Simplified Procedures and Charts for Rapid Estimation of Bending Frequencies of Rotating Beams', *NACA TN 3459*, (June 1955).

[25] Young, M. I., 'A Simplified Theory of Hingeless Rotors With Application to Tandem Helicopters', *Proceedings of the 18th Annual National Forum, American Helicopter Society*, pp. 38–45 (May 1962).

[26] Curtiss, H. C., Jr., 'Sensitivity of Hingeless Rotor Blade Flap-Lag Stability in Hover to Analytical Modelling Assumptions', *Princeton University Department of Aerospace and Mechanical Sciences Report 1236* (January 1975).

[27] Halley, D. H., 'ABC Helicopter Stability, Control and Vibration Evaluation on the Princeton Dynamic Model Track', *American Helicopter Society Preprint 744* (May 1973).

[28] Curtiss, H. C., Jr., 'Complex Coordinates in Near Hovering Rotor Dynamics', *Journal of Aircraft*, Vol. 10, No. 8 (May 1973).

[29] Berrington, D. K., 'Design and Development of the Westland Sea Lynx', *Journal of the American Helicopter Society*, Vol. 19, No. 1 (January 1974).

[30] Ormiston, R. A. and Bousman, W. G., 'A Study of Stall-Induced Flap-Lag Instability of Hingeless Rotors', *Journal of the American Helicopter Society*, Vol. 20, No. 1 (January 1975).

[31] Curtiss, H. C., Jr. and Putman, W. F., 'An Experimental Investigation of the Flap-Lag Stability of a Hingeless Rotor Blade with Comparable Levels of Hub and Blade Stiffness in Hovering Flight', *Princeton University, Department of Aerospace and Mechanical Sciences Report 1300* (June 1976).

[32] Gaffey, T. M., 'The Effect of Positive Pitch-Flap Coupling (Negative δ_3) on Rotor Blade Motion Stability and Flapping', *Journal of the American Helicopter Society*, Vol. 14, No. 2 (April 1969).

[33] Ormiston, R. A., 'Techniques for Improving the Stability of Soft Inplane Hingeless Rotors', *NASA TM X-62,390* (October 1974).

[34] Ham, N. D. and Garelick, M. S., 'Dynamic Stall Considerations in Helicopter Rotors', *Journal of the American Helicopter Society*, Vol. 13, No. 2 (April 1968).

[35] Liiva, J. and Davenport, F. J., 'Dynamic Stall of Airfoil Sections for High-Speed Rotors', *Journal of the American Helicopter Society*, Vol. 14, No. 2 (April 1969).

[36] Martin, J. M., et al., 'An Experimental Analysis of Dynamic Stall on an Oscillating Airfoil', *Journal of the American Helicopter Society*, Vol. 19, No. 1 (January 1974).

[37] McCroskey, W. J. and Fisher, R. K., Jr., 'Detailed Aerodynamic Measurements on a Model Rotor in the Blade Stall Regime', *Journal of the American Helicopter Society*, Vol. 17, No. 1 (January 1972).

[38] Johnson, W. and Ham, N. D., 'On the Mechanism of Dynamic Stall', *Journal of the American Helicopter Society*, Vol. 17, No. 4 (October 1972).

[39] Ericsson, L. E. and Reding, J. P., 'Dynamic Stall of Helicopter Blades', *Journal of the American Helicopter Society*, Vol. 17, No. 1 (January 1972).

7 Aeroelastic problems of Rotorcraft

[40] Farren, W. S., 'The Reaction on a Wing Whose Angle of Incidence is Changing Rapidly', *ARC R&M 1648*, 1935.

[41] Harris, F. D. *et al.*, 'Rotor High Speed Performance Theory vs. Test', *Journal of the American Helicopter Society*, Vol. 15, No. 3 (July 1970).

[42] Tarzanin, F. J., Jr., 'Prediction of Control Loads Due to Blade Stall', *Journal of the American Helicopter Society*, Vol. 17, No. 2 (April 1972).

[43] Coleman, R. P. and Feingold, A. M., 'Theory of Self-Excited Mechanical Oscillations of Helicopter Rotors with Hinged Blades', *NACA TN 3844* (February 1957).

[44] Brooks, G. W., *The Mechanical Instability and Forced Response of Rotors on Multiple Degree of Freedom Supports*, Ph.D. Thesis, Princeton University Department of Aeronautical Engineering (October 1961).

[45] Bielawa, R. L., 'An Experimental and Analytical Investigation of the Mechanical Instability of Rotors on Multiple Degree-of-Freedom Supports', *Princeton University Department of Aeronautical Engineering Report 612* (June 1962).

[46] Hohenemser, K. H. and Yin, S. K., 'Some Applications of the Method of Multi-Blade Coordinates', *Journal of the American Helicopter Society*, Vol. 17, No. 3 (July 1972).

[47] Lytwyn, R. T., *et al.*, 'Airborne and Ground Resonance of Hingeless Rotors', *Journal of the American Helicopter Society*, Vol. 16, No. 2 (April 1971).

[48] Briczinski, S. and Cooper, D. E., 'Flight Investigation of Rotor/Vehicle State Feedback', *NASA CR-132546*, 1974.

[49] Reed, W. H., III, 'Review of Propeller-Rotor Whirl Flutter', *NASA TR R-264* (July 1967).

[50] Hall, E. W., Jr., 'Prop-Rotor Stability at High Advance Ratios', *Journal of the American Helicopter Society*, Vol. 11, No. 2 (April 1966).

[51] Young, M. I. and Lytwyn, R. T., 'The Influence of Blade Flapping Restraint on the Dynamic Stability of Low Disc Loading Propeller-Rotors', *Journal of the American Helicopter Society*, Vol. 12, No. 4 (October 1967).

[52] Johnson, W., 'Analytical Modelling Requirements for Tilting Prop Rotor Aircraft Dynamics', *NASA TN D-8013* (July 1975).

[53] Shamie, J. and Friedmann, P. 'Aeroelastic Stability of Complete Rotors with Application to a Teetering Rotor in Forward Flight', *American Helicopter Society Preprint No. 1031* (May 1976).

[54] Kawakami, N., 'Dynamics of an Elastic Seesaw Rotor', *Journal of Aircraft*, Vol. 14, No. 3 (March 1977).

8

Aeroelasticity in turbomachines

The advent of the jet engine and the high performance axial-flow compressor toward the end of World War II focussed attention on certain aeroelastic problems in turbomachines.

The concern for very light weight in the aircraft propulsion application, and the desire to achieve the highest possible isentropic efficiency by minimizing parasitic losses led inevitably to axial-flow compressors with cantilever airfoils of high aspect ratio. Very early in their development history these machines were found to experience severe vibration of the rotor blades at part speed operation; diagnosis revealed that these were in fact stall flutter oscillations. The seriousness of the problem was underlined by the fact that the engine operating regime was more precisely termed the 'part corrected speed' condition, and that in addition to passing through this regime at ground start up, the regime could be reentered during high flight speed conditions at low altitude. In either flight condition destructive behavior of the turbojet engine could not be tolerated.

In retrospect it is probable that flutter had occurred previously in some axial flow compressors of more robust construction and in the latter stages of low pressure axial-flow steam turbines as well. Subsequently a variety of significant forced and self-excited vibration phenomena have been detected and studied in axial-flow turbomachinery blades.

8.1 Aeroelastic environment in turbomachines

Consider an airfoil or blade in an axial flow turbine or compressor which is running at some steady rotational speed. For reasons of steady aerodynamic and structural performance the blade has certain geometric properties defined by its length, root and tip fixation, possible mechanical attachment to other blades and by the chord, camber, thickness, stagger and profile shape which are functions of the radial coordinate. Furthermore, the blade may be constructed in such a manner that the line of centroids and the line of shear centers are neither radial nor straight, but are defined by schedules of axial and tangential coordinates as functions of radius. In fact, in certain cases, it may not be possible to define the elastic axis (i.e., the line of shear centers). The possibility of a build-up sheet metal and spar construction, a laid-up plastic laminate construction, movable or articulated fixations and/or supplemental damping devices attached to the blade would complicate the picture even further.

The blade under consideration, which may now be assumed to be completely defined from a geometrical and kinematical point of view, is capable of deforming* in an infinite variety of ways depending upon the loading to which it is subjected. In general, the elastic axis (if such can be defined) will assume some position given by axial and tangential coordinates which will be continuous functions of the radius (flapwise and chordwise bending). About this axis a certain schedule of twisting deformations may occur (defined, say by the angular displacement of a straight line between leading and trailing edges). Finally, a schedule of plate type bending deformations may occur as functions of radius and the chordwise coordinate. (Radial extensions summoned by centrifugal forces may further complicate the situation).

One has now to distinguish between steady and oscillatory phenomena. If the flow through the machine is completely steady in time and there are no mechanical disturbances affecting the blade through its connections to other parts of the machine, the blade will assume some deformed position as described above (and as compared to its manufactured shape) which is also steady in time. This shape or position will depend on the elastic and structural properties of the blade and upon the steady aerodynamic and centrifugal loading. (The centrifugal contribution naturally does not apply to a stator vane.)

Consider the ultimate situation, however, where disturbances may exist in the airstream, or may be transmitted through mechanical attachments from other parts of the structure. Due to the unsteadiness of the aerodynamic and/or the external forces the blade will assume a series of time-dependent positions. If there is a certain repetitive nature with time of the displacements relative to the equilibrium position, the blade is said to be executing vibrations, the term being taken to include those cases

---

* Deformations are reckoned relative to a steadily rotating coordinate system in the case of a rotor blade.

where the amplitude of the time-dependent displacements is either increasing, decreasing or remaining constant as time progresses.

It is the prediction and control of these vibrations with which the aeroelastician is concerned. In doing so it is found that a portion of the difficulty is involved in the fact that once the blade is vibrating the aerodynamic forces are no longer a function only of the airstream characteristics and the blade's angular position and velocity in the disturbance field, but depend in general upon the blade's vibratory position, velocity and acceleration as well. There is a strong interaction between the blade's time-dependent motion and the time-dependent aerodynamic forces which it experiences. It is appropriate at this point to note that in certain cases the disturbances may be exceedingly small, serving only to 'trigger' the unsteady motion, and that the vibration may be sustained or amplified purely by the aforementioned interdependence between the harmonic variation with time of the blade's position and the harmonic variation with time of the aerodynamic forces (the flutter condition).

8.2 The compressor performance map

The axial flow compressor, and its aeroelastic problems, are typical; the other major important variant being the axial flow turbine (gas or steam). In the compressor the angle of attack of each rotor airfoil at each radius r is compounded of the tangential velocity of the airfoil section due to rotor rotation and the through flow velocity as modified in direction by the upstream stator row. Denoting the axial component by V_x and the

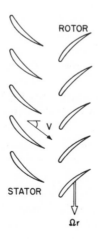

angular velocity by Ω, it is clear that the angle of attack will increase inversely with the ratio $\phi = V_x/(r\Omega)$. In the compressor, an increase in angle of attack (or an increase in 'loading') results in more work being done on the fluid and a greater stagnation pressure increment Δp_0 being imparted to it. Hence the general aspects of the single 'stage' (i.e., pair of fixed and moving blade rows) characteristics in the accompanying sketches are not without rational explanation. Note that the massflow through the stage equals the integral over the flow annulus of the product of V_x and fluid density.

When the various parameters are expressed in dimensionless terms, and the complete multistage compressor is compounded of a number of

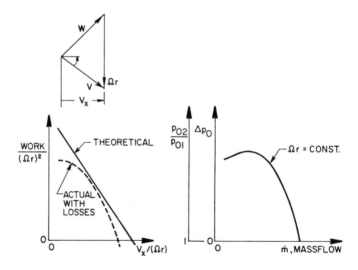

stages, the overall compressor 'map', or graphical representation of multistage characteristics, appears as in the following figure, (see next page) where $\dot{m}$ is massflow, k and R are the ratio of specified heats and gas constant, respectively; T_0 is stagnation temperature and A is a reference flow area in the compressor. Conventionally the constants k and R are omitted where the identity of the working fluid is understood (e.g., air). The quantity A is a scaling parameter relating the absolute massflow of geometrically similar machines and is also conventionally omitted. The tangential velocity of the rotor blade tip, Ωr_{tip}, is conventionally replaced by the rotational speed in rpm. The latter omission and replacement are justified when discussing a particular compressor.

An important property of the compressor map is the fact that to each point there corresponds theoretically a unique value for angle of attack

386

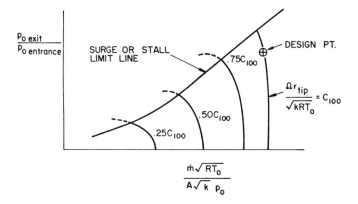

(or incidence) at any reference airfoil section in the compressor. For example, taking a station near the tip of the first rotor blade as a reference, contours of incidence may be superposed on the map coordinates. In the accompanying figure such angle of attack, α_i, contours have been shown for a specific machine. As defined here, α_i is the angle between the relative approach velocity W and the chord of the airfoil.

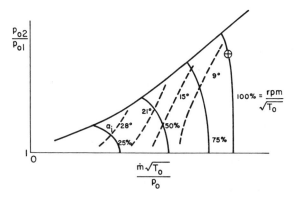

Here again axial velocity V_x (or massflow) is seen to display an inverse variation with respect to angle of attack as a line of constant rotational speed is traversed. The basic reason that such incidence contours can be established is that the two parameters which locate a point on the map, $\dot{m}\sqrt{T_0}/p_0$ and $\Omega r/\sqrt{T_0}$, are effectively a Mach number in the latter case and a unique function of Mach number in the former case. Thus the 'Mach number triangles' are established which yield the same 'angle of attack' as the velocity triangles to which they are similar.

387

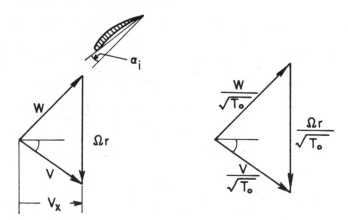

As a matter for later reference, contours of $V/(b\omega)$ for a particular stator airfoil, or else $W/(b\omega)$ for a particular rotor airfoil, can be superimposed on the same map, provided the environmental stagnation temperature, T_0, is specified. These contours are roughly parallel, though not exactly, to the constant rotational speed lines. The natural frequency of vibration, ω, tends to be constant for a rotor blade at a given rotational speed; and of course a stator blade's frequency does not depend directly on rotation. However, upon viewing the velocity triangles above, it is clear that if Ωr is kept constant and the *direction* of V is kept constant, the size of W may increase or decrease as V_x (or massflow) is changed. In fact, if the angle between V and W is initially close to 90 degrees, a not uncommon situation, the change in the magnitude of W will be minimal. For computing the stator parameter, $V/(b\omega)$, the direction of W *leaving* a rotor is considered to be virtually constant, and the corresponding changes in V (length and direction) as V_x is varied lead to similar conclusions with regard to angle of attack and magnitude of V experienced by the following stator. The values of $W/(b\omega)$ increase with increasing values of Ωr_{tip}, since the changes in W (or V) will dominate the somewhat smaller changes in the appropriate frequency, ω, at least in the first few stages of the compressor. Compressibility phenomena, when they become significant will sometimes alter these general conclusions.

8.3   Blade mode shapes and materials of construction

Flutter and vibration of turbomachinery blades occur with a wide variety of these beam-like structures and their degrees of end restraint. Rotor blades in use vary from cantilever with perfect root fixity all the way to a single pinned attachment such that the blade behaves in bending like a pendulum 'flying out' and being maintained in a more or less radial orientation by the centrifugal (rather than a gravity) field. Stator vanes may be cantilevered from the outer housing or may be attached at both ends, with degrees of fixity ranging from 'encastred' to 'pinned'.

The natural modes and frequencies of these blades, or blade-disc systems when the blades are attached to their neighbors in the same row or the discs are not effectively rigid, are obtainable by standard methods of structural mechanics. Usually twisting and two directions of bending are considered. If plate-type deformations are significant, the beam representation must be replaced by more sophisticated plate or shell models which recognize static twist and variable thickness.

In predicting the first several natural modes and frequencies of rotor blades it is essential to take into account the effect of rotor rotational speed. Although the description is not analytically precise in all respects, the effect of rotational speed can be approximately described by stating

$$\omega_n^2 = \omega_{0n}^2 + \kappa_n \Omega^2$$

where ω_{0n} is the static (nonrotating) frequency of the rotor blade and κ_n is a proportionality constant for any particular blade in the nth mode. The effect is most pronounced in the natural modes which exhibit predominantly bending displacements; the modes associated with the two gravest frequencies are usually of this type, and it is here that the effect is most important.

Materials of construction are conventionally aluminum alloys, steel or stainless steel (high Nickel and/or Chromium content). However, in recent applications Titanium and later Beryllium have become significant. In all these examples, considering flutter or else forced vibration in air as the surrounding fluid, the mass ratios are such that the critical mode and frequency may be taken to be one or a combination of the modes calculated, or measured, in a vacuum.

More recently there has been a reconsideration of using blades and vanes made of laminated materials such as glass cloth graphite or metal oxide fibers laid up in polymeric or metal matrix materials and molded under pressure to final airfoil contours. Determining the modes and frequencies of these composite beams is more exacting. However, once determined, these data may be used in the same manner as with conventional metal blades. It should also be noted that aeroelastic programs related to turbomachinery often make a great deal of practical use out of mode and frequency data determined experimentally from prototype and development hardware.

A major consideration in all material and mode of construction studies is the determination of mechanical damping characteristics. Briefly stated the damping may be categorized as material and structural. The former is taken to describe a volume-distributed property in which the rate of energy dissipation into heat (and thus removed from the mechanical system) is locally proportional to a small power of the amplitude of the local cyclical strain. The proportionality constant is determined by many factors, including the type of material, state of mean or steady strain, temperature and other minor determinants.

The structural damping will usually be related to interfacial effects, for example in the blade attachment to the disk or drum, and will depend on normal pressure across the interface, coefficient of friction between the surfaces, mode shape of vibration, and modification of these determinants by previous fretting or wear. Detailed knowledge about damping is usually not known with precision, and damping information is usually determined and used in 'lumped' or averaged fashion. Comparative calculations may be used to predict such gross damping parameters for a new configuration, basing the prediction on the known information for an existing and somewhat similar configuration. By this statement it is not meant to imply that this is a satisfactory state of affairs. More precise damping prediction capabilities would be very welcome in modern aeroelastic studies of turbomachines.

8.4   Nonsteady potential flow in cascades

Unwrapping an annulus of differential height dr from the flow passage of an axial turbomachine results in a two-dimensional representation of a cascade of airfoils and the flow about them. The airfoils are identical in shape, equally spaced, mutually congruent and infinite in number.

When a cascade is considered, as opposed to a single airfoil, the fact that the flexible blades may be vibrating means that the relative pitch and stagger may be functions of time and also position in the cascade. The flow, instead of being a uniform stream, will now undergo turning; large velocity gradients may occur in the vicinity of the blades and in the passages between them. These complications imply that the blade thickness and steady lift distribution must be taken into account for more complete fidelity in formulating the nonsteady aerodynamic reactions.

A fundamental complication which occurs is the necessity for treating the wakes of shed vorticity from *all* the blades in the cascade.

Assume the flow is incompressible. As a simple example of cascading effects consider only the steady lift distribution on each blade in the cascade and compute the disturbance velocity produced at the reference blade by a vibration of all the blades in the cascade.

Assume that standard methods of analyzing steady cascade performance provide the steady vorticity distribution common to all the blades, $\gamma_s(x)$, and its dependence on W_1 and β_1.

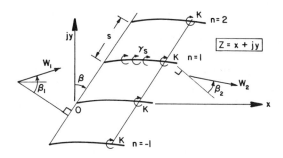

In what follows the imaginary index j for geometry and the imaginary index i for time variation (i.e., complex exponential) cannot be 'mixed'. That is $ij \neq -1$. Furthermore, it is convenient to replace the coordinate normal to the chord, z, by y and the upwash on the reference airfoil w_a by v. The velocities induced by an infinite row of vortices of equal strength, K, are given by

$$\delta[u(z) - jv(z)] = \frac{jK}{2\pi} \sum_{n=-\infty}^{\infty} \frac{1}{z - \zeta_n}$$

where the location ζ_n of the nth vortex

$$\zeta_n = \xi + jnse^{-j\beta} + jY_n(\xi_n, t) + X_n(t)$$

indicates small deviations from uniform spacing s, $(Y_n \ll s, X_n \ll c)$. The point Z is on the zeroth blade

$$Z = x + jY(x, t) + X(t)$$

and the location of the vortices will ultimately be congruent points on different blades so that

$$\xi_n = \xi + ns \sin \beta$$

(The subscript naught, indicating the zeroth blade, is conventionally omitted.) Finally, harmonic time dependence with time lag $-\tau$ between the motions of adjacent blades* is indicated by

$$Y_n(\xi_n, t) = e^{in\omega\tau} Y(\xi, t)$$

With these provisions the Cauchy kernel may be written

$$\frac{1}{Z - \zeta_n} = \frac{1}{x - \xi - jnse^{-j\beta} + j[Y(x, t) - Y_n(\xi_n, t)] + X(t) - X_n(t)}$$

$$\sum_{n=-\infty}^{n=\infty} \frac{1}{Z - \zeta_n} = \frac{1}{x - \xi + j[Y(x, t) - Y(\xi, t)]} + \sum_{n=-\infty}^{\infty\prime} \frac{1}{Z - \zeta_n}$$

where the primed summation indicates $n = 0$ is excluded. The first term on the RHS is a self-induced effect of the zeroth foil. The part $Y(x, t) - Y(\xi, t)$ is conventionally ignored in the thin-airfoil theory; it is small compared to $x - \xi$ and vanishes with $x - \xi$. Hence the first term supplies the single airfoil part of the steady state solution. Expanding the remaining term yields

$$\sum\prime \frac{1}{Z - \zeta_n} = \sum\prime \frac{1}{x - \xi - jnse^{-j\beta}} + i \sum\prime \frac{Y_n(\xi_n, t) - Y(x, t)}{(x - \xi - jnse^{-j\beta})^2}$$

$$+ \sum\prime \frac{X_n(t) - X(t)}{(x - \xi - jnse^{-j\beta})^2}$$

where the last two summations on the RHS are the *time-dependent portions*. The corresponding unsteady induced velocities may be expressed as follows

$$\delta[\bar{u}(x') - j\bar{v}(x')] \simeq -\frac{\gamma_s(\xi')\delta\xi'}{2\pi c} P^2 \left\{ \sum\prime \frac{e^{in\omega\tau} Y(\xi', t) - Y(x', t)}{(\chi - jn\pi)^2} \right.$$

$$\left. + \frac{1}{j} \sum\prime \frac{e^{in\omega\tau} X(t) - X(t)}{(\chi - jn\pi)^2} \right\}$$

where

$$P = \pi e^{j\beta} c / s$$

$$\chi = P(x' - \xi')$$

and $\bar{u}$, $\bar{v}$ are the time dependent parts of u, v. The local chordwise

---

* The so-called 'periodicity assumption' of unsteady cascade aerodynamics lends order, in principle and often in practice, to the processes of cascade aeroelasticity. The mode of every blade is assumed to be identical, with the same amplitude and frequency but with the indicated blade-to-blade phase shift. Absent this assumption, the cascade representing a rotor of n blades could have n distinct components (type of mode, modal amplitude, frequency).

distributed vortex strength $\gamma_s(\xi)\,d\xi$ has replaced K the discrete vortex strength in the last step. With the notation

$$q = 1 - \omega\tau/\pi$$

the above summations may be established in closed form. For example, when the blades move perpendicular to their chordlines with the same amplitude all along the chord the displacement function is a constant

$$Y = -\bar{h}e^{i\omega t} = -h$$

and there is obtained

$$\bar{u}(x') - j\bar{v}(x') = \frac{P^2}{2\pi c} \int_0^1 \gamma_s(\xi') \sum{}' \frac{e^{in\omega\tau}h(t) - h(t)}{(\chi - jn\pi)^2}\,d\xi'$$

or

$$\bar{u} = -\frac{h}{2\pi c} \int_0^1 \gamma_s(\xi')[F - iI]\,d\xi'$$

$$\bar{v} = \frac{h}{2\pi c} \int_0^1 \gamma_s(\xi')[G + iH]\,d\xi'$$

where

$$F + iG = P^2 \frac{q \sinh \chi \sinh q\chi - \cosh \chi \cosh q\chi + 1}{\sinh^2\chi}$$

$$H + iI = P^2 \frac{q \sinh \chi \cosh q\chi - \cosh \chi \sinh q\chi}{\sinh^2\chi}$$

Similar disturbance velocity fields can be derived for torsional motion, pure chordwise motion, etc. Another separate set of disturbance fields may be generated to take account of the blade thickness effects by replacing the steady vorticity distribution $\gamma(x)$ by, say $j\varepsilon(x)$, the steady source distribution, in the above development.

The net input to the computation of oscillatory aerodynamic coefficients is then obtained by adding the $\bar{v}$ of all the effects so described to the LHS of the integral equation which follows

$$\overbrace{v_1(x) + v_2(x) + v_3(x)}^{\text{on } y=0,\, 0<x<c} = \frac{1}{2\pi} \int_0^c [\gamma_1(\xi) + \gamma_2(\xi) + \gamma_3(\xi)]K(\xi - x)\,d\xi$$

$$+ \frac{1}{2\pi} \int_c^\infty [\gamma_1(\xi) + \gamma_2(\xi) + \gamma_3(\xi)]K(\xi - x)\,d\xi$$

In this formulation v_1 may be identified with the unsteady upwash, if any, convected as a gust with the mean flow and v_2 is the unsteady upwash attributable to vibratory displacement of all the blades in the cascade,

393

where each blade is represented by steady vortex and source/sink distribution. It is v_2 that was described for one special component in the derivation of $\tilde{v}$ previously.

The component v_3 may be identified with the unsteady upwash relative to the zeroth airfoil occasioned by it's harmonic vibration.

Since we are dealing here with a linear problem each of the subscripted sub-problems may be solved separately and independently of the others. It is also important to note that since the vortex distributions γ_1, γ_2 and γ_3 representing the lift distributions on the cascade chordlines are unsteady they must give rise to distributions of free vortices in the wake of each airfoil of the cascade. In other words vortex wakes emanate from the trailing edge of each airfoil and are convected downstream; at a point with fixed coordinates in the wake, the strength of the vortex element instantaneously occupying that point will vary with time. Hence, the integral equation will in general contain a term that is an integral over the wake $(c < \xi < \infty)$ to account for the additional induced velocities from the infinite number of semi-infinite vortex wakes. The kernel $\frac{1}{2}\pi K(\xi - x)$ accounts in every case for the velocity induced at $(x, 0)$ by a vortex element at the point $(\xi, 0)$ on the chord or wake of the reference, or zeroth, airfoil plus an element of equal strength located at the congruent point $(\xi + ns \sin \beta, ns \cos \beta)$ of every other profile of the cascade or its wake. The form of K may in fact be derived by returning to the previous derivation for $\tilde{v}$ and extracting the terms

$$\underbrace{\frac{1}{\xi - x}}_{\substack{\text{isolated}\\\text{airfoil}}} + \underbrace{\sum_{n=-\infty}^{\infty}{}' \frac{1}{\xi - x + jnse^{-j\beta}}}_{\text{cascade effect}}$$

In this expression the signs have been changed to imply calculations of positive v (rather than $-jv$) and with each term it is now necessary to associate a strength $\gamma_r(\xi) \exp(in\omega\tau)$ ($r = 1$, 2 or 3) since the inducing vortexes now pulsate rather than being steady in time. The kernel now appears as

$$\frac{1}{2\pi} K(\xi - x) = \frac{1}{2\pi} \sum_{n=-\infty}^{\infty} \frac{e^{in\omega\tau}}{\xi - x + jns \exp(-j\beta)}$$

which may be summed in closed form to yield

$$\frac{1}{2\pi} K(\xi - x) = \frac{e^{j\beta}}{2s}$$

$$\cdot \frac{\cosh\left[(1 - \sigma/\pi)\pi \exp(j\beta)(\xi - x)/s\right] + ij \sinh\left[(1 - \sigma/\pi)\pi \exp(j\beta)(\xi - x)/s\right]}{\sinh\left[\pi \exp(j\beta)(\xi - x)/s\right]}$$

where $\sigma = \omega\tau$ is known as the interblade phase angle, an assumed constant.

Taking the term for $n = 0$ in the previous summation yields

$$\frac{1}{2\pi} K_0(\xi - x) = \frac{1}{2\pi} \frac{1}{\xi - x}$$

which is the kernel for the isolated airfoil. Hence the added complexity of solving the cascaded airfoil problem is attributed to the additional terms giving the more complicated kernel displayed in the penultimate formula.

In contradistinction to the isolated airfoil case, solutions of the unsteady aerodynamics integral equation cannot be solved in closed form, or in terms of tabulated functions, for arbitrary geometry (β and s/c) and arbitrary interblade phase angle, σ. In fact, as noted previously, the thickness distribution of the profiles and the steady lift distribution become important when cascades of small space/chord ratio are considered to vibrate with nonzero interblade phasing. Consequently, solutions to the equation are always obtained numerically. It is found that the new parameters β, s/c and σ are strong determinants of the unsteady aerodynamic reactions. A tabular comparison of the effect of these variables on the lift due to bending* appears below. In this chart, the central stencil

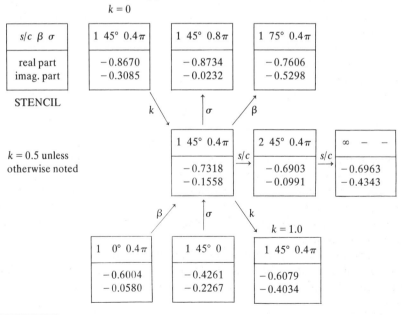

$k = 0$

s/c β σ	1 45° 0.4π	1 45° 0.8π	1 75° 0.4π
real part	-0.8670	-0.8734	-0.7606
imag. part	-0.3085	-0.0232	-0.5298

STENCIL

$k \searrow$ $\sigma \uparrow$ $\nearrow \beta$

$k = 0.5$ unless otherwise noted

	1 45° 0.4π		2 45° 0.4π		∞ $-$ $-$
	-0.7318	$\xrightarrow{s/c}$	-0.6903	$\xrightarrow{s/c}$	-0.6963
	-0.1558		-0.0991		-0.4343

$\beta \nearrow$ $\sigma \uparrow$ $k \searrow$

$k = 1.0$

1 0° 0.4π	1 45° 0	1 45° 0.4π
-0.6004	-0.4261	-0.6079
-0.0580	-0.2267	-0.4034

* Whitehead [1].

gives the lift coefficient for the reference values of $s/c = 1.0$, $\beta = 45°$, $\sigma = 0.4\pi$. Other values in the matrix give the coefficient resulting from changing one and only one of the governing parameters.

The effects of thickness and steady lift cannot be easily displayed, and are conventionally determined numerically for each application.

8.5 Compressible flow

The linearized problem of unsteady cascade flow in a compressible fluid may be conveniently formulated in terms of the acceleration potential, $-p/\rho$, where p is the perturbation pressure, i.e., the small unsteady component of fluid pressure. Using the acceleration potential as the primary dependent variable, a number of compact solutions have been obtained for the flat plate cascade at zero incidence. The most reliable in subsonic flow is that due to Smith,* and in supersonic flow the solution of Verdon† is most flexible (in the latter regime at the time of publishing there was intense development of competing methods of calculation).

Supersonic flow relative to the blades of a turbomachine is of practical importance in steam turbines and near the tips of transonic compressor blades. In these cases the axial component of the velocity remains subsonic; hence analytic solutions in this flow regime (the so-called subsonic leading edge locus problem) are of the most interest. It may be that in future applications the axial component will be supersonic. In this event the theory actually becomes simpler so that the present concentration on subsonic values of M_{axial} represents the most difficult problem. Currently efforts are underway to account for such complicating effects as changing back-pressure on the stage, flow turning, shock waves, etc.

To illustrate the effect of varying the Mach number from incompressible on up to supersonic, a particular aerodynamic coefficient has been graphed in the accompanying sketch as a function of the relative Mach number. It is seen that the variation of the coefficient in the subsonic regime is not great except in the immediate neighborhood of the so-called 'resonant' Mach number, or the Mach number at which 'aerodynamic resonance' occurs.

It is possible to generalize the situation with respect to compressibility by indicating that the small disturbance approximations are retained, but the velocities, velocity potential, acceleration potential, or pressure

---

* Smith [2].
† Verdon [3].

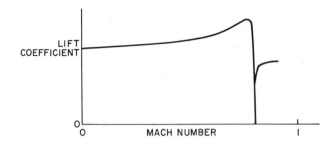

(in every case the disturbance component of these quantities) no longer satisfy the Laplace equation, but rather an equation of the following type.

$$(1 - M^2)\phi_{xx} + \phi_{yy} - \frac{1}{a^2}\phi_{tt} - 2\frac{M}{a}\phi_{xt} = 0$$

Here M is the relative Mach number and a is the sound speed. Note that the presence of time derivatives make this partial differential equation hyperbolic whatever the magnitude of M, a situation quite different from the steady flow equation.

Although the above equation is appropriate to either subsonic or supersonic flow, the resonance phenomenon occurs in the regime of subsonic axial component of the relative velocity when geometric and flow conditions satisfy a certain relationship.

Equating the time of propagation of a disturbance along the cascade to the time for an integral number of oscillations to take place plus the time lag associated with the interblade phase angle, σ, yields

$$\frac{S}{V_p^+} = \frac{2\pi\nu}{\omega} - \frac{\sigma}{\omega}$$

$$\frac{S}{V_p^-} = \frac{2\pi\nu}{\omega} + \frac{\sigma}{\omega}$$

where $V_p^\pm$, the velocity of propagation, has two distinct values associated with the two directions along the cascade, see sketch.

$$V_p^\pm = a[\sqrt{1 - M^2\cos^2\beta} \pm M\sin\beta]$$

These expressions can be reduced to the equation

$$\frac{\omega s}{a} = (2\nu\pi \pm \sigma)(\sqrt{1 - M^2\cos^2\beta} \mp M\sin\beta)$$

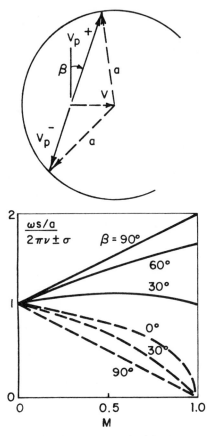

where ν may be any positive integer, and with the upper set of signs may also be zero.

The equation may be graphed and potential acoustic resonances discerned by plotting the characteristics of a given stage on the same sheet for possible coincidence. (It is convenient to take β as the parameter with axes $\omega s/a$ and M.) Acoustic resonances of the variety described above may be dangerous because they account for the vanishing, or near vanishing, of all nonsteady aerodynamic reactions including therefore the important aerodynamic damping. Although it is difficult to establish with certainty, several cases of large vibratory stresses have been presumably correlated with the acoustic resonance formulation. It should be recognized that the effects of blade thickness and nonconstant Mach number throughout the field are such as to render the foregoing formulation somewhat approximate.

The foregoing development may also be based more rigorously on the theoretically derived integral equation relating the harmonically varying downwash on the blade to the resulting harmonically varying pressure difference across the blade's thickness. Symbolically

$$\bar{v}_a(x) = \int_0^c K(\xi - x)\, \Delta \bar{p}_a(\xi)\, d\xi$$

and the acoustic resonance manifests itself by an infinite singularity appearing in the kernel K for special values of k, s, c, τ and β of which K is also a function. Under this circumstance the downwash v_a can only remain finite, as it must physically, by a vanishing of Δp_a as noted above. The previous development shows why the compressible flow solutions have received such an impetus and are so closely related to the acoustic properties of compressor and fan cascades.

The acoustic resonance phenomenon, as just described results from standing waves, albeit with impressed throughflow velocity, of the fluid filling the interblade passages.

8.6 Periodically stalled flow in turbomachines

Rotating, or propagating, stall are terms which describe a phenomenon of circumferentially asymmetric flow in axial compressors. Such a flow usually appears at rotationally part-speed conditions and manifests itself as one or more regions of reduced (or even reversed) through-flow which rotate about the compressor axis at a speed somewhat less than rotor speed, albeit in the same direction.

A major distinction between propagating stall and surge is that in the former case the integrated massflow over the entire annulus remains steady with time whereas in the latter case this is not true. The absolute propagation rate can be brought to zero or even made slightly negative by choosing pathological compressor design parameters.

If the instigation of this phenomenon can be attributed to a single blade row (it obviously must be in a single-stage compressor) then insofar as this blade row is concerned, it represents a periodic stalling and unstalling of each blade in the row. Later or preceding blade rows (i.e., half-stages) may or may not experience individual blade stall periodically, depending on the magnitude of the flow fluctuation at that stage, as well as the cascade stall limits in that stage.

The regions of stalled flow may extend across the annulus (full span)

399

or may be confined to the root or tip regions of the blades (partial-span stall). The number of such regions which may exist in the annulus at any one time varies from perhaps 1 to 10 with greater numbers possible in special types of apparatus.

The periodic loading and unloading of the blades may prove to be extremely harmful if a resonant condition of vibration obtains. Unfortunately the frequency of excitation cannot be accurately predicted at the present time so that avoidance of resonance is extremely difficult.

The results of various theories concerning propagating stall are all moderately successful in predicting the propagational speed. However the number of stall patches which occur (i.e., the wavelength of the disturbance) seems to be essentially unpredictable so that the frequency of excitation remains uncertain. Furthermore, the identification of the particular stage which is controlling the propagating stall, in the sense noted above, is often uncertain or impossible so that the application of any theory to practice in a multi-stage machine cannot be done.

Since this situation regarding propagating stall is one requiring additional research, no further attention will be devoted to the phenomenon of occurrence. Presumably if the fluctuations in flow direction and magnitudes can be predicted or described, the effect on blade vibratory response can be predicted using the gust functions similar to those developed for the single airfoil but rederived for the cascade situation and accounting for periodic stalling.

Although the precise classification of vibratory phenomena of an aeromechanical nature is often somewhat difficult in turbomachines because of the complication due to cascading and multistaging, it is nevertheless necessary to make such distinctions as are implied by an attempt at classification. The manifestation of stall flutter in turbomachines is a good example of what is meant. When a given blade row, or cascade, approaches the stalling incidence in some sense (i.e., stalling defined by rapid rise of relative total pressure loss, or defined by rapid increase in deviation angle, or defined by the appearance of flow separation from the suction surface of the blades, etc.) it is found experimentally that a variety of phenomena may exist. Thus the region of reduced throughflow may partially coalesce into discrete patches which propagate relative to the cascade giving rise to the type of flow instability previously discussed under rotating stall. There is no dependence on blade flexibility.

Under certain other overall operating conditions it is found that in the absence of, or even coexistent with, the previous manifestation, the blades vibrate somewhat sporadically at or near their own natural frequency. There is no immediately obvious correlation between the motions of adjacent blades, and the amplitudes of vibrations change with time in an apparently random manner. (We exclude here all vibration attributable to resonance with the propagating stall frequency, should the propagation phenomenon also be present.) This behavior is termed stall flutter or stalling flutter and the motion is often in the fundamental bending mode. Another term is random vibration. Since the phenomenon may be explained on the basis of nonlinear mechanics, (see the chapter on Stall Flutter) the sporadicity of the vibration can be attributed provisionally to the fact that the excitation has not been strong enough to cause 'entrainment of frequency', a characteristic of many nonlinear systems. Hence, each blade vibrates, *on the average*, as if the adjacent blades were not also vibrating. However, a careful analysis might very well demonstrate that the instantaneous amplitude of a particular blade was effected somewhat by the 'instantaneous phase difference' between its motion and the motion of the adjacent blade(s). One must also speak of 'instantaneous' frequency since a frequency modulation is also apparent. As a general statement it must be said that the frequency, amplitude and phase of adjacent blades are functionally linked in some complicated aeromechanical manner which results in modulations of all three quantities as functions of time. While the frequency modulation will normally be small (perhaps less than 1 or 2 percent) the amplitude and the phase modulations can be quite large. Here the term phase difference has been used rather loosely to describe the relationship between two motions of slightly different frequency. Since this aerodynamic coupling would also depend on the instantaneous amplitude of the adjacent blade(s), it is not surprising that the vibration gives a certain appearance of randomness. On the linear theory for identically tuned blades one would not expect to find sporadic behavior as described above. However, it is just precisely the failure to satisfy these two conditions that accounts for the observed motion; the average blade system consists of an assembly of slightly detuned blades (nonidentical frequencies) and furthermore the oscillation mechanism is nonlinear.

When the relative magnitudes of the nonsteady aerodynamic forces increase it may be expected that entrainment of frequency will occur. In certain nonlinear systems it can be shown that the 'normalized' frequency interval $(\omega - \omega_0)/\omega$ (where ω is the impressed frequency and ω_0 is the frequency of self-excitation) within which one observes entrainment, is proportional to h/h_0, where h is the amplitude of the impressed motion and h_0 is the amplitude of the self-excited oscillation. In case of entrainment one would expect to find a common phase difference between the motion of adjacent blades which implies also motion with a common flutter frequency. This latter phenomenon is also termed stalling flutter, although the term limit-cycle vibration is sometimes used to emphasize the constant-amplitude nature of the motion, which is often in the fundamental torsional mode.

Finally it should be noted that the distinction between blade instability (flutter) and flow instability (rotating stall) is not always perfectly distinct. When the sporadic stall flutter occurs it is clear that there is no steady tangentially propagating feature of the instability. Similarly, when propagating stall occurs with little or no vibration (stiff blades away from resonance) it is apparent that the instability is not associated with vibratory motion of the blade. However, the limit cycle type of behavior can be looked upon (due to the simultaneously observed constant interblade phase relationship) as the propagation of a disturbance along the cascade. Furthermore, the vorticity shed downstream of the blade row would have every aspect of a propagating stall region. For instance if the interblade phasing was 180° the apparent stall region would be on one blade pitch in tangential extent and each would be separated by one blade pitch of unstalled throughflow. Because of the large number of such regions, and the small tangential extent of each, this situation is still properly termed stall flutter since the blades are controlling and the blade amplitudes are constant. At the other extreme when one or two stall patches appear in the annulus it is obvious that the flow instability is controlling and then the phase relationships between adjacent blades' motions may appear to be rather sporadic. At any rate, in the middle ground between these extremes it is probable that a strong interaction between flow stability and blade stability exists and the two phenomenon cannot be easily separated.

Another distinction may be attempted to assist in understanding the operative phenomena. When a *single airfoil* is subjected to an increasing angle of attack an instability of the fluid may arise, related to the Karman vortex frequency or the extension of this concept to a distributed frequency spectrum. If the frequency of this fluid instability coincides with the natural frequency of the blade in any mode, the phenomenon is termed buffetting. If the dynamic moment coefficient (or force coefficient) attains a negative slope a self-excited vibration known as stall flutter occurs. The two phenomena may merge when airfoil vibration exerts some influence on the vortex shedding frequency. Stall-flutter is usually observed in the torsional mode and buffetting in the bending mode, but this distinction is not always possible. These concepts cannot be carried over directly to the cascade where steady bending amplitudes of the limit cycle variety have been observed. The explanation rests on the additional degrees of freedom present in the cascaded configuration.

8.7   Stall flutter in turbomachines

On account of the foregoing complications and the lack of quantitative theories noted in the Stall Flutter chapter (semi-empiricism seems to be the best that can be hoped for in the near term) it is not surprising that prediction for turbomachines has rested almost entirely on correlation of experimental data. The single most important parameter governing stalling is the incidence and the reduced frequency has been seen in all aeroelastic formulations to exert a profound influence. Hence it is not surprising that these variables have been used to correlate the data.

Typically stall flutter will occur at part-speed operation and will be confined to those rotor stages operating at higher than average incidence. With luck the region of flutter will be above the operating line on a compressor map and extend up to the surge line. Under less fortunate circumstances the operating line will penetrate the flutter region. The flutter boundary and contours of constant flutter stress (or tip amplitude) will have the appearance shown on the accompanying figure. Traditional parameters for this typical experimental correlation are reduced velocity, $W/b\omega$, (the inverse of reduced frequency) and incidence, at some characteristic radius such as 75% or 80% of the blade length for a cantilever blade. The curve is typical of data obtained in turbomachines or cascades; essentially a new correlation is required for each major change of any aerodynamic variable (Mach number, stagger, blade contour, etc.). The

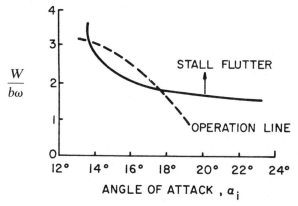

mode shape will usually be first torsion. The single contour shown in the previous figure is for that level of cyclic stress (or strain) in the blade material that is arbitrarily taken to represent some distinct and repeatable measurement attributable to the flutter vibration and discernible above the 'noise' in the strain measuring system. A typical number might be a stress of 2000 psi used to define the flutter boundary. However, small changes in relative airspeed, V, may increase the flutter stress substantially, or, in the case of 'hard' flutter, a small increase in incidence might have a similar effect. Hence, in keeping with the nonlinear behavior described in the Stall Flutter Chapter, the contours of constant flutter stress may be quite closely spaced in some regions of the correlation diagram.

Naturally, when considering three-dimensional effects it is the integrated energy passing from airstream to airfoil that determines whether flutter will occur, or not. The stalled tip of a rotor blade, for example, must extract more energy from the airstream than is put back into the airstream by airfoil sections at smaller radii and that is dissipated from the system by damping.

This total description of stall flutter in turbomachine rotor blades is consistent with the appearance of the stall flutter boundary as it appears on the following typical compressor performance map; the vibrations are usually confined to the first two or three stages. This figure may be viewed

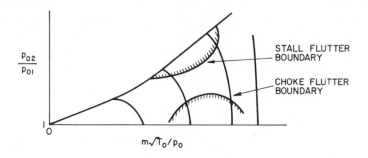

in conjunction with the performance map on page 387 which shows typical angles of attack for a rotor blade tip in the first stage of a compressor. Keeping in mind that the mass flow parameter $\dot{m}\sqrt{T_0}/p_0$ is virtually proportional to the throughflow velocity in the first few stages of a compressor, it is clear that any typical operating line as shown on the compressor map will traverse the flutter boundary somewhat as the dotted line on the $W/(b\omega)$ *versus* α_i plot shown previously.

This explains the general shape and location of the region of occurrence of stall flutter. Experimental determinations confirm increasing stresses as the region is penetrated from below and the specific behavior is a function of the aeroelastic properties of the individual machine, consistent with the broad principles enunciated here.

8.8   Choking flutter

In the middle stages of a multistage compressor it may be possible to discern another region on the compressor map wherein so-called choking flutter will appear. This will normally occur at part-speed operation and will be confined to those rotor stages operating at lower than average incidence (probably negative values are encountered). The region of flutter will normally lie below the usual operating line on a compressor map, but individual stages may encounter this type of instability without greatly affecting the operating line; this is particularly true when the design setting angle of a particular row of rotor blades had been arbitrarily changed from the average of adjacent stages through inadvertence or by a sequence of aerodynamic redesigns.

The physical manifestation of choking flutter is usually discriminated by a plot of a stage's operating line on coordinates of relative Mach number *vs.* incidence, as in the sketch. On these same coordinates the choke boundaries are shown; a coincidence or intersection of these graphs indicate the possible presence of choking flutter and is usually confined to a very small range of incidence values. The mechanism of choking flutter is not fully understood. It is related to compressibility phenomena in the fluid and separation of the flow is probably also involved. The graph labelled '2 × Loss$_{min}$' is a locus of constant aerothermodynamic loss coefficient (closely related to the drag coefficient of an airfoil); the interior of the nose-shaped region representing low values of loss, or efficient operation of the compressor stage. The curve labelled 'choke boundary' represents the combination of relative Mach number and flow angle at which the minimum flow area between adjacent blades (the throat) is passing the flow with the local sonic velocity. Presumably separation of the flow at the nose of each airfoil on the pressure surface, and the *relative* motion between adjacent blades as they vibrate, conspire to change the effective throat location in a time dependent manner. These

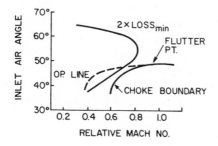

405

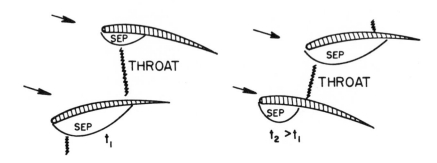

oscillatory changes effect the pressure distribution on each blade in such a fashion (including a phase angle) as to pump energy from the airstream into the vibration and thus sustain the presumed motion. No great certitude or fidelity should be attached to the accompanying sketches since the suggested mechanism has not been confirmed experimentally. Nor is the explanation sufficiently detailed to be subjected to a test of plausibility.

Choking flutter occurs in practice with sufficient frequency and destructive potential as to be an important area for current research efforts.

8.9   Supersonic torsional flutter

On the same fan or compressor map a flutter region has been identified in which the relative Mach numbers are supersonic and the vibratory mode is torsional. This remark is confined, of course, to those turbomachines which are designed for supersonic relative Mach numbers, or which can attain such an operating condition at off-design overspeed conditions.

The flutter boundary is more or less vertical on the map (sometimes termed the 'stone wall') and the selected vibratory motion, although predominantly torsion, has some bending component and also a specific interblade phase angle associated with a bladed-disk mode having two or more nodal diameters (i.e., a travelling wave about the periphery).

Analytically formulated low incidence predictions* are only moderately successful in predicting the occurrence of the problem; they do not match the type of data observed in practice, see figure. The predictions

---

* Snyder and Commerford [4].

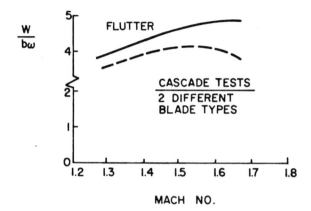

are lines of essentially constant reduced velocity in the range between 1 and 2 depending on the stagger and spacing of the cascade. More complete elucidation of this phenomenon depends on the acquisition of more accurate aerodynamic coefficients (including thickness and back pressure effects) correlated with experimental data of actual flutter conditions.

The need for these coefficients has already been noted in the previous discussion of the 'subsonic leading edge locus problem'.

8.10   Supersonic bending flutter

This is essentially a concatenation of high incidence and high Mach number effects; the flutter boundary is virtually horizontal on the compressor map and lies above the design point. This phenomenon is poorly understood but is probably closely related to high-speed aileron buzz of airplane wings. Strong shocks are formed which are relatively unstable; vibratory motion of the blades results in periodic movement of the shocks and oscillatory separation of the flow in the interblade channel.

Theoretical treatment has not been attempted, although some analytical models may possibly be borrowed from subsonic stall flutter and the aileron buzz formations which have not been presented. Fortunately the occurrence of this problem is not as important as the other three shown by shaded areas on the compressor map since it is usually possible to avoid the combination of high corrected speed and high incidence by judicious choice of operating schedules.

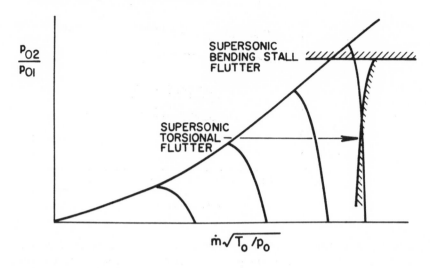

$$\dot{m}\sqrt{T_0}/p_0$$

8.11 Concluding remarks

The reader will readily appreciate the great complexity of this area of aeroelasticity and the demanding challenges which remain. Nevertheless research in this area has reached a new level of intensity motivated in part by several recent dramatic occurrences of flutter behavior in practice. Any extensive citation of the rapidly expanding literature runs the risk of obsolescence and the discussion here has concentrated on our present physical understanding of the phenomena. For a brief and readable overview of the recent research literature the article by Fleeter may be consulted.* For an entree to the companion literature on unsteady aerodynamics the survey article by Sisto is recommended.†

References for Chapter 8

[1] Whitehead, D. S., 'Force and Moment Coefficients for Vibrating Aerofoils in Cascade', *ARC R&M 3254, London*, 1960.

[2] Smith, S. M., 'Discrete Frequency Sound Generation in Axial Flow Turbo-machines', *ARC R&M 3709, London*, 1072.

[3] Verdon, J. M., 'Further Developments in the Aerodynamic Analysis of Unsteady Supersonic Cascades—Parts 1 and 2', *ASME Papers 77-GT-44* and *77-GR-45* (March 1977).

---

* Fleeter [5].
† Sisto [6].

[4] Snyder, L. E. and Commerford, G. L., 'Supersonic Unstalled Flutter in Fan Rotors; Analytical and Experimental Results' *ASME Trans.*, *Journal of Engineering for Power*, v. 96, Series A, n. 4 (1974) pp. 379–386.

[5] Fleeter, S., 'Aeroelasticity Research for Turbomachine Applications', *AIAA Paper 77–437, AIAA Dynamics Specialist Conference*, San Diego, California (March 24–25, 1977).

[6] Sisto, F., 'Review—A Review of the Fluid Mechanics of Aeroelasticity in Turbomachines', *Journal of Fluids Engineering, Transactions of the ASME* (March 1977) pp. 40–44.

A primer for structural response to random pressure fluctuations[*]

Nomenclature

a	Plate length
b	Plate width
D	$Eh^3/12(1-\nu^2)$, Plate bending stiffness
E	Modulus of elasticity
H_n	Plate transfer function
h	Plate thickness
$I_n; I$	Plate impulse function; see equation (I.21)
K_n^2	$=\dfrac{m\omega_n^2 a^4}{D}$
M_m	Plate generalized mass
m	Plate mass/area
n	normal
p	Pressure on plate
Q_m	Generalized force on plate
q_n	Generalized plate coordinate
R	Correlation function
t	Time
w	Plate deflection
x, y, z	Cartesian coordinates
∇^2	Laplacian
Φ	Power spectral density
ρ_m	Plate density
σ	stress
τ	Dummy time
ζ_m	Modal damping
ω_m	Modal frequency

---

[*] This Appendix is based upon a report by E. H. Dowell and R. Vaicaitis of the same title, Princeton University AMS Report No. 1220, April 1975.

Appendix I

Introduction

In this appendix we shall treat the response of a structure to a convecting-decaying random pressure field. The treatment follows along conventional lines after Powell [1] and others. That is, the pressure field is modelled as a random, stationary process whose correlation function (and/or power spectra) is determined from experimental measurements. Using this empirical description of the random pressure, the response of the structure is determined using standard methods from the theory of linear random processes [2, 3]. The major purpose of the appendix is to provide a complete and detailed account of this theory which is widely used in practice (in one or another of its many variants). A second purpose is to consider systematic simplifications to the complete theory. The theory presented here is most useful for obtaining analytical results such as scaling laws or even, with enough simplifying assumptions, explicit analytical formulae for structural response.

It should be emphasized that, if for a particular application the simplifying assumptions which lead to analytical results must be abandoned, numerical simulation of structural response time histories may be the method of choice [4, 5]. Once one is committed to any substantial amount of numerical work (e.g., computer work) then the standard power spectral approach loses much of its attraction.

Excitation-response relations for the structure

In the present section we derive the excitation-response relations for a flat plate. It will be clear, however, that such relations may be derived in a similar manner for any linear system.

The equation of motion for the small (linear) deformation of a uniform isotropic flat plate is

$$D \nabla^4 w + m \frac{\partial^2 w}{\partial t^2} = p \tag{I.1}$$

where w is the plate deflection, p the pressure loading and the other terms are defined in the Nomenclature. Associated with (I.1) are the natural modes and frequencies of the plate which satisfy

$$D \nabla^4 \psi_n - \omega_n^2 m \psi_n = 0 \tag{I.2}$$

where ω_n is the frequency and $\psi_n(x, y)$ the shape of the nth natural mode.

A primer for structural response to random pressure fluctuations

In standard texts it is shown that the ψ_n satisfy an orthogonality condition

$$\iint \psi_n \psi_m \, dx \, dy = 0 \quad \text{for} \quad m \neq n \tag{I.3}$$

If we expand the plate deflection in terms of the natural modes

$$w = \sum_n q_n(t) \psi_n(x, y) \tag{I.4}$$

then substituting (I.4) into (I.1), multiplying by ψ_m and integrating over the plate area we obtain

$$M_m[\ddot{q}_m + \omega_m^2 q_m] = Q_m \quad m = 1, 2, \ldots \tag{I.5}$$

where we have used (I.2) and (I.3) to simplify the result. M_m and Q_m are defined as

$$M_m \equiv \iint m\psi_m^2 \, dx \, dy$$

$$Q_m \equiv \iint p\psi_m \, dx \, dy \tag{I.6}$$

$$\cdot \equiv d/dt$$

For structures other than a plate the final result would be unchanged, (I.5) and (I.6); however, the natural modes and frequencies would be obtained by the appropriate equation for the particular structure rather than (I.1) or (I.2). Hence, the subsequent development, which depends upon (I.5) only, is quite general.

Before proceeding further we must consider the question of (structural) damping. Restricting ourselves to structural damping only we shall include its effect in a gross way by modifying (I.5) to read

$$M_m[\ddot{q}_m + 2\zeta_m \omega_m \dot{q}_m + \omega_m^2 q_m] \equiv Q_m \tag{I.7}$$

where ζ_m is a (nondimensional) damping coefficient usually determined experimentally. This is by no means the most general form of damping possible. However, given the uncertainty in our knowledge of damping from a fundamental theoretical viewpoint (see [6]) it is generally sufficient to express our meager knowledge. If damping is inherent in the material properties (stress-strain law) of the structure, the theory of viscoelasticity may be useful for estimating the amount and nature of the damping.

413

However, often the damping is dominated by friction at joints, etc., which is virtually impossible to estimate in any rational way.

Now let us turn to the principal aim of this section, the stochastic relations between excitation (pressure loading), and response (plate deflection or stress). We shall obtain such results in terms of correlation functions and power spectra.

The correlation function of the plate deflection w is defined as

$$R_w(\tau; x, y) \equiv \lim_{T \to \infty} \frac{1}{2T} \int_{-T}^{T} w(x, y, t) w(x, y, t + \tau) \, dt \tag{I.8}$$

Using (I.4) we obtain

$$R_w(\tau; x, y) = \sum_m \sum_n \psi_m(x, y) \psi_n(x, y) R_{q_m q_n}(\tau) \tag{I.9}$$

where

$$R_{q_m q_n}(\tau) \equiv \lim_{T \to \infty} \frac{1}{2T} \int_{-T}^{T} q_m(t) q_n(t + \tau) \, d\tau \tag{I.10}$$

is defined to be the cross-correlation of the generalized coordinates, q_m. Defining power spectra

$$\Phi_w(\omega; x, y) \equiv \frac{1}{\pi} \int_{-\infty}^{\infty} R_w(\tau; x, y) e^{i\omega\tau} \, d\tau \tag{I.11}$$

$$\Phi_{q_m q_n}(\omega) \equiv \frac{1}{\pi} \int_{-\infty}^{\infty} R_{q_m q_n}(\tau) e^{i\omega\tau} \, d\tau \tag{I.12}$$

we may obtain from (I.9) via a Fourier Transform

$$\Phi_w(\omega; x, y) = \sum_m \sum_n \psi_m(x, y) \psi_n(x, y) \Phi_{q_m q_n}(\omega) \tag{I.13}$$

(I.9) and (I.13) relate the physical deflection, w, to the generalized coordinates, q_m.

Consider next similar relations between physical load p and generalized force Q_m. Define the cross-correlation

$$R_{Q_m Q_n}(\tau) \equiv \lim_{T \to \infty} \frac{1}{2T} \int_{-T}^{T} Q_m(t) Q_n(t + \tau) \, dt \tag{I.14}$$

Using the definition of generalized force (I.6)

$$Q_m(t) \equiv \iint p(x, y, t) \psi_m(x, y) \, dx \, dy$$

$$Q_n(t + \tau) \equiv \iint p(x^*, y^*, t + \tau) \psi_n(x^*, y^*) \, dx^* \, dy^*$$

and substituting into (I.14) we obtain

$$R_{Q_m Q_n}(\tau) = \iiiint \psi_m(x, y)\psi_n(x^*, y^*)$$

$$\cdot R_p(\tau; x, y, x^*, y^*)\, dx\, dy\, dx^*\, dy^* \qquad (I.15)$$

where we define the pressure correlation

$$R_p(\tau; x, y, x^*, y^*) \equiv \lim_{T \to \infty} \frac{1}{2T} \int_{-T}^{T} p(x, y, t)p(x^*, y^*, t+\tau)\, dt \qquad (I.16)$$

Note that a rather extensive knowledge of the spatial distribution of the pressure is required by (I.16).

Again defining power spectra

$$\Phi_{Q_m Q_n}(\omega) \equiv \frac{1}{\pi} \int_{-\infty}^{\infty} R_{Q_m Q_n}(\tau)e^{i\omega\tau}\, d\tau \qquad (I.17)$$

$$\Phi_p(\omega; x, y, x^*, y^*) \equiv \frac{1}{\pi} \int_{-\infty}^{\infty} R_p(\tau; x, y, x^*, y^*)e^{i\omega\tau}\, d\tau \qquad (I.18)$$

we may obtain from (I.15)

$$\Phi_{Q_m Q_n}(\omega) = \iiiint \psi_m(x, y)\psi_n(x^*, y^*)$$

$$\cdot \Phi_p(\omega; x, y, x^*, y^*)\, dx\, dy\, dx^*\, dy^* \qquad (I.19)$$

Finally, we must relate the generalized coordinates to the generalized forces. From (I.7) we may formally solve (see [2], for example or recall Section 3.3)

$$q_n(t) = \int_{-\infty}^{\infty} I_n(t - t_1)Q_n(t_1)\, dt_1 \qquad (I.20)$$

where the 'impulse function' is defined as

$$I_n(t) \equiv \frac{1}{2\pi} \int_{-\infty}^{\infty} H_n(\omega)e^{i\omega t}\, d\omega \qquad (I.21)$$

and the 'transfer function' is defined as

$$H_n(\omega) \equiv \frac{1}{M_n[\omega_n^2 + 2\zeta_n i\omega_n\omega - \omega^2]}$$

Appendix I

Also

$$H_n(\omega) = \int_{-\infty}^{\infty} I_n(t)e^{-i\omega t}\, dt$$

which is the other half of the transform pair, cf (I.21). From (I.20) and (I.10)

$$R_{q_m q_n}(\tau) = \lim_{T\to\infty} \frac{1}{2T} \int\!\!\int\!\!\int_{-T}^{T} I_m(t-t_1)I_n(t+\tau-t_2)Q_m(t_1)Q_n(t_2)\, dt_1\, dt_2\, dt$$

Performing a change of integration variables and noting (I.14),

$$R_{q_m q_n}(\tau) = \int\!\!\int_{-\infty}^{\infty} I_m(\xi)I_n(\eta)R_{Q_m Q_n}(\tau-\eta+\xi)\, d\xi\, d\eta \qquad (I.22)$$

Taking a Fourier Transform of (I.22) and using the definitions of power spectra (I.12) and (I.17), we have

$$\Phi_{q_m q_n}(\omega) = H_m(\omega)H_n(-\omega)\Phi_{Q_m Q_n}(\omega) \qquad (I.23)$$

Summarizing, the relations for correlation functions are (I.9), (I.15), and (I.22) and for power spectra (I.13), (I.19) and (I.23). For example, substituting (I.19) into (I.23) and the result into (I.13) we have

$$\Phi_w(\omega; x, y) = \sum_m \sum_n \psi_m(x, y)\psi_n(x, y)H_m(\omega)H_n(-\omega)$$

$$\cdot \int\!\!\int\!\!\int\!\!\int \psi_m(x, y)\psi_n(x^*, y^*)$$

$$\cdot \Phi_p(\omega; x, y, x^*, y^*)\, dx\, dy\, dx^*\, dy^* \qquad (I.24)$$

This is the *desired final result* relating the physical excitation to the physical response in stochastic terms.

Sharp resonance or low damping approximation

Often (I.24) is approximated further. Two approximations are particularly popular and useful. The first is the 'neglect of off-diagonal coupling'. This means omitting all terms in the double sum except those for which $m = n$. The second is the 'white noise' approximation which assumes that Φ_p is essentially constant relative to the rapidly varying transfer functions

416

A primer for structural response to random pressure fluctuations

$H_m(\omega)$. Making both of these approximations in (I.24) we may obtain the mean square response

$$\bar{w}^2(x, y) \equiv R_w(\tau = 0; x, y) = \int_0^\infty \Phi_w(\omega; x, y)\, d\omega$$

$$\approx \frac{\pi}{4} \sum_m \frac{\psi_m^2(x, y)}{M_m^2 \omega_m^3 \zeta_m} \iiint \psi_m(x, y)\psi_m(x^*, y^*)$$

$$\cdot \Phi_p(\omega_m; x, y, x^*, y^*)\, dx\, dy\, dx^*\, dy^* \qquad (I.25)$$

Of course, only one or the other of these approximations may be made, rather than both. However, both stem from the same basic physical idea: The damping is small and hence, H_m has a sharp maximum near $\omega = \omega_m$. That is

$$H_m(\omega_m)H_n(-\omega_m) \ll |H_m(\omega_m)|^2$$
$$H_m(\omega_n)H_n(-\omega_n) \ll |H_n(\omega_n)|^2$$

and the 'neglect of off-diagonal coupling' follows. Also

$$\int \Phi_p |H_m(\omega)|^2\, d\omega \approx \Phi_p(\omega_m) \int |H_m(\omega)|^2\, d\omega$$

and (I.25) follows by simple integration.

Note that if we take the spatial mean square of (I.24) then using orthogonality (for a uniform mass distribution) one may show that the off-diagonal terms do not contribute (see Powell [1]).

Finally note that if we desire stress rather than deflection, then it may be shown that analogous to (I.25) one obtains

$$\bar{\sigma}^2 = \frac{\pi}{4} \sum_m \frac{\sigma_m^2(x, y)}{M_m^2 \omega_m^3 \zeta_m} \iiiint \psi_m(x, y)\psi_m(x^*, y^*)$$

$$\cdot \Phi_p(\omega_m; x, y, x^*, y^*)\, dx\, dy\, dx^*\, dy^* \qquad (I.26)$$

where σ_m is stress due to $w = \psi_m$.

References for Appendix I

[1] Powell, A., Chapter 8 in book, *Random Vibration*, edited by S. H. Crandall, Technology Press, Cambridge, Mass., 1958.
[2] Laning, J. H. and Battin, R. H., *Random Processes in Automatic Control*, McGraw-Hill, New York, N.Y., 1956.

Appendix I

[3] Lin, Y. K., *Probabilistic Theory of Structural Dynamics*, McGraw-Hill, New York, N.Y., 1967.

[4] Dowell, E. H., *Aeroelasticity of Plates and Shells*, Noordhoff International Publishing, Leyden, The Netherlands, 1974.

[5] Vaicaitis, R., Dowell, E. H. and Ventres, C. S., 'Nonlinear Panel Response by a Monte Carlo Approach', *AIAA Journal*, Vol. 12, No. 5 (May 1974) pp. 685–691.

[6] Lazan, B. J., *Damping of Materials and Members in Structural Mechanics*, Pergamon Press, New York, N.Y., 1968.

418

Some example problems

Problems such as these have been used successfully as homework assignments. When used as a text, the instructor may wish to construct variations on these problems.

Chapter 2

Questions

Typical section with control surface

1. Compute $q_{REVERSAL}$ for finite K_δ and show it is the same as computed in the text for $K_\delta \to \infty$.

2. Compute $q_{DIVERGENCE}$ explicitly in terms of K_α, K_δ, etc.

Beam-rod model

3. Compute $q_{DIVERGENCE}$ using one and two mode models with uniform beam-rod eigenfunctions.

Assume

$$GJ = GJ_0[1 - y/l]$$

How do these results compare to those for

$$GJ = GJ_0 \sim \text{constant?}$$

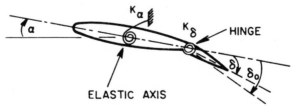

Answers 1. The two equations of static moment equilibrium are as follows:

Sum moments

$$eqS\left(\frac{\partial C_L}{\partial \alpha}\alpha + \frac{\partial C_L}{\partial \delta}\delta\right) + qSc\frac{\partial C_{MAC}}{\partial \delta}\delta - K_\alpha\alpha = 0 \quad \text{(about elastic axis)}$$

$$qSc\left(\frac{\partial C_H}{\partial \alpha}\alpha + \frac{\partial C_H}{\partial \delta}\delta\right) - K_\delta(\delta - \delta_0) = 0 \quad \text{(about hinge axis)}$$

These equations are given in matrix form as follows:

$$\begin{bmatrix} eqS\dfrac{\partial C_L}{\partial \alpha} - K_\alpha & eqS\dfrac{\partial C_L}{\partial \delta} + qSC\dfrac{\partial C_{MAC}}{\partial \delta} \\[2ex] qSC\dfrac{\partial C_H}{\partial \alpha} & qSc\dfrac{\partial C_H}{\partial \delta} - K_\alpha \end{bmatrix}\begin{bmatrix} \alpha \\ \delta \end{bmatrix} = \begin{bmatrix} 0 \\ -K_\delta \cdot \delta_0 \end{bmatrix}$$

Solving for α and δ, one obtains

$$\begin{cases} \alpha = \dfrac{\begin{vmatrix} 0 & eqS\dfrac{\partial C_L}{\partial \delta} + qSc\dfrac{\partial C_{MAC}}{\partial \delta} \\[2ex] -K_\delta \cdot \delta_0 & qSc\dfrac{\partial C_H}{\partial \delta} - K_\delta \end{vmatrix}}{\Delta} = \dfrac{K_\delta \cdot \delta_0 qS\left(e\dfrac{\partial C_L}{\partial \delta} + c\dfrac{\partial C_{MAC}}{\partial \delta}\right)}{\Delta} \quad (\text{II.1}) \\[8ex] \delta = \dfrac{\begin{vmatrix} eqS\cdot\dfrac{\partial C_L}{\partial \alpha} - K_\alpha & 0 \\[2ex] qSc\dfrac{\partial C_H}{\partial \alpha} & -K_\delta \cdot \delta_0 \end{vmatrix}}{\Delta} = \dfrac{-K_\delta \cdot \delta_0\left(eqS\dfrac{\partial C_L}{\partial \alpha} - K_\alpha\right)}{\Delta} \quad (\text{II.2}) \end{cases}$$

where

$$\Delta \equiv \begin{vmatrix} eqS\dfrac{\partial C_L}{\partial \alpha} - K_\alpha & eqS\dfrac{\partial C_L}{\partial \delta} + qSc\dfrac{\partial C_{MAC}}{\partial \delta} \\[2ex] qSc\dfrac{\partial C_H}{\partial \alpha} & qSc\dfrac{\partial C_H}{\partial \delta} - K_\delta \end{vmatrix} \quad (\text{II.3})$$

If control surface reversal occurs when $q = q_R$, then

$$L = qS\left[\frac{\partial C_L}{\partial \alpha}\alpha + \frac{\partial C_L}{\partial \delta}\delta\right] = 0$$

for $q = q_R$, i.e.,

$$\left[\frac{\partial C_L}{\partial \alpha}\alpha + \frac{\partial C_L}{\partial \delta}\delta\right] = 0$$

$$\text{at } q = q_R \quad (\text{II.4})$$

Substitution of (II.1) and (II.2) into (II.4) gives

$$0 = \frac{\partial C_L}{\partial \alpha} \cdot \frac{K_\delta \cdot \delta_0}{\Delta} q_R S \left(e \frac{\partial C_L}{\partial \delta} + c \frac{\partial C_{MAC}}{\partial \delta} \right) - \frac{\partial C_L}{\partial \delta} \cdot \frac{K_\delta \delta_0}{\Delta} \left(e q_R S \frac{\partial C_L}{\partial \alpha} - K_\alpha \right)$$

$$= \frac{K_\delta \cdot \delta_0}{\Delta} \left(e q_R S \frac{\partial C_L}{\partial \alpha} \frac{\partial C_L}{\partial \delta} + q_R S C \frac{\partial C_L}{\partial \alpha} \frac{\partial C_{MAC}}{\partial \delta} - e q_R S \frac{\partial C_L}{\partial \alpha} \frac{\partial C_L}{\partial \delta} + K_\alpha \frac{\partial C_L}{\partial \delta} \right)$$

$$= \frac{K_\sigma \cdot \delta_0}{\Delta} \left(q_R S c \frac{\partial C_L}{\partial \alpha} \frac{\partial C_{MAC}}{\partial \delta} + K_\alpha \frac{\partial C_L}{\partial \delta} \right)$$

Thus the reversal dynamic pressure q_R for finite K_δ is

$$q_R = \frac{\dfrac{K_\alpha}{Sc} \left(\dfrac{\partial C_L}{\partial \delta} \Big/ \dfrac{\partial C_L}{\partial \alpha} \right)}{-\dfrac{\partial C_{MAC}}{\partial \delta}}$$

which is identical with q_R when $K_\delta \to \infty$!

2. The divergence dynamic pressure is determined by $\Delta = 0$. That is,

$$\left(eqS \frac{\partial C_L}{\partial \alpha} - K_\alpha \right) \left(qSc \frac{\partial C_H}{\partial \delta} - K_\delta \right) - qSc \frac{\partial C_H}{\partial \alpha} \left(eqS \frac{\partial C_L}{\partial \delta} + qSc \frac{\partial C_{MAC}}{\partial \delta} \right) = 0$$

$$\therefore \quad q^2 S^2 c^2 \left(\bar{e} \cdot \frac{\partial C_L}{\partial \alpha} \frac{\partial C_H}{\partial \delta} - \bar{e} \frac{\partial C_H}{\partial \alpha} \frac{\partial C_L}{\partial \delta} - \frac{\partial C_{MAC}}{\partial \delta} \right)$$

quadratic

$$- qSc \left(K_\alpha \cdot \frac{\partial C_H}{\partial \delta} + K_\delta \cdot \bar{e} \frac{\partial C_L}{\partial \alpha} \right) + K_\alpha K_\delta = 0$$

$(\bar{e} \equiv e/c)$ (II.5)

If $A \neq 0$ (A is defined below),

$$q = \frac{1}{Sc} \cdot \frac{B \pm \sqrt{B^2 - 4AC}}{2A}$$

where

$$\begin{cases} A \equiv \bar{e} \left(\dfrac{\partial C_L}{\partial \alpha} \dfrac{\partial C_H}{\partial \delta} - \dfrac{\partial C_H}{\partial \alpha} \dfrac{\partial C_L}{\partial \delta} \right) - \dfrac{\partial C_{MAC}}{\partial \delta} \\[3mm] B \equiv K_\alpha \dfrac{\partial C_H}{\partial \delta} + K_\delta \bar{e} \dfrac{\partial C_L}{\partial \alpha} \\[3mm] C \equiv K_\alpha K_\alpha \end{cases}$$

Then divergence occurs when $AB > 0$ and $B^2 - 4AC \geq 0$ (for which e.g., (II.5) has two positive roots), and the divergence dynamic pressure q is

$$q_D = \min\left[\frac{1}{Sc} \cdot \frac{B + \sqrt{B^2 - 4AC}}{2A}, \frac{1}{Sc} \frac{B - \sqrt{B^2 - 4AC}}{2A} \right]$$

If $A = 0$, then divergence occurs when $B > 0$, and the divergence dynamic pressure q_D is

$$q_D = \frac{1}{Sc} \cdot \frac{C}{B}$$

To sum up, divergence occurs when
(a)

$$\begin{cases} \left\{ \bar{e}\left(\frac{\partial C_L}{\partial \alpha} \frac{\partial C_H}{\alpha \delta} - \frac{\partial C_H}{\partial \alpha} \frac{\partial C_L}{\partial \delta} \right) - \frac{\partial C_{MAC}}{\partial \delta} \right\} \left(K_\alpha \frac{\partial C_H}{\partial \delta} + K_\delta \bar{e} \frac{\partial C_L}{\partial \alpha} \right) > 0 \\ \text{and} \\ \left(K_\alpha \frac{\partial C_H}{\partial \delta} - K_\delta \bar{e} \frac{\partial C_L}{\partial \alpha} \right)^2 + 4 K_\alpha K_\delta \left(\bar{e} \frac{\partial C_H}{\partial \alpha} \frac{\partial C_L}{\partial \delta} + \frac{\partial C_{Mac}}{\partial \delta} \right) \geq 0 \end{cases}$$

and the divergence dynamic pressure q_D is

$$q_D = \frac{K_\alpha \dfrac{\partial C_H}{\partial \delta} + K_\delta \bar{e} \dfrac{\partial C_L}{\partial \alpha} - \sqrt{\left(K_\alpha \dfrac{\partial C_H}{\partial \delta} - K_\delta \bar{e} \dfrac{\partial C_L}{\partial d} \right)^2 + 4 K_\alpha K_d \left(\bar{e} \dfrac{\partial C_H}{\partial \alpha} \dfrac{\partial C_L}{\partial \delta} + \dfrac{\partial C_{MAC}}{\partial \delta} \right)}}{2Sc \left\{ \bar{e} \left(\dfrac{\partial C_L}{\partial \alpha} \dfrac{\partial C_H}{\partial \delta} - \dfrac{\partial C_H}{\partial \alpha} \dfrac{\partial C_L}{\partial \delta} \right) - \dfrac{\partial C_{MAC}}{\partial \delta} \right\}}$$

when

$$\bar{e}\left(\frac{\partial C_L}{\partial \alpha} \frac{\partial C_H}{\partial \delta} - \frac{\partial C_H}{\partial \alpha} \frac{\partial C_L}{\partial \sigma} \right) - \frac{\partial C_{MAC}}{\partial \delta} > 0$$

and

$$q_D = \frac{K_\alpha \dfrac{\partial C_H}{\partial \delta} + K_\delta \bar{e} \dfrac{\partial C_L}{\partial \alpha} + \sqrt{\left(K_\alpha \dfrac{\partial C_H}{\partial \delta} - K_\delta \bar{e} \dfrac{\partial C_L}{\partial \alpha} \right)^2 + 4 K_\alpha K_\delta \left(\bar{e} \dfrac{\partial C_H}{\partial \alpha} \dfrac{\partial C_L}{\partial \delta} + \dfrac{\partial C_{MAC}}{\partial \delta} \right)}}{2Sc \left\{ \bar{e} \left(\dfrac{\partial C_L}{\partial \alpha} \dfrac{\partial C_H}{\partial \delta} - \dfrac{\partial C_H}{\partial \alpha} \dfrac{\partial C_L}{\partial \delta} \right) - \dfrac{\partial C_{MAC}}{\partial \delta} \right\}}$$

when

$$\bar{e}\frac{\partial C_L}{\partial \alpha}\frac{\partial C_H}{\partial \delta} - \frac{\partial C_H}{\partial \alpha}\frac{\partial C_L}{\partial \delta} - \frac{\partial C_{MAC}}{\partial \delta} < 0$$

or,

(b)

$$\begin{cases} \bar{e}\left(\frac{\partial C_L}{\partial \alpha}\frac{\partial C_H}{\partial \delta} - \frac{\partial C_H}{\partial \alpha}\frac{\partial C_L}{\partial \delta}\right) - \frac{\partial C_{MAC}}{\partial \delta} = 0 \\ \text{and} \\ K_\alpha\frac{\partial C_H}{\partial \delta} + K_\delta\bar{e}\frac{\partial C_L}{\partial \alpha} > 0 \end{cases}$$

and the divergence dynamic pressure q_D is

$$q_D = \frac{K_\alpha K_\delta}{Sc\left(K_\alpha\dfrac{\partial C_H}{\partial \delta} + K_\delta\bar{e}\dfrac{\partial C_L}{\partial \alpha}\right)}$$

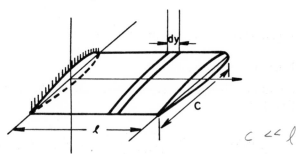

$$c \ll \ell$$

3. The equation of static torque equilibrium for a beam rod is

$$\frac{d}{dy}\left(GJ\frac{d\alpha_e}{dy}\right) + M_y = 0 \tag{II.6}$$

where

$$M_y = M_{AC} + L_e$$
$$= qc^2 C_{MAC_0} + eqc\frac{\partial C_L}{\partial \alpha}(\alpha_0 + \alpha_e) \tag{II.7}$$

If we put $\gamma = [1 - y/l]$ and $y = l\bar{y}$, then, from (II.6) and (II.7), we have

$$\frac{d}{d\bar{y}}\left(\gamma\frac{d\alpha_e}{d\bar{y}}\right) + \frac{qcel^2\dfrac{\partial C_L}{\partial \alpha}}{GJ_0}\alpha_e = -\frac{qcl^2}{GJ_0}\left(c \cdot C_{MAC_0} + e\frac{\partial C_L}{\partial \alpha}\alpha_0\right) \tag{II.8}$$

423

Appendix II

(1) Eigenvalues and functions for constant wing properties. Putting

$$\lambda^2 \equiv \frac{qcel^2 \dfrac{\partial C_L}{\partial \alpha}}{GJ_0}$$

we have the characteristic equation as follows

$$(\text{II.7}) \rightarrow \frac{d^2\alpha_e}{d\bar{y}^2} + \lambda^2\alpha_e = 0 \qquad (\gamma = 1 \text{ for constant wing properties})$$

Hence, $\alpha_e = A \sin \lambda\bar{y} + B \cos \lambda\bar{y}$.
As boundary conditions are

$$\begin{cases} \alpha_e = 0 \text{ at } \bar{y} = 0 \rightarrow B = 0 \\ \dfrac{d\alpha_e}{d\bar{y}} = 0 \text{ at } \bar{y} = 1 \rightarrow A\lambda \cos \lambda - B\lambda \sin \lambda \rightarrow \cos \lambda = 0 \end{cases}$$

(If $A\lambda = 0$ then $\alpha_e \equiv 0$, which is of no interest.)
So

$$\begin{cases} \text{Eigenvalues:} \quad \lambda_m = (2m-1)\dfrac{\pi}{2}, \ m = 1, 2, \dots \\ \text{Eigenfunctions:} \ \alpha_m = \sin \lambda_m\bar{y} \end{cases}$$

We first find the divergence dynamic pressure for the wing with constant properties. Let

$$\alpha_e = \sum_m a_n\alpha_n, \qquad K \equiv -\frac{qcl^2}{(GJ)_0}\left(cC_{MAC_0} + e\frac{\partial C_L}{\partial \alpha}\alpha_0\right) = \sum_n A_n\alpha_n$$

Then

$$\sum_n a_n\left(\frac{d^2\alpha_n}{d\bar{y}^2} + \lambda^2\alpha_n\right) = K$$

As

$$\frac{d^2\alpha_n}{d\bar{y}^2} = -\lambda_n^2\alpha_n$$

so

$$\sum_n a_n(\lambda^2 - \lambda_n^2)\alpha_n = K$$

$$\sum_n a_n\int_0^1 (\lambda^2 - \lambda_n^2)\alpha_n\alpha_m \, d\bar{y} = \int_0^1 K\alpha_m \, d\bar{y} = \tfrac{1}{2}A_m$$

since

$$\int_0^1 \alpha_n \alpha_m \, \mathrm{d}\bar{y} = \tfrac{1}{2}\delta_{mn}$$

$$= \tfrac{1}{2}(m = n)$$

$$= 0(m \neq n)$$

Hence

$$\frac{a_m}{2}(\lambda^2 - \lambda_m^2) = \tfrac{1}{2}A_m$$

$$a_m = \frac{A_m}{\lambda^2 - \lambda_m^2}$$

Thus

$$\alpha_e = \sum_n \frac{A_n}{\lambda^2 - \lambda_n^2} \cdot \alpha_n$$

$\alpha_e \to \infty$ when

$$\lambda = \lambda_m = (2m - 1)\frac{\pi}{2}$$

hence, the divergence dynamic pressure q_D, corresponds to the minimum value of λ_m, i.e., $\pi/2$. Thus

$$q_D = \frac{GJ_0}{cel^2 \dfrac{\partial C_L}{\partial \alpha}} \frac{\pi^2}{4}$$

for constant wing properties.

(2) $GJ = GJ_0(1 - y/l) = GJ_0(1 - \bar{y})$, variable wing properties. We assume for simplicity that only the torsional stiffness varies along span and that other characteristics remain the same.

Putting

$$\alpha_e = \sum_n b_n \cdot \alpha_n, \qquad K \equiv -\frac{qcl^2}{GJ_0}\left(cC_{MAC_0} + e\frac{\partial C_L}{\partial \alpha}\alpha_0\right) = \sum_n A_n \alpha_n$$

and

$$\lambda^2 \equiv \frac{qcel^2}{GJ_0}\frac{\partial C_L}{\partial \alpha}$$

425

we get from (II.8)

$$\sum_n b_n \left\{ \frac{d}{d\bar{y}} \left(\gamma \frac{d\alpha_n}{d\bar{y}} \right) + \lambda^2 \alpha_n \right\} = K$$

$$\therefore \quad \sum_n b_n \int_0^1 \left\{ \frac{d}{d\bar{y}} \left(\gamma \frac{d\alpha_n}{d\bar{y}} \right) + \lambda^2 \alpha_n \right\} \alpha_m \, d\bar{y} = \int_0^1 K\alpha_m \, dy = \tfrac{1}{2} A_m$$

$$\therefore \quad [C_{mn}]\{b_n\} = \tfrac{1}{2} A_m \text{ (for finite } n)$$ (II.9)

where

$$C_{mn} = \int_0^1 \left\{ \frac{d}{d\bar{y}} \left(\gamma \frac{d\alpha_n}{d\bar{y}} \right) + \lambda^2 \alpha_n \right\} \alpha_m \, d\bar{y}$$

$$= -\int_0^1 \gamma \frac{d\alpha_n}{d\bar{y}} \frac{d\alpha_m}{d\bar{y}} \, d\bar{y} + \frac{\lambda^2}{2} \delta_{mn}$$

$\left(\gamma \dfrac{d\alpha_n}{d\bar{y}} \alpha_m = 0 \text{ at } \bar{y} = 0 \text{ and } 1 \text{ because of the boundary conditions for eigenfunctions.} \right)$

(1) One mode model. The assumed mode is as follows:

$$\alpha_1 = \sin \lambda_1 \bar{y} = \sin \frac{\pi}{2} \bar{y} \rightarrow \frac{d\alpha_1}{d\bar{y}} = \frac{\pi}{2} \cos \frac{\pi}{2} \bar{y}$$

Equation (II.9) is

$$C_{11} b_1 = \frac{A_1}{2}$$ (II.10)

where

$$C_{11} = -\int_0^1 (1 - \bar{y}) \left(\frac{d\alpha_1}{d\bar{y}} \right)^2 d\bar{y} + \frac{\lambda^2}{2}$$

$$\therefore \quad C_{11} = \frac{\lambda^2}{2} - \frac{\pi^2 + 4}{16}$$

From (II.10),

$$b_1 = \frac{A_1}{\lambda^2 - \dfrac{\pi^2 + 4}{8}}$$

Then divergence occurs when

$$\lambda^2 = \frac{\pi^2 + 4}{8}$$

and

$$q_D = \frac{GJ_0}{cel^2 \dfrac{\partial C_L}{\partial \alpha}} \frac{\pi^2 + 4}{8} = (q_D)_{\substack{\text{const.} \\ \text{wing prop.}}} \times 0.703$$

(2) Two mode model. Assumed modes are

$$\begin{cases} \alpha_1 = \sin \lambda_1 \bar{y} = \sin \dfrac{\pi}{2} \bar{y} \Rightarrow \dfrac{d\alpha_1}{d\bar{y}} = \dfrac{\pi}{2} \cos \dfrac{\pi}{2} \bar{y} \\[3mm] \alpha_2 = \sin \lambda_2 \bar{y} = \sin \dfrac{3}{2} \pi \bar{y} \Rightarrow \dfrac{d\alpha_2}{d\bar{y}} = \dfrac{3}{2} \pi \cos \dfrac{3}{2} \pi \bar{y} \end{cases}$$

Equation (II.9) is as follows:

$$\begin{bmatrix} C_{11} & C_{12} \\ C_{21} & C_{22} \end{bmatrix} \begin{bmatrix} b_1 \\ b_2 \end{bmatrix} = \frac{1}{2} \begin{bmatrix} A_1 \\ A_2 \end{bmatrix} \tag{II.11}$$

where

$$C_{11} = -\int_0^1 (1 - \bar{y}) \left(\frac{d\alpha_1}{d\bar{y}} \right)^2 d\bar{y} + \frac{\lambda^2}{2} = \frac{\lambda^2}{2} - \frac{\pi^2 + 4}{16}$$

$$C_{12} = -\int_0^1 (1 - \bar{y}) \frac{d\alpha_2}{d\bar{y}} \frac{d\alpha_1}{d\bar{y}} d\bar{y} = -\frac{3}{4}$$

$$C_{21} = -\int_0^1 (1 - \bar{y}) \frac{d\alpha_1}{d\bar{y}} \frac{d\alpha_2}{d\bar{y}} d\bar{y} = C_{12} = -\frac{3}{4} \qquad \underline{\textit{Integration by parts}}$$

$$C_{22} = -\int_0^1 (1 - \bar{y}) \left(\frac{d\alpha_2}{d\bar{y}} \right)^2 d\bar{y} + \frac{\lambda^2}{2} = \frac{\lambda^2}{2} - \frac{9\pi^2 + 4}{16}$$

Then equation (II.11) is as follows:

$$\begin{bmatrix} \lambda^2 - \dfrac{\pi^2 + 4}{8} & -\dfrac{3}{2} \\[3mm] -\dfrac{3}{2} & \lambda^2 - \dfrac{9\pi^2 + 4}{8} \end{bmatrix} \begin{bmatrix} b_1 \\ b_2 \end{bmatrix} = \begin{bmatrix} A_1 \\ A_2 \end{bmatrix}$$

427

Appendix II

Thus divergence occurs when

$$\begin{vmatrix} \lambda^2 - \dfrac{\pi^2+4}{8} & -\dfrac{3}{2} \\[4mm] -\dfrac{3}{2} & \lambda^2 - \dfrac{9\pi^2+4}{8} \end{vmatrix} = 0$$

$$\therefore \quad \lambda^2 = \frac{5\pi^2+4}{8} \pm \frac{1}{2}\sqrt{\pi^4+9}$$

q_D is given by the smaller value of λ^2, i.e.,

$$q_D = \frac{GJ_0}{cel^2 \dfrac{\partial C_L}{\partial \alpha}} \times \left(\frac{5\pi^2+4}{8} - \frac{1}{2}\sqrt{\pi^4+9} \right)$$

$$= (q_D)_{\substack{\text{const.} \\ \text{wing prop.}}} \times 0.612$$

Question

Beam-rod model

4. For a constant GJ, etc. wing, use a two 'lumped element' model and compute the divergence dynamic pressure. Neglect rolling. Compare your result with the known analytical solution. How good is a one 'lumped element' solution?

Answer

4.

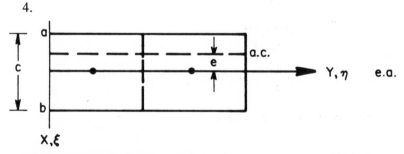

(a) Two lumped element model

$$\alpha(y) = \int_0^l C^{\alpha M}(y, \eta) M(\eta)\, d\eta \tag{II.12}$$

where

$C^{\alpha M}(y, \eta)$: twist about y axis at y due to unit moment at η

$$M(\eta) = \int_a^b p(\xi, \eta)\xi \, d\xi$$

Equation (II.12) in matrix form is

$$\{\alpha\} = [C^{\alpha M}]\{M\} \Delta \eta \qquad \text{(II.12)}$$

where from structural analysis,

$$[C^{\alpha M}] = \begin{bmatrix} \dfrac{l/4}{GJ} & \dfrac{l/4}{GJ} \\ \dfrac{l/4}{GJ} & \dfrac{3l/4}{GJ} \end{bmatrix} \qquad \text{(II.13)}$$

and $C^{\alpha M}(i, j)$ is the twist at i due to unit moment at j. Using an aerodynamic 'strip theory' approximation, the aerodynamic moment may be related to the twist,

$$\{M\} = qce\frac{\partial C_L}{\partial \alpha}\begin{bmatrix} 1 & 0 \\ 0 & 1 \end{bmatrix}\{\alpha\} = qce\frac{\partial C_L}{\partial \alpha}\{\alpha\} \qquad \text{(II.14)}$$

From (II.12) and (II.14), one has

$$\{\alpha\} = [C^{\alpha M}]\{M\}\Delta\eta = qCe\frac{\partial C_L}{\partial \alpha}[C^{\alpha M}] \, \Delta\eta\{\alpha\} \quad \leftarrow \; (2,4,5b) \;\; p.39$$

or rewritten, using $\Delta\eta = l/2$,

$$\left[\begin{bmatrix} 1 & 0 \\ 0 & 1 \end{bmatrix} - \frac{l}{2}qce\frac{\partial C_L}{\partial \alpha}\frac{l}{4GJ}\begin{bmatrix} 1 & 1 \\ 1 & 3 \end{bmatrix}\right]\{\alpha\} = \begin{Bmatrix} 0 \\ 0 \end{Bmatrix} \qquad \text{(II.15)}$$

w

analogous to $(2,4,11)$

Setting the determinant of coefficients to zero, gives

$$|\quad| = 0 \rightarrow 2Q^2 - 4Q + 1 = 0 \qquad \text{(II.16)}$$

where

$$Q \equiv \frac{l^2}{8GJ}qce\frac{\partial C_L}{\partial \alpha}$$

Solving (II.16), one obtains

$$Q = \frac{2 \pm \sqrt{2}}{2}$$

The smaller Q gives the divergence q_D.

$$q_D = 4(2 - \sqrt{2}) \frac{GJ/l}{(lc)e \dfrac{\partial C_L}{\partial \alpha}}$$

$$\doteq 2.34 \frac{GJ/l}{(lc)e \dfrac{\partial C_L}{\partial \alpha}}$$

(b) One lumped element model

$$\alpha = qce \frac{\partial C_L}{\partial \alpha} C^{\alpha M} \Delta \eta \alpha$$

where

$$\Delta \eta = l, \qquad C^{\alpha M} = \frac{l/2}{GJ}$$

$$\left(1 - qCe \frac{\partial C_L}{\partial \alpha} \frac{l/2}{GJ}\right)\alpha = 0$$

$$\therefore \quad q_D = 2 \frac{GJ/l}{(lc)e \dfrac{\partial C_L}{\partial \alpha}}$$

Recall that the analytical solution is (cf. Section 2.2)

$$q_D = \left(\frac{\pi}{2}\right)^2 \frac{GJ}{l} /(lc)e \frac{\partial C_L}{\partial \alpha} \qquad see\ top\ p.18$$

$$= 2.46 \cdots$$

A comparison of the several approximations is given below. In the two element model the error is about 5%.

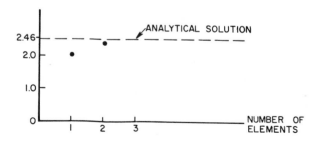

Question

5. Consider a thin cantilevered plate of length l and width b which represents the leading edge of a wing at supersonic speeds. See sketch.

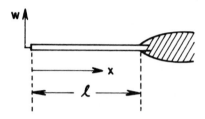

The aerodynamic pressure loading (per unit chord and per unit span) at high speeds is given by (Sections 3.4 and 4.2)

$$p = \frac{2\rho U^2}{(M^2-1)^{\frac{1}{2}}} \frac{\partial w}{\partial x} \qquad \begin{array}{c} \textit{Sign convention} \\ p \quad \text{down} \\ w \quad \text{up} \end{array}$$

where M is mach number and w is transverse deflection (not downwash!). Compute the divergence speed.

(1) Work out a formal mathematical solution, without numerical evaluation, using classical differential equation methods.

(2) How would you use Galerkin's method with an assumed mode of the form

$$w = a\{2(1-x/l)^2 - \tfrac{4}{3}(1-x/l)^3 + \tfrac{1}{3}(1-x/l)^4\}$$

to obtain a numerical answer? What boundary conditions on w does the assumed mode satisfy?

Answer

(1) Governing equilibrium equation is:

$$EI\frac{\partial^4 w}{\partial x^4} = -p = -\frac{2\rho U^2}{(M^2-1)^{\frac{1}{2}}} \frac{\partial w}{\partial x}$$

Define

$$K \equiv \left[\frac{2\rho U^2}{(M^2-1)^{\frac{1}{2}}}\right]\left[\frac{1}{EI}\right]$$

the equation above becomes

$$\frac{\partial^4 w}{\partial x^4} + K\frac{\partial w}{\partial x} = 0 \qquad\qquad\text{(II.17)}$$

431

Appendix II

The boundary conditions are:

$$w(l) = \frac{\partial w}{\partial x}(l) = \frac{\partial^2 w}{\partial x^2}(0) = \frac{\partial^3 w}{\partial x^3}(0) = 0 \tag{II.18}$$

The characteristic equation of differential equation (II.17) is

$$\gamma^4 + K\gamma = 0 \tag{II.19}$$

The roots are $\gamma_1 = 0$ and γ_2, γ_3, γ_4 such that $\gamma^3 = -K$. Now

$$(-K)^{\frac{1}{3}} = K^{\frac{1}{3}} e^{i\frac{1}{3}(\pi + 2n\pi)}, \qquad n = 0, 1, 2$$

and defining $K^1 = K^{\frac{1}{3}}$ the roots γ_2, γ_3, γ_4 become

$$\gamma_2 = K^1 e^{i\pi/3} = K^1[\cos \pi/3 + i \sin \pi/3] = K^1 \left[\frac{1}{2} + i \frac{\sqrt{3}}{2} \right]$$

$$\gamma_3 = K^1 e^{i\pi} = K^1[\cos \pi + i \sin \pi] = K^1[-1]$$

$$\gamma_4 = K^1 e^{i5\pi/3} = K^1 \left[\cos \frac{5\pi}{3} + i \sin \frac{5\pi}{3} \right] = K^1 \left[\frac{1}{2} - i \frac{\sqrt{3}}{2} \right]$$

Therefore $w(x)$ has the form:

$$w(x) = b_1 + b_2 e^{-K^1 x} + e^{K^1(x/2)} \left[b_3 \cos \left(K^1 \frac{\sqrt{3}}{2} x \right) + b_4 \sin \left(K^1 \frac{\sqrt{3}}{2} x \right) \right]$$

$$w'(x) = -b_2 K^1 e^{-K^1 x} + \frac{K^1}{2} e^{K^1(x/2)} \left[(b_3 + b_4\sqrt{3}) \cos \left(K^1 \frac{\sqrt{3}}{2} x \right) \right.$$
$$\left. + (b_4 - b_3\sqrt{3}) \sin \left(K^1 \frac{\sqrt{3}}{2} x \right) \right]$$

$$w''(x) = b_2 K^{1^2} e^{-K^1 x} + \left(\frac{K^1}{2} \right)^2 e^{K^1(x/2)} \left[(2\sqrt{3}b_4 - 2b_3) \cos \left(K^1 \frac{\sqrt{3}}{2} x \right) \right.$$
$$\left. + (2\sqrt{3}b_3 + 2b_4) \sin \left(K^1 \frac{\sqrt{3}}{2} x \right) \right]$$

$$w'''(x) = -b_2 K^{1^3} e^{-K^1 x} - K^{1^3} e^{K^1(x/2)} \left[b_3 \cos \left(K^1 \frac{\sqrt{3}}{2} x \right) + b_4 \sin \left(K^1 \frac{\sqrt{3}}{2} x \right) \right]$$

$$\tag{II.20}$$

Using boundary conditions (II.18), we obtain from (II.20),

$$w(l) = 0 = b_1 + b_2 e^{-K^1 l} + e^{K^1 \frac{l}{2}} \left[(b_3 \cos\left(K^1 \frac{\sqrt{3}}{2} l\right) + b_4 \sin\left(K^1 \frac{\sqrt{3}}{2} l\right)\right]$$

$$\frac{\partial w}{\partial x}(l) = 0 = -b_2 K^1 e^{-K^1 l} + \frac{K^1}{2} e^{K^1 \frac{l}{2}} \left[(b_3 + b_4\sqrt{3}) \cos\left(K^1 \frac{\sqrt{3}}{2} l\right)\right.$$

$$\left. + (b_4 - b_3\sqrt{3}) \sin\left(K^1 \frac{\sqrt{3}}{2} l\right)\right]$$

$$\frac{\partial^2 w}{\partial x^2}(0) = 0 = b_2 K^{1^2} + \left(\frac{K^1}{2}\right)^2 (2\sqrt{3}b_4 - 2b_3)$$

$$\frac{\partial^3 w}{\partial x^3}(0) = 0 = -b_2 K^{1^3} - b_3 K^{1^3} \tag{II.21}$$

The condition for nontrivial solutions is that the determinant of coefficients of the system of linear, algebraic equations given by (II.21) be zero. This leads to

$$e^{-\frac{3}{2}K''} = -2 \cos\left(\frac{\sqrt{3}}{2} K''\right) \tag{II.22}$$

where

$$K'' \equiv K^1 l$$

In order to find the solution to equation (II.22), one would plot on the same graph as a function of K'' the right and left sides of this equation and note the points (if any) of intersection. The first intersection for $K'' > 0$ is the one of physical interest. Knowing this particular K'', call it K_D'', one may compute

$$U_D^2 = \frac{K_D''^3 (M^2 - 1)^{\frac{1}{2}} EI}{2\rho l^3}$$

to find the speed U at which divergence occurs.

(2) This is left as an excercise for the reader.

Section 3.1

Question

Starting from

$$U = \frac{1}{2} \iiint [\sigma_{xx}\varepsilon_{xx} + \sigma_{xy}\varepsilon_{xy} + \sigma_{yx}\varepsilon_{yx} + \sigma_{yy}\varepsilon_{yy}]\, dx\, dy\, dz$$

planar or 2-D
form of
(3.1.13) p 51

433

and

$$\varepsilon_{xx} = -z\frac{\partial^2 w}{\partial x^2}$$

$$\varepsilon_{yy} = -z\frac{\partial^2 w}{\partial y^2}$$

$$\varepsilon_{xy} = -z\frac{\partial^2 w}{\partial x\,\partial y}$$

$$\sigma_{xx} = \frac{E}{(1-\nu^2)}[\varepsilon_{xx} + \nu\varepsilon_{xy}]$$

$$\sigma_{yy} = \frac{E}{(1-\nu^2)}[\varepsilon_{yy} + \nu\varepsilon_{xx}]$$

$$\sigma_{xy} = \frac{E}{(1+\nu)}\varepsilon_{xy} = \sigma_{yx}$$

$w = w(x, y)$ only

1. Compute $U = U(w)$
2. For $w = h(y) + \alpha(y)x$

Compute $U = U(h, \alpha)$
3. Using a kinetic energy expression

$$T = \frac{1}{2}\iiint \rho\left(\frac{\partial w}{\partial t}\right)^2 \mathrm{d}x\,\mathrm{d}y\,\mathrm{d}z \qquad \text{form of } (3.1.10)\ p.50$$

Compute $T = T(h, \alpha)$
4. Assume $h(y) = af(y)$

$$\alpha(y) = bg(y)$$

where f, g are specified.
Determine equations of motion for a, b using Lagrange's Equations, where the virtual work done by aerodynamic pressure, p, is given by

$$\delta W = \iint p\delta w\,\mathrm{d}x\,\mathrm{d}y$$

5. Return to 1; now assume

$$w = \sum_m q_m\psi_m(x, y)$$

Handwritten annotations:

$*$ If $\vec{r} = u\vec{i} + v\vec{j} + w\vec{k}$

$\frac{d\vec{r}}{dt}\cdot\frac{d\vec{r}}{dt} = \left(\frac{du}{dt}\right)^2 + \left(\frac{dv}{dt}\right)^2 + \left(\frac{dw}{dt}\right)^2$

Eq. 3, p.377 of Kreyzig 4th Ed.

where

ψ_m is specified.
Determine equations of motion for q_m.

Answer

1. Potential energy U:

$$U = \frac{1}{2} \iiint (\sigma_{xx}\varepsilon_{xx} + \sigma_{xy}\varepsilon_{xy} + \sigma_{yx}\varepsilon_{yx} + \sigma_{yy}\varepsilon_{yy}) \, dx \, dy \, dz$$

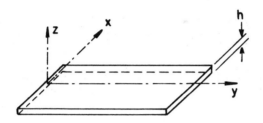

where

$$
\begin{cases}
\varepsilon_{xx} = -z\dfrac{\partial^2 w}{\partial x^2} \\[2mm]
\varepsilon_{xy} = -z\dfrac{\partial^2 w}{\partial x \, \partial y} = \varepsilon_{yx} \\[2mm]
\varepsilon_{yy} = -z\dfrac{\partial^2 w}{\partial y^2}
\end{cases}
$$

$$
\begin{cases}
\sigma_{xx} = \dfrac{E}{1-\nu^2}(\varepsilon_{xx} + \nu\varepsilon_{yy}) \\[2mm]
\sigma_{xy} = \dfrac{E}{1+\nu}\varepsilon_{xy} \\[2mm]
\sigma_{yy} = \dfrac{E}{1-\nu^2}(\varepsilon_{yy} + \nu\varepsilon_{xx})
\end{cases}
$$

Thus

$$\sigma_{xx}\varepsilon_{xx} + \sigma_{xy}\varepsilon_{xy} + \sigma_{yx}\varepsilon_{yx} + \sigma_{yy}\varepsilon_{yy}$$

$$= \frac{E}{1-\nu^2}(\varepsilon_{xx} + \nu\varepsilon_{yy})\cdot\varepsilon_{xx} + 2\frac{E}{1+\nu}\varepsilon_{xy}^2 + \frac{E}{1-\nu^2}(\varepsilon_{yy} + \nu\varepsilon_{xx})\varepsilon_{yy}$$

$$= \frac{E}{1-\nu^2}(\varepsilon_{xx}^2 + 2\nu\varepsilon_{xx}\varepsilon_{yy} + \varepsilon_{yy}^2) + \frac{2E(1-\nu)}{1-\nu^2}\varepsilon_{xy}^2$$

$$= \frac{E}{1-\nu^2}\left\{ z^2\left(\frac{\partial^2 w}{\partial z^2}\right)^2 + 2\nu\cdot z^2\left(\frac{\partial^2 w}{\partial x^2}\right)\left(\frac{\partial^2 w}{\partial y^2}\right) + z^2\left(\frac{\partial^2 w}{\partial y^2}\right)^2 \right.$$

$$\left. + 2(1-\nu)z^2\left(\frac{\partial^2 w}{\partial x\,\partial y}\right) \right\}$$

$$= \frac{Ez^2}{1-\nu^2}\left\{ \left(\frac{\partial^2 w}{\partial x^2}\right) + \left(\frac{\partial^2 w}{\partial y^2}\right)^2 + 2\nu\left(\frac{\partial^2 w}{\partial x^2}\right)\left(\frac{\partial^2 w}{\partial y^2}\right) + 2(1-\nu)\left(\frac{\partial^2 w}{\partial x\,\partial y}\right)^2 \right\}$$

$$\therefore\quad U = \frac{1}{2}\iiint \frac{Ez^2}{1-\nu^2}\left[\left(\frac{\partial^2 w}{\partial x^2}\right)^2 + \left(\frac{\partial^2 w}{\partial y^2}\right)^2 + 2\nu\left(\frac{\partial^2 w}{\partial x^2}\right)\left(\frac{\partial^2 w}{\partial y^2}\right) + 2(1-\nu) \right.$$

$$\left. \times\left(\frac{\partial^2 w}{\partial x\,\partial y}\right)^2 \right]\mathrm{d}x\,\mathrm{d}y\,\mathrm{d}z$$

$$= \frac{1}{2}\iint D\left[\left(\frac{\partial^2 w}{\partial x^2}\right)^2 + \left(\frac{\partial^2 w}{\partial y^2}\right)^2 + 2\nu\left(\frac{\partial^2 w}{\partial x^2}\right)\left(\frac{\partial^2 w}{\partial y^2}\right) + 2(1-\nu)\left(\frac{\partial^2 w}{\partial x\,\partial y}\right)^2 \right]\mathrm{d}x\,\mathrm{d}y$$

$$(\text{II}.23)$$

where

$$D \equiv \frac{E}{1-\nu^2}\int z^2\,\mathrm{d}z$$

2. For $w = h(y) + \alpha(y)\cdot x$

$$\begin{cases} \dfrac{\partial^2 w}{\partial y^2} = \dfrac{\partial^2 h}{\partial y^2} + \dfrac{\partial^2 \alpha}{\partial y^2}\cdot x \\[2mm] \dfrac{\partial^2 w}{\partial x\,\partial y} = \dfrac{\partial\alpha}{\partial y} \\[2mm] \dfrac{\partial^2 w}{\partial x^2} = 0 \end{cases}$$

Hence, from (II.23), we have

$$U = \frac{1}{2} \iint D\left[\left(\frac{\partial^2 h}{\partial y^2}\right) + \left(\frac{\partial^2 \alpha}{\partial y^2} x\right)^2 + 0^2 + 2\nu \cdot \left(\frac{\partial^2 h}{\partial y^2} + \frac{\partial^2 \alpha}{\partial y^2} \cdot y\right) \cdot 0 \right.$$

$$\left. + 2(1-\nu)\left(\frac{\partial \alpha}{\partial y}\right)^2 \right] dx\, dy$$

$$= \frac{1}{2} \iint D\left[\left(\frac{\partial^2 h}{\partial y^2}\right)^2 + 2\left(\frac{\partial^2 h}{\partial y^2}\right)\left(\frac{\partial^2 \alpha}{\partial y^2}\right) \cdot x + \left(\frac{\partial^2 \alpha}{\partial y^2}\right)^2 x^2 + 2(1-\nu)\left(\frac{\partial \alpha}{\partial y}\right)^2 \right] dx\, dy$$

$$\tag{II.24}$$

3. $w(z, y, t) = h(y, t) + \alpha(y, t) \cdot x$

$$= a(t) \cdot f(y) + b(t)g(y)x$$

$$\therefore \quad \frac{\partial w}{\partial t} = \dot{a}f(y) + \dot{b}g(y)x \qquad \cdot \equiv \frac{d}{dt}$$

$$T = \frac{1}{2} \iiint \rho\left(\frac{\partial w}{\partial t}\right)^2 dx\, dy\, dz$$

$$= \frac{1}{2} \iiint \rho(\dot{a}f(y) + \dot{b}g(y)x)^2 \, dx\, dy\, dz$$

$$= \frac{1}{2} \iiint \rho(\dot{a}^2\{f(y)\}^2 + 2\dot{a}\dot{b}f(y)g(y)x + \dot{b}^2\{g(y)\}^2 x^2) \, dx\, dy\, dz$$

$$= \frac{1}{2}\left[\dot{a}^2 \iiint \rho\{f(y)\}^2 dx\, dy\, dz + 2\dot{a}\dot{b} \iiint \rho f(y)g(y)x\, dx\, dy\, dz \right.$$

$$\left. + \dot{b}^2 \iiint \{g(y)\}^2 x^2 dx\, dy\, dz \right]$$

$$= \tfrac{1}{2}(\dot{a}^2 A_{ff} + 2\dot{a}\dot{b}A_{fg} + \dot{b}^2 A_{gg}) \tag{II.25}$$

where

$$A_{ff} \equiv \iiint \rho\{f(y)\}^2 \, dx\, dy\, dz$$

$$A_{fg} \equiv \iiint \rho f(y)g(y)x \, dx\, dy\, dz \tag{II.26}$$

$$A_{gg} \equiv \iiint \rho\{g(y)\}^2 x^2 dx\, dy\, dz$$

Appendix II

4. For $w = a(t)f(y) + b(t)g(y)x$ the potential energy is given as follows:

$$' \equiv \frac{\mathrm{d}}{\mathrm{d}y}$$

$$\left(\frac{\partial^2 h}{\partial y^2}\right) = af''(y), \frac{\partial \alpha}{\partial y} = bg'(y), \frac{\partial^2 \alpha}{\partial y^2} = bg''(y) \text{ into (II.24)}$$

$$U = \frac{1}{2}\iint D[\{af''(y)\}^2 + 2af''(y)bg''(y)x + \{bg''(y)\}^2 x^2$$

$$+ 2(1-\nu)\{bg'(y)\}^2]\,\mathrm{d}x\,\mathrm{d}y$$

$$= \frac{1}{2}\iint D\{a^2\{f''(y)\}^2 + 2abf''(y)g''(y)x + b^2[\{g''(y)\}^2 y^2$$

$$+ 2(1-\nu)\{g'(y)\}^2]\,\mathrm{d}x\,\mathrm{d}y$$

$$= \frac{1}{2}\left[a^2\iint D\{f''(y)\}^2\mathrm{d}x\,\mathrm{d}y + 2ab\iint Df''(y)g''(y)x\,\mathrm{d}x\,\mathrm{d}y\right.$$

$$\left. + b^2\iint D\{g''(y)\}^2 x^2 + 2(1-\nu)\{g'(y)\}^2\right]\mathrm{d}x\,\mathrm{d}y$$

$$= \tfrac{1}{2}[a^2 B_{ff} + 2abB_{fg} + b^2 B_{gg}] \qquad\qquad\qquad\text{(II.27)}$$

where

$$\begin{cases} B_{ff} \equiv \iint D\{f''(y)\}^2\mathrm{d}x\,\mathrm{d}y \\[2em] B_{fg} \equiv \iint Df''(y)g''(y)x\,\mathrm{d}x\,\mathrm{d}y \qquad\qquad \text{(II.28)} \\[2em] B_{gg} \equiv \iint D[\{g''(y)\}^2 x^2 + 2(1-\nu)\{g'(y)\}^2]\,\mathrm{d}x\,\mathrm{d}y \end{cases}$$

Virtual work

$$\delta W = \iint p\delta w\,\mathrm{d}x\,\mathrm{d}y$$

where

$$\delta w = \delta h + \delta \alpha x = f(y)\, \delta a + g(y)x\, \delta b$$

$$\therefore \quad \delta W = \iint p(f(y)\, \delta a + g(y)x\, \delta b)\, \mathrm{d}x\, \mathrm{d}y$$

$$= \delta a \iint pf(y)\, \mathrm{d}x\, \mathrm{d}y + \delta b \iint pg(y)x\, \mathrm{d}x\, \mathrm{d}y$$

$$= Q_a\, \delta a + Q_b \cdot \delta b$$

where

$$\begin{cases} Q_a \equiv \iint pf(y)\, \mathrm{d}x\, \mathrm{d}y \\[2mm] Q_b \equiv \iint pg(y) \cdot x\, \mathrm{d}x\, \mathrm{d}y \end{cases} \tag{II.29}$$

Lagrangian, $L \equiv T - U$, may be written

$$= \tfrac{1}{2}(\dot{a}^2 A_{ff} + 2\dot{a}\dot{b}A_{fg} + \dot{b}^2 A_{gg})$$

$$- \tfrac{1}{2}(a^2 B_{ff} + 2abB_{fg} + b^2 B_{gg})$$

$$\therefore \quad \begin{cases} \dfrac{\partial L}{\partial \dot{a}} = \dot{a}A_{ff} + \dot{b}A_{fg}, \dfrac{\partial L}{\partial a} = -a \cdot B_{ff} - bB_{fg} \\[3mm] \dfrac{\partial L}{\partial \dot{b}} = \dot{a}A_{fg} + \dot{b}A_{gg}, \dfrac{\partial L}{\partial b} = -aB_{fg} - bB_{gg} \end{cases}$$

Then Lagrange's equations of motion are

$$\begin{cases} \dfrac{\mathrm{d}}{\mathrm{d}t}\left(\dfrac{\partial L}{\partial \dot{a}}\right) - \dfrac{\partial L}{\partial a} = Q_a \rightarrow A_{ff}\ddot{a} + A_{fg}\ddot{b} + B_{ff}a + B_{fg} \cdot b = Q_a \\[3mm] \dfrac{\mathrm{d}}{\mathrm{d}t}\left(\dfrac{\partial L}{\partial \dot{b}}\right) - \dfrac{\partial L}{\partial b} = Q_b \rightarrow A_{fg}\ddot{a} + A_{gg}\ddot{b} + B_{fg}a + B_{gg}b = Q_b \end{cases} \tag{II.30}$$

where A_{ff}, A_{fg}, A_{gg}, B_{ff}, B_{fg}, B_{gg}, Q_a and Q_b are given in (II.26), (II.28) and (II.29).

5. When

$$w(x, y, t) = \sum_m q_m(t)\Psi_m(x, y)$$

$$\begin{cases} \dfrac{\partial^2 w}{\partial x^2} = \sum_m q_m \dfrac{\partial^2 \psi_m}{\partial x^2} \\[2mm] \dfrac{\partial^2 w}{\partial y^2} = \sum_m q_m \dfrac{\partial^2 \psi_m}{\partial y^2} \\[2mm] \dfrac{\partial^2 w}{\partial x\, \partial y} = \sum_m q_m \dfrac{\partial^2 \psi_m}{\partial x\, \partial y} \end{cases}$$

$$\therefore \begin{cases} \left(\dfrac{\partial^2 w^2}{\partial x^2}\right) = \sum_m \sum_n q_m q_n \dfrac{\partial^2 \psi_m}{\partial x^2} \dfrac{\partial^2 \psi_n}{\partial x^2} \\[2mm] \left(\dfrac{\partial^2 w^2}{\partial y^2}\right) = \sum_m \sum_n q_m q_n \dfrac{\partial^2 \psi_m}{\partial y^2} \dfrac{\partial^2 \psi_n}{\partial y^2} \\[2mm] \left(\dfrac{\partial^2 w^2}{\partial x^2}\right)\left(\dfrac{\partial^2 w}{\partial y^2}\right) = \sum_m \sum_n q_m q_n \dfrac{\partial^2 \psi_m}{\partial x^2} \dfrac{\partial^2 \psi_n}{\partial y^2} \\[2mm] \left(\dfrac{\partial^2 w^2}{\partial x\, \partial y}\right) = \sum_m \sum_n q_m q_n \dfrac{\partial^2 \psi_m}{\partial x\, \partial y} \dfrac{\partial^2 \psi_n}{\partial x\, \partial y} \end{cases}$$

Then from (II.23) the potential energy is

$$U = \frac{1}{2} \iint D\left[\sum_m \sum_n q_m q_n \frac{\partial^2 \psi_m}{\partial x^2} \frac{\partial^2 \psi_n}{\partial x^2} + \sum_m \sum_n q_m q_n \frac{\partial^2 \psi_m}{\partial y^2} \frac{\partial^2 \psi_n}{\partial y^2} \right.$$

$$\left. + 2\nu \sum_m \sum_n q_m q_n \frac{\partial^2 \psi_m}{\partial x^2} \frac{\partial^2 \psi_n}{\partial y^2} + 2(1-\nu) \sum_m \sum_n q_m q_n \frac{\partial^2 \psi_m}{\partial x\, \partial y} \frac{\partial^2 \psi_n}{\partial x\, \partial y} \right] dx\, dy$$

$$= \frac{1}{2} \sum_m \sum_n q_m q_n \iint D\left[\frac{\partial^2 \psi_m}{\partial x^2} \frac{\partial^2 \psi_n}{\partial x^2} + \frac{\partial^2 \psi_m}{\partial y^2} \frac{\partial^2 \psi_n}{\partial y^2} + 2\nu \frac{\partial^2 \psi_m}{\partial x^2} \frac{\partial^2 \psi_n}{\partial y^2} \right.$$

$$\left. + 2(1-\nu) \frac{\partial^2 \psi_m}{\partial x\, \partial y} \frac{\partial^2 \psi_n}{\partial x\, \partial y} \right] dx\, dy$$

$$= \frac{1}{2} \sum_m \sum_n q_m q_n \cdot B_{mn} \tag{II.31}$$

where

$$B_{mn} \equiv \iint D \left[\frac{\partial^2 \psi_m}{\partial x^2} \frac{\partial^2 \psi_n}{\partial x^2} + \frac{\partial^2 \psi_m}{\partial y^2} \frac{\partial^2 \psi_n}{\partial y^2} + 2\nu \frac{\partial^2 \psi_m}{\partial x^2} \frac{\partial^2 \psi_n}{\partial y^2} \right.$$

$$\left. + 2(1-\nu) \frac{\partial^2 \psi_m}{\partial x \, \partial y} \frac{\partial^2 \psi_n}{\partial x \, \partial y} \right] \mathrm{d}x \, \mathrm{d}y$$

Note $B_{mn} \neq B_{nm}$!

Kinetic energy

$$\frac{\partial w}{\partial t} = \sum_m \dot{q}_m \psi_m (x, y)$$

$$\therefore \left(\frac{\partial w}{\partial t} \right)^2 = \sum_m \sum_n \dot{q}_m \dot{q}_n \psi_m \psi_n$$

$$T = \frac{1}{2} \iiint \rho \left(\sum_m \sum_n \dot{q}_m \dot{q}_n \psi_m \psi_n \right) \mathrm{d}x \, \mathrm{d}y \, \mathrm{d}z$$

$$= \frac{1}{2} \sum_m \sum_n \dot{q}_m \dot{q}_n \iiint \rho \psi_m \psi_n \, \mathrm{d}x \, \mathrm{d}y \, \mathrm{d}z$$

$$= \frac{1}{2} \sum_m \sum_n \dot{q}_m \dot{q}_n A_{mn} \qquad\qquad (\text{II.32})$$

where

$$A_{mn} \equiv \iiint \rho \psi_m \psi_n \, \mathrm{d}x \, \mathrm{d}y \, \mathrm{d}z$$

Virtual work

$$\delta W = \iint pw \, \mathrm{d}x \, \mathrm{d}y$$

$$= \iint p \left(\sum_m \delta q_m \Psi_m \right) \mathrm{d}x \, \mathrm{d}y$$

$$\delta W = \sum_m \delta q_m \iint p \Psi_m \, \mathrm{d}x \, \mathrm{d}y$$

$$= \sum_m Q_m \delta q_m \qquad\qquad (\text{II.33})$$

Appendix II

where

$$Q_m \equiv \iint p\Psi_m \, \mathrm{d}x \, \mathrm{d}y$$

Lagrangian:

$$L = T - U = \frac{1}{2} \sum_m \sum_n \dot{q}_m \dot{q}_n A_{mn} - \frac{1}{2} \sum_m \sum_n q_m q_n B_{mn}$$

$$\frac{\partial L}{\partial \dot{q}_j} = \frac{1}{2} \sum_n \dot{q}_n A_{jn} + \frac{1}{2} \sum_m \dot{q}_m A_{mj} = \frac{1}{2} \sum_m \dot{q}_m (A_{jm} + A_{mj})$$

$$= \sum_m \dot{q}_m A_{mj} \qquad (A_{mj} = A_{jm})$$

$$\frac{\partial L}{\partial q_j} = -\frac{1}{2} \left(\sum_n q_n B_{jn} + \sum_m q_m B_{mj} \right) = -\frac{1}{2} \sum_m (B_{mj} + B_{jm})$$

Lagrange's equations of motion

$$\frac{\mathrm{d}}{\mathrm{d}t}\left(\frac{\partial L}{\partial \dot{q}_j}\right) - \frac{\partial L}{\partial q_j} = \sum_m \ddot{q}_m A_{mj} + \frac{1}{2} \sum_m q_m (B_{mj} + B_{jm}) = Q_j \qquad (j = 1, 2, \ldots)$$

(II.34)

Note: $B_{mj} + B_{jm} = B_{jm} + B_{mj}$, i.e., coefficient symmetry is preserved in final equations.

Section 3.3

Question. Use the vertical translation of and angular rotation about the center of mass of the typical section as generalized coordinates.
a. Derive equations of motion.
b. Determine the flutter dynamic pressure and show that it is the same as discussed in text. Use steady or quasi-static aerodynamic theory.

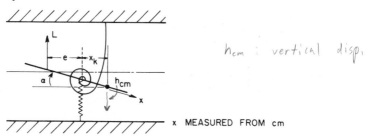

442

Answer

a)

$$T = \frac{m}{2} \dot{h}_{cm}^2 + \frac{I_{cm}}{2} \dot{\alpha}^2$$

$$U = \frac{1}{2} K_h (h_{cm} - \alpha x_k)^2 + \frac{1}{2} K_\alpha \alpha^2$$

$$\delta W = \int p \delta w \, dx, \quad w = -h_{cm} - x\alpha, \qquad \text{not needed}$$

vertical displacement of a point on airfoil

$$= \int p(-\delta h_{cm} - x\delta\alpha) \, dx$$

$$= \delta h_{cm}\left(-\int p \, dx\right) + \delta\alpha\left(-\int px \, dx\right)$$

$$= \delta h_{cm}(-L) + \delta\alpha(M_y), \quad M_y \text{ is moment around c.m}$$

$$Q_{h_{cm}} = -L \equiv -\int p \, dx$$

$$Q_\alpha = M_y \equiv -\int px \, dx$$

$$T - U = \frac{m}{2} \dot{h}_{cm}^2 + \frac{I_{cm}}{2} \dot{\alpha}^2 - \frac{K_h}{2}(h_{cm} - \alpha x_k)^2 - \frac{K_\alpha}{2} \alpha^2$$

From Lagrange's equations,

$$-m\ddot{h}_{cm} - K_h(h_{cm} - \alpha x_K) - \int p \, dx = 0$$

$$-I_{cm}\ddot{\alpha} + K_h x_K(h_{cm} - \alpha x_K) - K_\alpha \alpha - \int px \, dx = 0$$

(II.35)

Substituting

$$\int p \, dx = qS\frac{\partial C_L}{\partial \alpha} \alpha, \qquad \int px \, dx = -qS(e + x_k)\frac{\partial C_L}{\partial \alpha} \alpha, \qquad h = \bar{h}e^{pt}$$

and $\alpha = \bar{\alpha}e^{pt}$ into the above equations, we obtain

$$\begin{bmatrix} (mp^2 + K_h) & -K_h x_k + qS\frac{\partial C_L}{\partial \alpha} \\ -K_h x_K & I_{cm}p^2 + K_h x_k^2 + K_\alpha - qS\left(e + x_k\frac{\partial C_L}{\partial \alpha}\right) \end{bmatrix} \left\{ \begin{matrix} \bar{h}e^{pt} \\ \bar{\alpha}e^{pt} \end{matrix} \right\} = \left\{ \begin{matrix} 0 \\ 0 \end{matrix} \right\}$$

The condition that the coefficient matrix is zero gives

$$Ap^4 + Bp^2 + C = 0 \tag{II.36}$$

where

$$A = mI_{cm} = mI_\alpha - S_\alpha^2 \qquad (I_{cm} = I_\alpha - mx_k^2, \quad S_\alpha = mx_k)$$

$$B = m\left[K_h x_k^2 + K_\alpha - qS(e + x_k)\frac{\partial C_L}{\partial \alpha} \right] + K_h I_{cm}$$

$$= m\left[K_\alpha - qSe\frac{\partial C_L}{\partial \alpha} \right] + K_h \alpha - S_\alpha qS\frac{\partial C_L}{\partial \alpha}$$

$$C = K_h\left[K_h x_k^2 + K_\alpha - qS(e + x_k)\frac{\partial C_L}{\partial \alpha} \right] + K_h x_k\left(-K_h x_k + qS\frac{\partial C_L}{\partial \alpha} \right)$$

$$= K_h\left[K_\alpha - qSe\frac{\partial C_L}{\partial \alpha} \right]$$

These *A*, *B* and *C* are the same as in equation (3.3.51), Section 3.3, in the text. Thus we have the same flutter boundary.

$$p^2 = \frac{-B + [B^2 - 4AC]^{\frac{1}{2}}}{2A}$$

(a) $B > 0$ ($A > 0$, $C > 0 \leftarrow$ divergence free). If p^2 is complex (not real), then instability occurs.

$\therefore$ $B^2 - 4AC = 0$ gives flutter boundary, i.e.,

$$Dq_F^2 + Eq_F + F = 0$$

or

$$q_F = \frac{-E \pm [E^2 - 4DF]}{2D}$$

where

$$D \equiv \left\{ (me + S_\alpha)S\frac{\partial C_L}{\partial \alpha} \right\}^2$$

$$E \equiv \left\{ -2(me + S_\alpha)[mK_\alpha + K_h I_\alpha] + 4[mI_\alpha - S_\alpha^2]eKS\frac{\partial C_L}{\partial \alpha} \right\}$$

$$F \equiv [mK_\alpha + K_h I_\alpha]^2 - 4[mI_\alpha - S_\alpha^2]K_h K_\alpha$$

The smaller, real, and positive q_F is the flutter dynamic pressure.

(b) $B < 0$. Note that $B = 2\sqrt{AC}$ before $B = 0$ as q increases. Hence flutter always occurs for $B > 0$.

Question. Prove that

1.

$$\phi_{hF}(\tau) = \phi_{Fh}(-\tau)$$

and

2.

$$\Phi_{hF}(\omega) = H_{hF}(-\omega)\Phi_{FF}(\omega)$$

This is a useful exercise to confirm one's facility with the concepts of correlation function and power spectral density.

Answer
1. Prove that $\phi_{hF}(\tau) = \phi_{Fh}(-\tau)$. We start with the definition of the cross-correlation function:

$$\phi_{hF}(\tau) = \lim_{T \to \infty} \frac{1}{2T} \int_{-T}^{+T} h(t)F(t+\tau)\, \mathrm{d}t \tag{II.37}*$$

The response $h(t)$ is given by

$$h(t) = \int_0^t I_{hF}(t - \tau_1)F(\tau_1)\, \mathrm{d}\tau_1 \tag{II.38}$$

Here we have taken $h(t)$ in dimensional form and $I_{hF}(t)$ represents the response to an impulse. Substituting (II.38) into (II.37),

$$\phi_{hF}(\tau) = \lim_{T \to \infty} \frac{1}{2T} \int_{-T}^{+T} \int_{-\infty}^{+\infty} I_{hF}(t - \tau_1)F(\tau_1)F(t+\tau)\, \mathrm{d}\tau_1\, \mathrm{d}t$$

(One may change the limit $(0, t)$ in the inner integral to $(-\infty, +\infty)$ since the impulse will be zero for $(t - \tau_1) < 0$.) Let $t' \equiv t - \tau_1 \Rightarrow \tau_1 = t - t'$ and interchange the order of integration. Then $\mathrm{d}\tau_1 = -\mathrm{d}t'$ and

$$\phi_{hF}(\tau) = -\int_{+\infty}^{-\infty} I_{hF}(t') \lim_{T \to \infty} \frac{1}{2T} \int_{-T}^{+T} F(t - t')F(t+\tau)\, \mathrm{d}t\, \mathrm{d}t'$$

$$= -\int_{+\infty}^{-\infty} I_{hF}(t')\phi_{FF}(\tau + t')\, \mathrm{d}t'$$

---

* A short proof goes as follows. Define $\eta \equiv t - \tau$. Then $\mathrm{d}\eta = \mathrm{d}t$ and $t = \eta - \tau$; using these and (II.37) the proof follows by inspection.

Thus

$$\phi_{hF}(\tau) = + \int_{-\infty}^{+\infty} I_{hF}(\lambda)\phi_{FF}(\tau + \lambda)\, d\lambda \tag{II.39}$$

where $\lambda \equiv t' =$ dummy variable.
We follow the same procedure for $\phi_{Fh}(\tau)$.

$$\phi_{Fh}(\tau) = \lim_{T\to\infty} \frac{1}{2T} \int_{-T}^{+T} F(t)h(t+\tau)\, dt$$

$$= \lim_{T\to\infty} \frac{1}{2T} \int_{-T}^{+T} F(t)\left\{\int_{-\infty}^{+\infty} I_{hF}(t+\tau-t_2)F(t_2)\, dt_2\right\} dt$$

let $t'' = t + \tau - t_2. \Rightarrow dt'' = -d\tau_2,\ \tau_2 = \tau + t - t''$

$$\phi_{Fh}(\tau) = -\int_{+\infty}^{-\infty} I_{hF}(t'')\left\{\lim_{T\to\infty} \frac{1}{2T} \int_{-T}^{+T} F(t - t'' + \tau)F(t)\, dt\right\} dt''$$

$$= -\int_{+\infty}^{-\infty} I_{hF}(t'')\phi_{FF}(\tau - t'')\, dt''$$

$$= \int_{-\infty}^{+\infty} I_{hF}(\lambda)\phi_{FF}(\tau - \lambda)\, d\lambda$$

Let $\tau \to -\tau$:

$$\phi_{Fh}(-\tau) = + \int_{-\infty}^{+\infty} I_{hF}(\lambda)\phi_{FF}(-\tau - \lambda)\, d\lambda$$

but $\phi_{FF}(\tau) = \phi_{FF}(-\tau)$ and hence

$$\phi_{Fh}(-\tau) = + \int_{-\infty}^{+\infty} I_{hF}(\lambda)\phi_{FF}(+\tau + \lambda)\, d\lambda \tag{II.40}$$

Comparing (II.39) and (II.40) we see that

$$\phi_{hF}(\tau) = \phi_{Fh}(-\tau)$$

2. Prove that $\Phi_{hF}(\omega) = H_{hF}(-\omega)\Phi_{FF}$. By definition, the spectral density function is the Fourier transform of the correlation function.

We apply this to the cross correlation function defined by (II.39).

$$\Phi_{hF}(\omega) \equiv \frac{1}{\pi} \int_{-\infty}^{+\infty} \phi_{hF}(\tau) e^{-i\omega\tau} \, d\tau$$

$$\Phi_{hF}(\omega) = \frac{1}{\pi} \int_{-\infty}^{+\infty} \int_{-\infty}^{+\infty} I_{hF}(t) \phi_{FF}(\tau + t) e^{-i\omega\tau} \, dt \, d\tau$$

$$= \frac{1}{\pi} \int_{-\infty}^{+\infty} \int_{-\infty}^{+\infty} I_{hF}(t) e^{-i\omega t} \phi_{FF}(\tau + t) e^{-i\omega\tau + i\omega t} \, dt \, d\tau$$

$$= \int_{-\infty}^{+\infty} I_{hF}(t) \left\{ \frac{1}{\pi} \int_{-\infty}^{+\infty} \phi_{FF}(\tau + t) e^{-i\omega(\tau + t)} \, d\tau \right\} e^{+i\omega t} \, dt$$

By definition

$$\frac{1}{\pi} \int_{-\infty}^{+\infty} \phi_{FF}(\tau') e^{-i\omega\tau'} \, d\tau' = \Phi_{FF}(\omega)$$

Let $\tau' \equiv \tau - t$, and substitute in RHS of equation for Φ_{hF}. Then

$$\Phi_{hF}(\omega) = \int_{-\infty}^{+\infty} I_{hF}(t) e^{+i\omega t} \Phi_{FF}(\omega) \, dt$$

Now, since

$$H_{hF}(\omega) = \int_{-\infty}^{+\infty} I_{hF}(t) e^{-i\omega t} \, dt \text{ it follows that}$$

$$\Phi_{hF}(\omega) = H_{hF}(-\omega) \Phi_{FF}(\omega)$$

Section 3.6

Typical section flutter analysis using piston theory aerodynamics

Pressure: $\quad p = \rho a \left[\dfrac{\partial z_a}{\partial t} + \dfrac{\partial z_a}{\partial x} \right]$ $\qquad U\frac{\partial z_a}{\partial x}$ *see top p.86 / p.81*

Motion: $\quad z_a = -h - \alpha(x - x_{EA})^*$ $\qquad (3.4.2) \; p.82$ *p.180 / p.174*

Upper surface: $\quad p_U = \rho a[-\dot{h} - \dot{\alpha}(x - x_{EA}) - U\alpha]$

Lower surface: $\quad p_L = -\rho a[-\dot{h} - \dot{\alpha}(x - x_{EA}) - U\alpha]$

last eq. p.82

---

*x is measured from airfoil leading edge; b is half-chord of airfoil.

$\dfrac{\partial z_a}{\partial t} =$ *normal vel. of the body surface*

normal vel. of surface

fluid vel. comp. normal to the surface must equal this for an inviscid fluid

447

Appendix II

Net pressure:
$$p_L - p_U = \frac{4\rho U^2}{2M}\left[\frac{\dot{h}}{U} + \frac{\dot{\alpha}}{U}\cdot(x - x_{EA}) + \alpha\right]$$

Lift:
$$L \equiv \int_0^{2b}(p_L - p_U)\,dx$$

$$= \frac{4\rho U^2}{2M}\left\{\left[\frac{\dot{h}}{U} - \frac{\dot{\alpha}x_{EA}}{U} + \alpha\right]2b + \frac{\dot{\alpha}}{U}\frac{(2b)^2}{2}\right\}$$

Moment:
$$M_y = -\int_0^{2b}(p_L - p_U)(x - x_{EA})\,dx$$

$$= x_{EA}L - \frac{4\rho U^2}{2M}\left[\frac{\dot{h}}{U} - \dot{\alpha}\frac{x_{EA}}{U} + \alpha\right]\frac{(2b)^2}{2}$$

$$- \frac{4\rho U^2}{2M}\frac{\dot{\alpha}}{U}\frac{(2b)^3}{3} \tag{II.41}$$

Assume simple harmonic motion,

$$h = \bar{h}e^{i\omega t}$$

$$\alpha = \bar{\alpha}e^{i\omega t}$$

$$L = \bar{L}e^{i\omega t}$$

$$M_y = \bar{M}_y e^{i\omega t}$$

$$\bar{L} = \frac{4\rho U^2}{2M}\left\{\frac{i\omega}{U}2b\bar{h}\right.$$

$$\left. + \left[\frac{-i\omega x_{EA}}{U} + 1 + \frac{i\omega}{U}\frac{(2b)}{2}\right]2b\bar{\alpha}\right\}$$

$$\equiv 2\rho b^2\omega^2(2b)\left\{(L_1 + iL_2)\frac{\bar{h}}{b} + [L_3 + iL_4]\bar{\alpha}\right\}$$

Thus from equation (3.6.3) *p.106* in Section 3.6,

$$L_1 + iL_2 = \frac{\dfrac{2\rho U^2}{M}\dfrac{i\omega\rho 2b}{U}}{2\rho b^2\omega^2(2b)\dfrac{1}{b}} = \frac{i}{M}\frac{U}{\omega b} \tag{II.42}$$

448

and

$$L_3 + iL_4 = \frac{\dfrac{2\not{\rho}U^2}{M}\left[\dfrac{-i\omega x_{EA}}{U} + 1 + \dfrac{i\omega(2b)}{U^2}\right]2\not{b}}{2\not{\rho}b^2\omega^2(2\not{b})}$$

$$= \frac{1}{M}\left(\frac{U}{b\omega}\right)^2\left[\frac{-i\omega b}{U}\frac{x_{EA}}{b} + 1 + \frac{i\omega b}{U}\right] \tag{II.43}$$

Questions

(1) Derive similar equations for

$M_1 + iM_2$ and $M_3 + iM_4$

(2) Fix $\dfrac{\omega_h}{\omega_\alpha} = 0.5$, $r_\alpha = 0.5$, $x_\alpha = 0.05$

$\dfrac{x_{ea}}{b} = 1.4$, $M = 2$

Choose several k, say $k = 0.1$, 0.2, 0.5, and solve for

$\left(\dfrac{\omega}{\omega_\alpha}\right)^2$ and $\dfrac{m}{2\rho_\infty bS} \equiv \mu$ $(S \equiv 2b)$

from (3.6.4) using the method described on pp. 107 and 108. Plot k vs μ and ω/ω_α vs μ.

Finally plot $\dfrac{U}{b\omega_\alpha} \equiv \dfrac{\omega/\omega_\alpha}{k}$ vs μ. This is the flutter velocity as a function of mass ratio.

Answers

Recall equation (3.6.3) again from Section 3.6,

$$\bar{M}_y = -2\rho b^3\omega^2(2b)\left\{[M_1 + iM_2]\frac{h}{b} + [M_3 + iM_4]\bar{\alpha}\right\}$$

Comparing the above and (II.41), one can identify

$$M_1 + iM_2 = \frac{iU}{Mb\omega}\left[1 - \frac{x_{ea}}{b}\right]$$

$$M_3 + iM_4 = \frac{1}{M}\left(\frac{U}{b\omega}\right)^2\left[1 - \frac{x_{ea}}{b}\right] + i\frac{1}{M}\frac{U}{b\omega}\left\{\left[1 - \frac{x_{ea}}{b}\right]^2 + \frac{1}{3}\right\} \tag{II.44}$$

449

Appendix II

Recall the method described in Section 3.6 for determining the flutter boundary.

1. Evaluate real and imaginary parts of equation (3.6.4) and set each individually to zero.

2. Solve for $(\omega_\alpha/\omega)^2$ in terms of themass ratio, μ, from the imaginary part of the equation.

3. Substituting this result into the real part of the equation, obtain a quadratic in μ. Solve for possible values of μ for various k. To be physically meaningful, μ must be positive and real.

4. Return to 2. to evaluate ω/ω_α

5. Finally determine $\dfrac{U}{b\omega_\alpha} = \dfrac{U}{b\omega}\dfrac{\omega}{\omega_\alpha} = \dfrac{1}{k}\dfrac{\omega}{\omega_\alpha}$.

In detail these steps are given below.

1. *Real part*

$$\mu^2\left\{\left[1-\left(\frac{\omega_\alpha}{\omega}\right)^2\left(\frac{\omega_h}{\omega_\alpha}\right)^2\right]r_\alpha^2\left[1-\left(\frac{\omega_\alpha}{\omega}\right)^2\right]-x_\alpha^2\right\}$$

$$+\mu\left\{\frac{-1}{k^2M}\left(1-\frac{x_{ea}}{b}\right)\left[1-\left(\frac{\omega_\alpha}{\omega}\right)^2\left(\frac{\omega_h}{\omega_\alpha}\right)^2\right]+\frac{x_\alpha}{k^2M}\right\}-\frac{1}{3k^2M^2}=0 \qquad \text{(II.45)}$$

Imaginary part

$$\mu r_\alpha^2\left[1-\left(\frac{\omega_\alpha}{\omega}\right)^2\right]-\frac{1}{k^2M}\left(1-\frac{x_{ea}}{b}\right)$$

$$+\left[\left(1-\frac{x_{ea}}{b}\right)^2+\frac{1}{3}\right]\left[1-\left(\frac{\omega_\alpha}{\omega}\right)^2\left(\frac{\omega_h}{\omega_\alpha}\right)^2\right]\mu$$

$$-\left[1-\frac{x_{ea}}{b}\right]\mu x_\alpha-\left[1-\frac{x_{ea}}{b}\right]\left[\mu x_\alpha-\frac{1}{k^2M}\right]=0 \qquad \text{(II.46)}$$

2. Solving for $(\omega_\alpha/\omega)^2$ from (II.46),

$$\left(\frac{\omega_\alpha}{\omega}\right)^2=\frac{r_\alpha^2+\left(1-\frac{x_{ea}}{b}\right)^2+\frac{1}{3}-2x_\alpha\left(1-\frac{x_{ea}}{b}\right)}{r_\alpha^2+\left(\frac{\omega_h}{\omega_\alpha}\right)^2\left[\left(1-\frac{x_{ea}}{b}\right)^2+\frac{1}{3}\right]} \qquad \text{(II.47)}$$

Note (II.47) is independent of μ and k; this is a consequence of using piston theory aerodynamics and would not be true, in general, for a more elaborate (and hopefully more accurate) aerodynamic theory.

450

Substituting the various numerical parameters previously specified into (II.47) gives

$$\left(\frac{\omega_\alpha}{\omega}\right)^2 = 2.099 \quad \text{or} \quad \frac{\omega}{\omega_\alpha} = 0.69 \tag{II.48}$$

3. Using (II.48) in (II.45) along with the other numerical parameters gives

$$-0.133\,\mu^2 + \frac{0.121}{k^2}\,\mu - \frac{1}{12k^2} = 0 \tag{II.49}$$

Solving for μ,

$$\mu_{1,2} = \frac{0.45}{k^2} \pm \frac{1}{k}\left[\frac{0.21}{k^2} - 0.63\right]^{\frac{1}{2}} \tag{II.50}$$

Note that there is a maximum value of k possible, $k_{max} = [0.21/0.63]^{\frac{1}{2}}$. Larger k give complex μ which are physically meaningless. Also note that $\mu \to 0.67, \infty$ as $k \to 0$.

4. ω/ω_α is evaluated in (II.48) and for these simple aerodynamics does not vary with μ or k.

5. From (II.48) and a knowledge of k, $U/b\omega_\alpha$ is known.

The above results are tabulated below.

k	μ_1	μ_2	$U/b\omega_\alpha$
0.0	0.67	∞	∞
0.1	0.69	89.6	6.9
0.2	0.72	22	3.45
0.3	0.75	9.28	2.3
0.4	0.81	4.3	1.73
0.5	0.937	2.66	1.38
0.57	1.39	1.39	1.21

From the above table (as well as equation (II.50)) one sees that for $\mu < 0.67$, no flutter is possible. This is similar to the flutter behavior of the typical section at incompressible speeds. At these low speeds mass ratios of this magnitude may occur in hydrofoil applications. Although no such applications exist at high supersonic speeds, it is of interest at least from a fundamental point of view that this somewhat surprising behavior at small μ occurs there as well.

451

Appendix II

Section 4.1

Questions.

1. Starting from Bernoulli's equation, show that

$$\frac{\hat{a}}{a_\infty} \sim M_\infty^2 \frac{\hat{u}}{U_\infty}$$

2. Previously we had shown that the boundary condition on a moving body is (within a linear approximation)

$$\frac{\partial \hat{\phi}}{\partial z}\bigg|_{z=0} = \frac{\partial z_a}{\partial t} + U_\infty \frac{\partial z_a}{\partial x}$$

What is the corresponding boundary condition in terms of $\hat{p}$?

3. Derive approximate formulae for the perturbation pressure over a two-dimensional airfoil at supersonic speeds for very low and very high frequencies.

Answers.

1. Bernoulli's equation is

$$\frac{\partial \phi}{\partial t} + \frac{\nabla \phi \cdot \nabla \phi}{2} + \int_{p_\infty}^{p} \frac{dp_1}{\rho^1(p^1)} = \frac{U_\infty^2}{2}$$

Since

$$a^2 \equiv \frac{\partial p}{\partial \rho} \quad \text{and} \quad \frac{p}{\rho^\gamma} = \text{constant}$$

we may evaluate integral in the above to obtain

$$\frac{U_\infty^2}{2} \frac{\partial \phi}{\partial t} + \frac{u^2}{2} = \frac{a^2 - a_\infty^2}{\gamma - 1}, \qquad u \equiv |\nabla \phi|$$

Assume

$$a = a_\infty + \hat{a}$$
$$u = U_\infty + \hat{u}$$
$$\phi = U_\infty x + \hat{\phi}$$

where $\hat{a} \ll a_\infty$, etc. To first order

$$-M_\infty^2 \frac{\hat{u}}{U_\infty} - \frac{1}{a_\infty^2} \frac{\partial \hat{\phi}}{\partial t} = \frac{2}{\gamma - 1} \frac{\hat{a}}{a_\infty} + \text{terms } (\hat{a}^2, \text{etc})$$

452

This means that $M_\infty^2(\hat{u}/U_\infty)$ and $\hat{a}/a_\infty$ are quantities of the same order, at least for steady flow where $\partial\hat{\phi}/\partial t \equiv 0$.

   2.

$$\left.\frac{\partial\hat{\phi}}{\partial z}\right|_{z=0} = Dz_a; \qquad D \equiv \frac{\partial}{\partial t} + U_\infty\frac{\partial}{\partial x} \tag{II.51}$$

By the linearized momentum equation

$$\frac{-\partial p}{\partial x} = \rho_\infty D\hat{u}$$

but

$$\hat{u} = \nabla_x\hat{\phi}$$

$$\therefore \quad \hat{p} = -\rho_\infty D\hat{\phi}$$

$$\therefore \quad -\frac{\partial\hat{p}}{\partial z} = \rho_\infty\frac{\partial}{\partial z}(D\hat{\phi})$$

$$= \rho_\infty\frac{\partial}{\partial z}(D(\hat{\phi})) = \rho_\infty D\left(\frac{\partial}{\partial z}\hat{\phi}\right)$$

From (II.51) and the above

$$\left.\frac{-\partial p}{\partial z}\right|_{z=0} = \rho_\infty D^2 z_a$$

$$\frac{\partial p}{\partial z} = -\rho_\infty D^2 z_a \quad \text{at} \quad z = 0 \tag{II.52}$$

   3.

$$\nabla^2\phi - \frac{1}{a_\infty^2}\left[\frac{\partial}{\partial t} + U_\infty\frac{\partial}{\partial x}\right]^2\phi = 0$$

where

$$\left.\frac{\partial\phi}{\partial z}\right|_{z=0} = \frac{\partial}{\partial t}z_a + U_\infty\frac{\partial}{\partial x}z_a \equiv w$$

off wing $\left.\dfrac{\partial\phi}{\partial z}\right|_{z=0} = 0$ thickness case   This does not matter here, because there are no disturbances ahead of wing in supersonic flow.

$$\phi|_{z=0} = 0 \quad \text{lifting case,}$$

453

Appendix II

For a two dimensional solution, let $\phi(x, z, t) = \bar{\phi}(x, z)e^{i\omega t}$ and $w = \bar{w}e^{i\omega t}$. Thus

$$\frac{\partial^2 \bar{\phi}}{\partial x^2} + \frac{\partial^2 \bar{\phi}}{\partial z^2} - \frac{1}{a_\infty^2}\left[-\omega^2\bar{\phi} + 2i\omega U_\infty \frac{\partial \bar{\phi}}{\partial x} + U_\infty^2 \frac{\partial^2 \bar{\phi}}{\partial x^2}\right] = 0$$

Recall $u, v, w = 0$ for $x \le 0$ (leading edge) in supersonic flow. Taking Laplace transform (quiescent condition at $x = 0$)

$$\Phi \equiv \int_0^\infty \bar{\phi}e^{-px}\,dx$$

then

$$p^2\Phi + \frac{\partial^2 \Phi}{\partial z^2} - \frac{1}{a_\infty^2}[-\omega^2\Phi + 2i\omega pU_\infty\Phi + p^2 U_\infty^2\Phi]$$

or

$$\frac{d^2\Phi}{dz^2} = \left[-p^2 - \frac{\omega^2}{a_\infty^2} + \frac{2i\omega pM}{a_\infty} + p^2 M^2\right]\Phi \equiv \mu^2\Phi$$

Thus

$$\Phi = Be^{-\mu z}$$

Now

$$\frac{d\Phi}{dz}\bigg|_{z=0} = W, \qquad W \equiv \int_0^\infty \bar{w}e^{-px}\,dx$$

Thus

$$\frac{\partial\Phi}{\partial z}\bigg|_{z=0} = -\mu B, \qquad B = -\frac{W}{\mu}$$

Hence

$$\Phi = -\frac{W}{\mu}e^{-\mu z}$$

so

$$\bar{\phi}\big|_{z=0} = \int \mathscr{L}^{-1}\left\{-\frac{1}{\mu}\right\}\bar{w}(\xi, \omega)\,d\xi$$

For low frequencies, we can ignore ω^2 terms, so

$$\mu^2 \cong (M^2 - 1)\left\{ p + \frac{iM\omega}{a_\infty(M^2 - 1)} \right\}^2$$

$$-\frac{1}{\mu} = \frac{-1}{\sqrt{M^2 - 1}} \left(\frac{1}{p + \dfrac{iM\omega}{a_\infty(M^2 - 1)}} \right)$$

$$\mathscr{L}^{-1}\left(\frac{-1}{\mu}\right) = \frac{-1}{\sqrt{M^2 - 1}} \exp\left[-iM\omega x/a_\infty(M^2 - 1)\right]$$

and

$$\bar{\phi}\big|_{z=0} = \frac{-1}{\sqrt{M^2 - 1}} \int_0^x \exp\left[-iM\omega(x - \xi)/a_\infty(M^2 - 1)\right]\bar{w}(\xi, \omega)\, d\xi$$

and if we select our coordinate system such that $w(0) = 0$, for low frequencies the perturbation pressure, $\hat{p}$, is from Bernoulli's equation

$$\hat{p} = \frac{\rho_\infty e^{i\omega t}}{\sqrt{M^2 - 1}} \left[-\frac{i\omega \exp\left[i\omega(t - Mx/a_\infty(M^2 - 1))\right]}{(M^2 - 1)} \right.$$

$$\left. \times \int \exp\left[iM\omega\xi/a_\infty(M^2 - 1)\right]\bar{w}(\xi, \omega)\, d\xi + U_\infty \bar{w}(x, \omega) \right]$$

$$\cong \frac{\rho_\infty e^{i\omega t}}{\sqrt{M^2 - 1}} U_\infty \bar{w}(x) \qquad \textit{for very low frequencies} \qquad (\text{II.53})$$

For high frequencies,

$$\frac{d^2\Phi}{dz^2} = \left[\frac{-\omega^2}{a_\infty^2} + \frac{2i\omega pM}{a_\infty} + (M^2 - 1)p^2 \right]\Phi$$

$$\cong \left[\frac{i\omega}{a_\infty} + pM \right]^2 \Phi$$

when we ignore the $(-p^2)$ term compared to those involving ω. Then,

$$\frac{-1}{\mu} \cong \frac{-1}{pM + \dfrac{i\omega}{a_\infty}}$$

and

$$\bar{\phi}\big|_{z=0} = \int \mathscr{L}^{-1}\left(\frac{-1}{\mu}\right)\bigg|_{x-\xi} \bar{w}(\xi, \omega)\, d\xi$$

455

by the convolution theorem. Now

$$\mathscr{L}^{-1}\left[\frac{-1}{pM+\dfrac{i\omega}{a}}\right] = \frac{-1}{M}\exp\left(-i\omega x/a_\infty M\right)$$

so

$$\bar{\phi}\big|_{z=0} = -\frac{1}{M}\int_0^x \exp\left[-i\omega(x-\xi)/a_\infty M\right]\bar{w}(\xi,x)\,\mathrm{d}\xi$$

and from Bernoulli's equation

$$\therefore\ \hat{p} = \frac{\rho_\infty}{M}i\omega\exp\left[i\omega(x-x/a_\infty M)\right]\int_0^x \exp\left(i\omega\xi/a_\infty M\right)\bar{w}(\xi,\omega)\,\mathrm{d}\xi$$

$$+\frac{\rho_\infty U_\infty}{M}\exp\left[i\omega(t-x/a_\infty M)\right]\exp\left(i\omega x/a_\infty M\right)\bar{w}(x,\omega)$$

$$-\frac{\rho_\infty U_\infty}{M}\frac{i\omega}{a_\infty M}\exp\left[i\omega(t-x/a_\infty M)\right]\int_0^x \exp\left(i\omega\xi/a_\infty M\right)\bar{w}(\xi,\omega)\,\mathrm{d}\xi$$

$$\hat{p} \cong \frac{\rho_\infty U_\infty}{M}\bar{w}(x,\omega)e^{i\omega t}\quad \textit{for high frequencies.}$$

This is known as the (linearized, small perturbation) piston theory approximation. It is a useful and interesting exercise (also a rather straight forward one) to determine pressure distributions, lift and moment for translation and rotation of a flat plate using the piston theory.* The low frequency approximation considered earlier is also useful in this respect.

---

* Ashley, H. and Zartarian, G., 'Piston Theory—A New Aerodynamic Tool for the Aeroelastician', *J. Aero. Sciences*, **23** (December 1956) pp. 1109–1118.

Subject index

457

Subject index

Subject index

Lumped element *see* Lumped element method
Rayleigh-Ritz *see* Rayleigh-Ritz method
Speed of sound 165
Stability *see* Divergence, flutter
Stall *see* Aerodynamics, stall; also see Flutter, stall
Stiffness
  bending 42
  nonlinear 12
  plate 52
  spring 5, 123
  torsional 15
Strain *see* Energy, strain
Structural
  Damping *see* Damping, structural
  Influence coefficients and functions 23, 131, 282
  Modes *see* Modes, structural
  Stiffness *see* Stiffness
Structural theory
  beam 42, 45, 52
  beam-rod 15, 52, 126, 340
  plate 40, 52
  rod 15, 52, 125
Subsonic *see* Aerodynamics, subsonic
Supersonic *see* Aerodynamics, supersonic

T

Theodorsen's function *see* Aerodynamics, Theodorsen function

Transfer function
  aerodynamic 72, 84
  aeroelastic 72, 274
  structural 60, 71, 415
Transform
  Fourier *see* Fourier transform
  Laplace *see* Laplace transform
Transonic *see* Aerodynamics, transonic
Transient motion *see* Motion, arbitrary and aerodynamics, transient
Tubes *see* Pipes
Turbomachines ix, 104, 110, 383
Typical section viii, 3, 55

V

Velocity potential *see* Aerodynamics, velocity potential
Virtual displacement 48
Virtual work 48
Vortex shedding *see* Aerodynamics, vortex shedding

W

Wagner function *see* Aerodynamics, Wagner function
Wave equation 168

Author index

461

Author index

Author index